MUSEUM WITHOUT WALLS

MUSEUM WITHOUT WALLS

Jonathan Meades

unbound

Museum Without Walls
First published 2012
Reprinted 2012
by Unbound
32–38 Scrutton Street, Shoreditch, London EC2A 4RQ
www.unbound.co.uk

Text design and typesetting by Palindrome
Cover design by Mark Ecob

A CIP record for this book
is available from the British Library

ISBN 978-1-908717-18-4

Printed in England by Clays Ltd, Bungay, Suffolk

To Robin Coates and his parents

CONTENTS

CONTENTS

Contents

PUBLISHER'S NOTE

THE PUBLISHER WOULD like to thank Faber and Faber and the Estate of T. S. Eliot for permission to quote from T. S. Eliot's essay *The Sacred Wood*; Faber and Faber for permission to quote from W. H. Auden's *Letters from Iceland*; Weidenfeld & Nicholson and the Estate of Vladimir Nabokov for permission to quote from *Pale Fire*, and Henry Holt & Company and the Estate of Robert Frost for permission to quote from 'Of a Winter Evening' (later published as 'Questioning Faces').

ACKNOWLEDGEMENTS

THE CONTENTS OF THIS BOOK appeared in various forms, in various media. They were published or otherwise disseminated by the following: The Academy of Urbanism, BBC2 and BBC4, *Building*, Dillington House, *The Independent on Sunday*, *Intelligent Life*, Liverpool School of Architecture, the Lowdham Festival, the National Gallery, *openDemocracy*, the Prince's Foundation, RIBA, *The Times*, the Victoria and Albert Museum, the Victorian Society, the Woodchester Mansion Trust.

Scripts are merely blueprints. Those included here were brought to life by a group of people with whom I have amiably collaborated over many years: Frank Hanly, Martha Wailes, Rob Payton, Tim Niel, Colin Murray, Russell England, Allan Campbell, Andrea Miller, Anthony Wall, John Whiston, Simon Holland, Jim Allison, Michael Fenton Stevens, Simon Mulligan, Emma Matthews, the late Nick Rossiter, Riaz Meer, Paul Ingvasson, Angela Slaven, Kirsty Hanly, Lesley Smith-Glasgow, Rachel Mackenzie, Phil Molloy, Pete Stanton, Christopher Biggins, Alastair McCormick, Luke Cardiff, Jamie Flynn, Martin Rowson, Clem Cecil and John Pritchard (Mr).

I'm hugely indebted to my wife Colette Forder who selected and edited this collection.

INTRODUCTION

WE ARE SURROUNDED by the greatest of free shows. Places. Most of them made by man, remade by man. Deserted streets, seething boulevards, teeming beaches, empty steppes, black reservoirs, fields of agricultural scrap, cute villages and disappearing points which have an unparalleled capacity to promote hope (I am thinking of the aspect north up rue Paradis in Marseille).

This book is the product of an obsessive preoccupation with places, mainly British places, with their ingredients, with how and why they were made, with their power over us, with their capacity to illumine the societies that inhabit them and, above all, with the ideas that they foment.

Much of it evidently concerns buildings, the gaps between them, their serendipitous conjunctions and grotesque collisions. High architecture comes into it, but so too do inspired bricolage, plutocratic boasts, arid estates, mighty sprawl. The catastrophic cock-ups of grandiloquent visionaries are as grimly appealing as the imaginatively bereft efforts of volume builders. There is an emphasis on buildings which are overlooked or which, should they be noticed, are unthinkingly despised, just as there is on topographies which are routinely neglected: the banal is a thing of joy. Everything is fantastical if you stare at it for long enough, everything is interesting. There is no such thing as a boring place.

These lectures, essays, polemics, squibs and telly scripts are intended to entertain, to instruct, to inform and to question the orthodoxies of the architectural, heritage and construction industries, to draw attention to the rich oddness of what we take for granted. But before that they are written because I want to read them, to watch them. If that sounds selfish and immodest so be it. But it is surely more honest to write for an audience of one whose peccadillos and limitations

I understand than for an inchoate mass of opinionated individuals whose multiple and conflicting tastes I can only guess at and which I have, above everything else, to be indifferent to. Régis Jauffret got it right when he said that he was disgusted by writers who think of their readers.

This is a pretty basic point which the cretinocracy that has seized control of television cannot begin to understand. There is much that it cannot understand: the unknown, alien opinions, intelligence. And what it cannot understand it seeks to quash. One becomes inured to censorship by dolts: mocking the INLA can cause the impressionable to take up terrorism; Muslims must be treated with a respect that is not accorded to other delusionists (who are less sensitive, less heavily armed); the sight of dead rabbits may offend – who? Living rabbits? I am inured, too, to being accused of bigotry and bias by persons who are blind to their own bigotry and bias, which they perceive as unexceptionable opinions or even The Truth. The scripts collected here include a number of passages deemed unfit for human consumption as well as the omission of some half-witted health warnings to the effect that my observations do not concur with the BBC's *pensée unique*. That organisation is, incidentally, waiting for the French government to come round to its way of thinking.

In the name of populism or 'accessibility' the cretinocracy has all but destroyed a medium which was for thirty or so years an instrument of beneficent cultural diffusion. There exists among telly executives and managers a conviction that everyone is as crass as they are, that everyone is preoccupied by football and its moronic overpaid pundits. Still, I can't say I wasn't warned. When I started to do telly in the mid-'80s Richard Williams, now a celebrated sports writer, then features editor of *The Times*, told me: 'You're going to meet a lot of very stupid people.' And so I have. But the directors and crews with whom I have spent years on the road in often questionable hotels are anything but stupid. Some of them are among my closest friends. One has to conclude that television is a medium in which it is the scum which rises inexorably to the top.

A priori the subjects here are: the cross-party tradition of governmental submission to the construction industry; architectural epiphanies – Marsh Court, Arc-et-Senans, l'Unité d'Habitation's roof; what to do with Anglican churches; Hadid and Legorreta; the folly of pedestrianisation; the hierarchy of land- and cityscapes; the hierarchy of building types; Birmingham's beauty; Bremen and the Hanseatic League; the futile vanity of 'landmark' buildings; why buildings are better unfinished; the congruence of the 1860s and 1960s; Letchworth's dreary legacy; the chasm between Hitler's architecture and Stalin's; the regeneration gravy train; the picturesque as an English disease; shopping malls; the Isle of Sheppey; the Isle of Rust; the Dome and domes; post-war churches; Pevsner and Nairn.

They're what I thought I was writing about. Revisiting them it is evident that I suffer recurrent devotions and, equally, a gamut of seldom submerged antipathies. Beaverbrook observed that his father taught him to hate, to hate. I enjoyed no such tuition. I've had to teach myself. It comes easily enough when one is presented with such objects as good taste, Georgian timidity and the nasty bland synthetic-modern 'legacy' of New Labour, made in the image of the grinning Tartuffish war criminal himself – but the happy Christmas Day will come when our Christian bomber and his gurning hag magically mutate into the Ceaucescus of Connaught Square. (The house has a basement.)

I guess that willing victims of Abrahamic systems of gross superstition may be offended by my mild mockery but these poor dupes surely have their 'faith' and their dietary idiocies and their intolerant belligerent paedophiliac 'prophet' to fortify them.

Such systems come and go. Two thousand years is not a long time: after two thousand years in the grave a Sicilian hasn't even decided on the form his revenge will take. Physical structures outlast dangerously frivolous religious pathologies. Whatever Carnac, Callanish, Stonehenge stood for is forgotten. These sites may have had nothing anyway to do with cosy supernaturalism or worship of coarse fictions. Their material survives, just as words remain when shorn of didactic nagging. The valley of the shadow of death is a potent construction which signifies to me a specific place rather than a metaphor for our apprisal of mortality (and god's playful tendency to visit that state on us). One case for architecture rather than mere building, for art rather than utility, is that a structure's purpose is ultimately provisional: the mosque is turned and becomes a cathedral, the warehouse is transformed into a skating rink, the old abattoir is now a boathouse. What remains is creativity, resourcefulness, the impetus to make.

It is the fruit of that impetus which has delighted and fascinated and even succoured me for as many years as I have been sentient. No doubt the fact that the only object I've ever managed to carpenter was a wrist-watch stand (BRG, 1959) inflates my admiration for those who work in three dimensions, who work virtuously, energetically, aggressively: Vanbrugh and Pilkington, Butterfield and Gordon, and, above all, Anon and Unattrib, whose oeuvre is vast, diverse and massively appreciated. Their names are forgotten but their shacks and leets, their lean-tos and footbridges will live forever more. For a while yet anyway.

I

FREE SHOW

Just Looking

HAMAS & KIBBUTZ

I'VE SPENT A lifetime writing about place – as it's now called, with neither a definite nor an indefinite article – writing about it in different media and in different ways: polemically, analytically, essayistically, fictively. I have also written about and filmed architecture and buildings – which are no more than components of place: the distinction between what is architecture and what is building strikes me, incidentally, as absolutely bogus.

The fiction I've written has almost invariably been triggered by a fascination with particular technical devices, by incident, and by place – rather than by character or by one of those grim plonkers called abstract ideas. Thus: the Surrey speciality of the private estate where light entertainers' quaintly kitschy houses are set among rhododendrons, pines, broom and golf courses; the New Forest where my mother taught the children of inbred families who came to school in winter sewn into untreated rabbit furs; the Isle of Portland whose landscape of abandoned quarries, prisons, wide village streets and baroque churches so obsessed me that I had to find a way of making it the mute protagonist of a novella – the memory of an overheard bus queue conversation did the rest; the precipitous south-east London suburbs suggested a particular kind of life . . .

Places render me suggestible in a way that all dance, all musicals and most theatre don't. I was driving on the M27 when I looked up to Portsdown Hill and realised what was wrong with it: it didn't have a 200-metre-tall neon cross on its escarpment. Why would such a cross exist? That was how a novel began.

Of course, amending a landscape is an architectural rather than a writerly aspiration. The films I make are fictions without dramatis personae, plays without players. They are, unequivocally, not about places but about the ideas that places foment. I have often written them having paid only the most cursory visit to the places in question. The directors I work with then conspire with me to create what may or may not be the truth, or a truth, about them. Ultimately they are expressions of an incurable topophilia, of a love of places – which are what you find when you're on the way to somewhere else.

The downlands of southern England have such a thin layer of top-soil that they are useless for cultivation of crops – all that grows is grass and blackthorns. But they are adequate for the grazing of sheep which used to be tended by necessarily solitary shepherds. In the valleys beneath the downs, meadows could be floated – i.e. irrigated by a system of narrow canals called leets and rudimentary sluices. The goal was to achieve maximum lushness which would in turn produce rich milk in the cows that fed off that pasture. On the downs one finds tiny isolated churches which are akin to hermitages and refuges, they speak of mysticism or, at least, of reflective contemplation. In the valleys nonconformism flourished in chapels where observance was collective and social rather than ritualistic. Two conjoining topo-graphies, two different ways of life. The bleak downs are formed of chalk. The end product of the rich cow's milk in the populous valleys was cheese.

What we have here is comparative synecdoche: in each case a characteristic part stands for the whole – that, however, is merely incidental. What we have, more pertinently, is a rudimentary example of classification. It happens to serve also as an instance of that most potentially toxic of divisions, between them and us. Us here, in this village in the valley, we do our wibs with beech nuts and tallow. Them in the hamlets on the downs, them are two miles distant – foreign: they call it wibbing and they use acorns and horse fat. They're wrong – they should be doing it our way. We're as different as chalk and cheese.

Classification or taxonomy or pigeon-holing or whatever we may call it may not be as fundamental a trait as seeking shelter, food, sex and narcosis, but it's universal. And it's not one of those characteristics of which it can be said: it's what makes us human.

A buzzard, for instance, rides on thermals. It looks good, that lazy wind-blown life. But what it is actually doing is keeping on the *qui vive*. Which means classifying what it sees from up there. It distinguishes between quarry and non-quarry. Then makes further distinctions. Between quarry such as a mouse or a duckling – which are just snacks. And a rabbit – which is a proper sit-down meal. And between non-quarry that is unthreatening – cars, say; buzzards sit on their posts quite unfazed as cars hurry past – and non-quarry which is, possibly, hostile: human beings walking or combine harvesters. Is it a New Holland, a John Deere, a Claas or a Massey Ferguson which most incites buzzardly suspicion? We may never know. Were we, in the best of all possible worlds, to ask a buzzard, I suspect we would be disappointed. Not because of the buzzard's assumption that Claas are the most threatening combines due to their being multicoloured. We would be disappointed because of the buzzard's voice. Buzzards are astonishingly shrill, fluting, piping. There is a chasmic incongruity between the magnificence of the vaunting serial killer and the sound it emits. Chasmic and comic.

A buzzard is an avian analogue of the even squeakier David Beckham. I'd seen Beckham play, I'd even been at Old Trafford the day that the Spice Girls came on pitch to do a club lottery draw, that most historic of historic days when Becks met Posh. But I had never heard him speak. He can be classified in a variety of conventional ways: as a fine footballer with a marked resemblance to the 1970s Leeds winger Peter Lorimer, as a charming clothes-horse in the same box as Rupert Everett, as a former Adonis who goes on the same shelf as, say, Alain Delon? Or Terence Stamp? As a man whose fame eclipsed his wife's – like Roger Moore? But when I did eventually hear him speak – well, he'll always be a wrong voice, a treble in a tenor's body. That for me is his defining characteristic. He is to be filed, ultimately, alongside Alan Ball and Emlyn Hughes in the Sportsmen Posing As Castrati category.

Two more instances from football after which I promise not to mention the game again, sorry – not to mention the *beautiful* game again, not to mention the *national* game again – if it is the national game how come we always lose? It would, of course, be discomfiting to win. Losing is what we are familiar with; we are thus reassured by loss as we are reassured by football's always being described in the same stunted vocabulary, with the same conventionalised thoughts – if thoughts is the word which it isn't: *criminal defending, great role model for any youngsters watching, there were twenty-one heroes out there but only one legend.* These brain-dead boobies are paid to maul language. But there is nothing so cosy as the third rate which we are already aware of, as the mediocre which we know so well that it's the sett we huddle into for mental hibernation.

Now, those two instances: I saw a game on telly in which the ones wearing blue played against the ones wearing a different colour. Can't remember what colour. And I don't know the name of the teams. Equally I don't know the score. Or who the players were. What I do know is that four of the blue players had ginger hair. That's 36.5 per cent of a team in a country where the ginger-haired constitute about 3 per cent of the population: even in Scotland where it's 12 per cent it would have been remarkable. It was both visually astonishing and uplifting; it was as though surrealism had at last found a way to leave its mark on soccer. It goes without saying that the commentators didn't see it that way, they didn't remark on the gross statistical anomaly or on the almost otherworldly sight it provided. Even had they noticed it, to have done so would have been to trespass beyond the conventions of their hackneyed trade and to have looked at a game of football from a different angle, as a festival of genetically recessive pigmentation.

Second instance: again, an otherwise unremembered match, against unremembered opposition when a team whose name I do know, Sunderland, in red vertical stripes with an array of hair colours representational of the national norm, included nine players whose name was Steve. That's an 81 per cent Steve-ness. There were possibilities here. No, probabilities! And opportunity! This was nomenclatural comedy. But there was, all too predictably, a muteness on the part of robotically programmed commentators who failed to realise that our experiences are improved, enriched and made joyful by being considered outside their obvious and intended context, by being viewed tangentially with a louche gleg – that's to say, an oblique squint.

Even when broaching the most familiar subjects – *especially* when broaching the most familiar subjects – we have always to be looking to illumine them in ways that make them seem unfamiliar, fresh, in ways that make us believe we have not seen them before. This is akin to the actor's primary task. The actor must initially convince him- or herself, then the audience, that the words spoken are being spoken for the first time, that they are the issue of spontaneity rather than of honing, chamfering and perfecting in rehearsal. What acting shows us – the same goes in writing or musical composition or indeed any art – is that the only kind of spontaneity that's worthwhile is rehearsed spontaneity. What *improvised* acting shows us is that genuine spontaneity has humans grabbing at the familiar, resorting to the usual, consoling themselves with cliché.

Here are three dicta which are not quite as straightforward as they seem, they require that we shed our usual mindset.

> JEAN-FRANÇOIS REVEL There are no styles, just talents.
>
> VLADIMIR NABOKOV There is only one school of writing, the school of talent.
>
> DUKE ELLINGTON The question is not whether it's jazz music or whether it's classical music but whether it's good music.

The essayist, the novelist and the composer are saying exactly the same thing. Look beyond aspiration to achievement. Do not judge by genre but by accomplishment. Indeed we should bear in mind that anything which is any good creates its own genre. It exists in the interstices of the already extant. We are not buzzards.

If we cannot act on these counsels and suppress our instinct for what might be termed prospective classification we can at least control it and, so to speak, do more than differentiate between the edible and the non-edible, the snack and the meal, the threatening and the harmless. We are capable of reclassifying classifications. The more specificities we heap into the mix the more we bury cliché, which is off-the-peg locution signifying off-the-peg thought – it is hardly surprising that religious faiths and ideological systems should be bolstered by clichés in the form of rote-repeated catechisms, mantras, easily memorised slogans and superstitious admonitions learnt by heart. These are forms of subjugation, and so, in a lesser way, is secular cliché itself.

There is a certain irony in the fact that cliché and idle classification can be deflected by what are almost set stratagems.

For instance:

1) *The simile* which is both accurate, and which lends the object described an entirely new facet. Craig Raine's description of a leg of pork as *a poison bouquet*, is precise, and so visually acute that, once we are apprised of it, it is difficult to see a leg of pork otherwise.

2) *The neologism* or coinage which while not visually precise has a figurative or emotional potency: so let us call sexual congress . . . *rubbing offal.*

3) *Oxymoron.* Robert Browning wrote against classification itself using oxymoron, or something close to it.

> *One's interest's on the dangerous edge of things,*
> *The honest thief, the tender murderer,*
> *The superstitious atheist, demirep*
> *Who loves and saves her soul in new French books –*
> *We watch while these in equilibrium keep*
> *The giddy line midway: one step aside,*
> *They're classed and done with.*

Classed meaning, of course, classified – whether they are characters of primary reality or of fiction they are done with because they are stripped of nuance, thus of interest.

A few lines later in the same monologue:

> *Fool or knave?*
> *Why needs a bishop be a fool or knave*
> *When there's a thousand diamond weights between.*

4) *Invention.* When, over half a century ago, Alain Robbe-Grillet was in Paris writing *Le Voyeur*, which is set on an island, presumably Ouessant, off the Brittany coast, he was attempting with difficulty to capture a seagull's flight. He thought he'd take the opportunity of a visit to his parents who lived in the port of Brest to study that flight.

He sat in his car and watched seagulls. They disappointed him. The seagulls let him down. They simply didn't fly as he wanted them to. He reasoned, then, that since it was *his* book the gulls in it were his too and they could be made to behave any way he decreed. So he invested seagulls with properties they don't possess but which he convinced the reader they do possess – so the reader detects those properties when he or she next watches a seagull flying. So a real seagull seems to mimic an imagined seagull. Seagulls may not be transformed by art but our perception of them is. And if seagulls, why not the physical world?

The very word cliché originally signified the metal plate which a printer used to transfer a woodblock to paper, to transfer it repetitively, over and over again. Cliché is thus as much visual as it is verbal. Visual cliché is all around us. Product design – that mobile, this door handle; the colours of cars, the cut of clothes, of course, advertising's idioms, typefaces, spectacles, buildings – most especially buildings. There is no area of creative endeavour which is as clichéd, as enslaved to fashion, as prone to plagiarism, as the built environment.

Plagiarism derives from a Latin word for kidnapping. Architects may not be kidnappers but they have, let us say, an enthusiastically flock-like mentality. For every goat there are 10,000 profoundly dependent sheep, gratefully filching mannerisms, ripping off devices, stealing shapes, clinging parasitically to those goats, those rare goats and committing grand larceny in public – architecture is incontestably very public.

In Belgium between the two World Wars a number of suits were prosecuted for what would now be called theft of intellectual property. It must be a great relief to the sheep that this never happened elsewhere. There is a very evident lack of *cunning* at work here. They must surely realise that they can hardly fail to be had bang to rights even if they are unlikely to suffer sanctions. Perhaps being an architect means having no shame, no pride. But then the financial stakes are high. Those for most artists are low going on nonexistent. Reputation is its own reward.

Eliot famously wrote that:

8

*Immature poets imitate; mature poets steal; bad poets deface
what they take, and good poets make it into something better,
or at least something different.*

He less famously added:

*A good poet will usually borrow from authors remote in time,
or alien in language, or diverse in interest.*

In other words a good poet is one who doesn't get caught out. A good
poet is sly enough to draw on recondite sources – that's sources, plural
– which may not be recognised. Not quite the perfect crime perhaps
but, equally, one that is unlikely to be discovered.

This is Robert Frost's *Of a Winter Evening* published in 1958:

*The winter owl banked just in time to pass
And save herself from breaking window glass
And her wide wings, stretched suddenly at spread,
Caught colour from the last of evening red
In a display of underdown and quill
To glassed-in children at the window sill*

And this is Nabokov again, the first two couplets of *Pale Fire*
published in 1962:

*I was the shadow of the waxwing slain
By the false azure in the window pane
I was the smudge of ashen fluff and I
Lived on, flew on, in the reflected sky.*

The congruence between the two verses or, at least, between the
two avian images was pointed out by the American scholar Abraham
Socher. There is no copying, no plagiarism: it's more a question of
taking on the baton of inspiration. Frost's owl survives. Nabokov's
waxwing does too – but only provisionally, in some sort of afterlife
entered through the de facto mirror of a window. What can also be
discerned here is the allusion to the parallel life of Alice in *Through
the Looking-Glass*. Nabokov had been Lewis Carroll's first Russian

translator and his oeuvre is haunted by Carroll.

The point is: to draw on one source is plagiarism, to draw on two or more sources is research – and, possibly, to attempt to begin to make something new. Through exogamy, admixtures, métissage, mongrelism, alloys. Above all not through purity – which must be shunned.

Architecture astonishingly doesn't do this. Astonishingly, because every building which is not prefabricated has the potential to be a one-off. It could be unique. But architects fail to see that the self-congratulatory admission that they are Miesian or Corbusian or Hadidian or Koolhaasian is actually an admission of intellectual and aesthetic barrenness, that they are merely admitting joining up to one or other flock. It strikes me that architects are seldom told this. No one bothers. Architecture can be – and indeed is – represented in a multitude of ways.

This most public of endeavours is practised by people who inhabit a smugly hermetic milieu which is cultish. If this sounds far-fetched just consider the way that initiates of this cult describe outsiders as the *lay* public, *lay* writers and so on: it's the language of the priesthood. And like all cults its primary interest is its own interests, that is to say its survival, and in the triumph of its values – which means building. Architects, architectural critics, architectural theorists, the architectural press (which is little more than a deferential PR machine operated by sycophants whose tongue can injure a duodenum) – the entire quasi-cult is cosily conjoined by mutual dependence and by an ingrown, verruca-like jargon which derives from the more dubious end of American academe. For instance:

> *Emerging from the now-concluding work on single-surface organisations, animated form, data-scapes, and box-in-box organisations are investigations into the critical consequences of complex vector networks of movement and specularity . . .*

There is incidentally no such word as specularity: but so what, nothing wrong in devising a new word – what *is* wrong is the dishonesty of the coinage, the way it is designed to con the reader into believing that this guff is the product of scrupulous research, of laboratory endeavours, when what is being referred to are merely methods of making buildings or thinking about making buildings. This is the dangerous cant of pseudo-science – self-referential, self-important, inelegant, obfuscatingly exclusive, occluded: it attempts to elevate architecture yet makes a mockery of it.

The boorish half-wit who wrote those words is the author of several monographs on big-name American architects which are far from critical – indeed they are often undertaken in collaboration with their subjects. They belong not to the history of reasoned analysis but to the history of vanity publishing, self-advertisement and over-produced calling cards. The product is intended to validate the supposed subject's oeuvre with a gloss of objectivity that fools no one. No one save, paradoxically, those in the know, those who subscribe to the architectural cult. Any sentient human, any despised lay person, will recognise that the prose is treacly jargon with all the rigour of copywriting and that the photography is mendacious.

From very early on in its history, photography was adopted by architects as a means of idealising their buildings. As beautiful and heroic, as tokens of their ingenuity and mankind's progress, et cetera. This debased tradition evidently continues to thrive. At its core lies the imperative to show the building out of context, as a monument. To represent it as pristine, untouched by its immediate surroundings unless those surroundings are part of the composition, more usually to separate it from streetscape, from awkward neighbours, from untidiness, to persuade the spectator of its stand-alone perfection. A vast institutional lie is being told in the pages of architectural magazines the world over, in the pages of newspapers and in countless television films. It's also being told on the web – which is significant, and depressing, for it demonstrates how thoroughly the conventions of professional architectural photography, that is of architectural propaganda, have seeped into the collective, how they are imitated, how we have learned to look at buildings through those biased deceiving eyes, just as we have learned to look at cute villagescapes through the eyes of postcard photographers.

The mediation of buildings, as of anything else, can never be neutral. The mediation has to a degree suppressed its subject. As long ago as the 1930s Harry Goodhart-Rendel observed that: *The modern architectural drawing is interesting, the photograph is magnificent, the building is an unfortunate but necessary stage between the two.* Goodhart-Rendel was an architect who belonged to no school, could not be pigeonholed and is thus regarded as peripheral. He was also a writer. A rare combination of talents. One can think of any number of teachers, soldiers, lawyers, actors, scientists, priests and layabouts who have excelled as writers, and as for doctors – Schiller, Keats, Chekhov, Conan Doyle, Maugham, Celine, William Carlos Williams, Dannie Abse. But architects . . . Vanbrugh was a good playwright who became a great if not the greatest of British architects. Sergei Eisenstein began to train as an architect at his father's insistence but gave it up. Thomas Hardy, to judge from Max Gate, the house he designed for himself in Dorchester, made the right choice when he elected to abandon architecture in favour of writing. Arthur Heygate Mackmurdo, another artist who belonged to no school, also abandoned architecture in favour of writing. He composed utopian tracts, among them 'The Human Hive', advising that human society should model itself on the beehive.

The two activities seem antithetical. There may be a neuro-psychological explanation: you know, the left hemisphere is verbal, sequential, linear while the right is pictorial, spatial, three-dimensional. The atypically high incidence of dyslexia and left-handedness among architects is always cited at this point. There are equally probable cultural explanations for the peculiarly straitened paucity of writing inside the architectural cult. Habit, custom, tradition – those practices which are passed down from one generation to the next. Jargon is after all only the coded slang of a particular professional group. And this group, this cult, is further characterised by its cosmopolitan complexion. English is its lingua franca. People whose first language is Urdu or Italian or Portuguese or Tagalog learn the language of architectural discourse in architectural schools and architectural practices without realising how debased it is: that may account for its perpetuation. For them *it is English*, an English stripped of its richness and infinite capacity for mood, nuance, prosaic extravagance and poetic ordinariness.

But there's also, as I say, the matter of exclusivity – perhaps *exclusion* is more apt. Architecture, the cult, talks about architecture as though it is disconnected from all other endeavours, it treats architecture as an autonomous discipline which is an end in itself, something watertight, something which can be understood only by architects and their acolytes: any criticism from without is invalid. Now it would be acceptable to discuss opera or sawmill technology or athletics or numismatics or the refinement of lard in such a way. They can be justifiably isolated for they don't impinge on anyone outside, say, the lard community – the notoriously factional lard community. To isolate architecture is blindness and an abjuration of responsibility.

If we want to understand the physical environment we should not ask architects about it. After all if we want to understand charcuterie we don't seek the opinion of pigs. We must ignore what architects say even if we can't ignore what they do. They make the error of confusing that physical environment with what they impose on it. Wrong. What is going on around us is the product of innumerable forces – some are, of course, created by design, in both senses of that word. But they are the exceptions. Accidents – chance juxtapositions, fortuitous collisions – some happy, some not – clashes of scale and material, harmonious elisions of contrasting idioms, stylistic hostilities, municipal idiocies and corporate boasts, the whimsical expressions of individuality made by the patronisingly named *ordinary man in the street* or by Jane Doe or by Anon – these are some of the more salient determinants of our urban and suburban and extra-urban environments – they are accidents. Buildings are, of course, the major component of these environments. Some of those buildings will be the work of architects. But with the exception of those places where they have been granted the licence to do they what they yearn to do – that is, to start from zero – architects have less influence than they believe. Their works are frequently compromised by the accidents I've just mentioned. It is as though a new orchestral work cannot be heard without tango music coming from next door, a military band playing upstairs and zydeco in the back yard.

The places where those accidents don't occur are salutary. The places where architects indeed had the opportunity to start from zero. Where architecture has enjoyed the sort of primacy it believes it deserves. Where a vision has been realised. A sort of homogeneity anyway. We think of Tourny's creation beside the Garonne of which Victor Hugo said: *Prenez Versailles, mêlez-y Anvers, vous avez Bordeaux. (Take Versailles, mix with Antwerp and you get Bordeaux.)* We think of Bath's crescents and circuses, of the successive Edinburgh new towns, of the exiled Polish court's rebuilding of Nancy. Of course, we also remind ourselves that an appealing facet of Bath is its closeness to Bristol just as Nancy is close to Metz; and the Edinburgh new towns are a walk away from the Old Town and the extravagantly Victorian southern suburbs. At a higher level Ledoux's Arc-et-Senans and Le Corbusier's l'Unité d'Habitation both instruct us in what genius is. Indeed the roof of l'Unité is a transcendent work: it is as though Odysseus is beside you. In a few gestures it summons the entirety of the Mediterranean's mythic history. It is exhilarating and humbling, it occasions aesthetic bliss. It demonstrates the beatific power of great art, great architecture. Even so it is unlikely to persuade us of the probity of Le Corbusier's Parisian wheeze – knock down the city and replace it with tower blocks – which he quaintly believed were immune to aerial attack.

L'Unité is absolutely atypical. So are the other places I mention. They are the exceptions to the rule that planned towns, tied towns, new towns, garden cities, garden villages, cottage estates, communist utopias, national socialist utopias, socialist utopias, one-nation utopias, comprehensive developments and wholesale regenerations lurch between the mediocre and the disastrous *irrespective of the style adopted.* From Letchworth to Marne la Vallée, from New Lanark to Welwyn – the first provincial town in Britain, incidentally, to develop a serious smack habit – from the Aylesbury Estate in south London to Seaside, Florida, from Possilpark in north Glasgow to Celebration, Florida, from Thamesmead to La Muette, from Canberra to Port

Sunlight, from Peterlee to Poundbury, from cuteness to high modernism, from concrete to tile-hanging, from beaux arts to new urbanism. It doesn't matter what *idiom* is essayed – we all know that the flock will jump on any passing bandwagon – it is the business of attempting to create places rather than create buildings, which are just a component of place, that defeats architects. Architects – or rather architecture – cannot make places. Architects cannot devise analogues for what has developed over centuries, for the strata of collective imprints, for generation upon generation of amendments. They cannot understand the appeal of untidiness and randomness and even if they could they wouldn't know how to replicate it. The notion of planning the unplanned has been mooted since the 1960s; so too has planning to allow for the unplanned which isn't quite the same. But planning is planning even if it is intended to result in its antithesis. And we are all products of our time, even those – perhaps most especially those – who try to swim against the tide and deny current fashion. Whatever an architect designs is bound by definition to be new, of its age. Today's restoration of a medieval cathedral, today's Georgian mansion or Gothic folly, a lavishly bogus Los Angeleno Bavarian schloss – these are just as new as a synthetic modern IT hub or an IKEA modern R&D trepan. When the gorgeously named Blunden Shadbolt made houses in the 1920s from the remnants of demolished Jacobean mansions, Tudor manors and yeoman farmhouses, the bricolaged results belonged nonetheless to the Jazz Age. Borges's Pierre Menard copies *Don Quixote* word for word and in so doing composes a brand-new fiction because he is writing in the mid-1930s, not over 300 years previously. The fact that it's kidnapping doesn't mitigate its newness. Borges might equally have conceived of Pierre Menard architect of the Parthenon – but then his paradox would have been less fresh for the western world was already littered with replicas of the Parthenon.

They are the easy bit. New buildings are simple: imagination and engineering. New places are not. Indeed it seems impossible to achieve by artifice the parts with no name, to mimic the bits in between, the pavement's warts and the avenue's lesions, the physical consequences of changed uses, the waste ground, the apparently purposeless plots, the tracts without name. It shouldn't be impossible. One cause of this failure is architects' lack of empathy, their failure to cast themselves as non-architects: Yona Friedman long ago observed that architecture entirely forgets those who use its products. Another cause of failure is their bent towards aesthetic totalitarianism – a trait Pevsner approved of incidentally. There was no work he more admired than St Catherine's College, Oxford: *a perfect piece of architecture*. And it is indeed impressive in an understated way. But it is equally an example of nothing less than micro-level totalitarianism. Arne Jacobson designed not only the building but every piece of furniture and every item of cutlery that would be used in the refectory. There is no escape from the will of the god of the drawing board. At macro-level a so-called master planner will attend to the details of countless streets, closes, avenues, drop-in centres, houses, offices, bridges. The master planner is almost certainly an architect. Even though planning and architecture are contrasting disciplines. There are evidently countless differences between a suburb and, say, a shopping mall in that suburb. We are all familiar with the hubristic pomp that often results when actors direct themselves. Appointing architects to conceive places rather than just sticking to buildings is like appointing foxes to advise on chicken security, like getting Hamas to babysit a kibbutz.

The architectural ideal is to fabricate topographical perfection: the immaculate, that is unstained, conception, a creation untroubled by context, by anything so messy as life, by what is already there – hence the covert enthusiasm for gated communities, which, of course, are not communities but civilian fortresses, expressions of separateness. Hence too the enthusiasm for demolition – which does not destroy just buildings, it destroys the sentiments we attach to buildings, it destroys a little bit of us. There is often a good aesthetic case for demolition. Ricardo Bofil declared in the early 1980s that he would only enter the competition for the National Gallery extension in London on condition that he could pull down the Wilkins building and start from scratch: we might, of course, have got a building preferable to Venturi's offensively apologetic aberration, but that's hardly the point. What was at stake was Trafalgar Square – and even if Wilkins's gallery is thoroughly feeble – it is not high enough to dominate the square as it should – it is nonetheless an integral part of the square, an integral part of our conception of London. Bofil – never falsely modest; he is after all an architect – then announced that he would only work in Britain if he could build an entire city – and even petitioned the government to this end.

The human, as opposed to architectural, ideal is to revel in urbanistic richness, in layers of *im*perfection, in the flesh that is attached to the architectural skeleton. I got sick of Rome when I worked there: too much perfection, too constant a diet of masterpieces – the lumbering sod-you-ness of Basil Spence's British Embassy was peculiarly attractive. The only town in the Cotswolds that attracts me is Stroud where the tyranny of oolitic limestone is ruptured by brick and slate.

Utopia is boring. Skeletons are lifeless. Place is composed of more than an armature and good intentions. It must do what architects are disinclined to do – that is to leave space for those who frequent place, or places. I'm not referring to polluters equipped with spray cans or lorryloads of York stone cladding who are the visual equivalent of reeking burger joint operatives. And I'm not referring to space in the sense that architects and urbanists use it: private space, public space and so on – physical volumes and voids. I mean mental space. The space that a creator leaves for his spectator or reader or audience to imagine in. Place, like art, affects us through a private compact, a conspiracy of insinuation. The avoidance of explicit meaning allows the spectator to become complicit, almost to enter into the creative process: the words on a page, the buildings in a street, the marks on a canvas are – if they're any good – electric triggers, synapse prods. They speak to a combination of faculties: brain, ears, eyes, spine. The most technically accomplished, intellectually appealing and well-meant work is bereft unless it is provokes joy or grated nerves, unless it is delightful – or emotionally harrowing.

Now, the failure to make place is surely allied to the way in which place is represented – that is, how it is first seen and then shown. The likelihood, as I've said, is that the sine qua non of architectural prowess is the ability to imagine in three dimensions, to conceive of a space by means of planar, isometric, dimetric, trimetric, parallel and orthographic projections. By means of axonometrics. These tools or devices, rendered infinitely more potent by computing, have gradually become ends rather than the mere means: method and product coalesce. Virtual buildings – developer willing – mutate into real buildings. I appreciate that we should always put quotes round *real* but you know what I mean: physical constructs composed of glass, poured concrete, Corten, breeze-block, steel, brick and, in special cases, mafiosi chunks. Certain of these metamorphoses from the virtual to the real will succeed, others won't.

They will be enclosed interior spaces. They will be confining and limiting. Although we move through them and are moved through them by lifts, escalators, walkways, and although they may be very large – superstores, airports, stadia, corporate HQs – we experience them as finite entities. Our movements, too, are finite. We are within them. We are at the mercy of the architectural disposition of space and of routes predetermined by utility. Buildings regulate our behaviour. For instance: baggage reclaim, passports, customs – or is it customs, passports? – shopping mall, smart signage which is also dysfunctional signage, rank of disobliging ashtrays called taxis. We are constantly aware of a building's purpose. We are, usually, within it as a result of that purpose. We do not go to a dentist's surgery for a coffee. We do not go to a café to get our teeth fixed. We are controlled by buildings. The majority of them are curtailments. They are monolithic. Provisional gaols.

Places are, on the other hand, heterogeneous and multipartite. Liberating. Places are feasts for the spirit. Or can be, should be. They resist classification. I was standing across the street from Richard Rogers's law courts in Bordeaux the other day – you know, award-winning, sustainable, *gesticulatoire*, *un bâtiment phare*, gestural engineering, an icon, a – yes – iconic landmark that is so achingly iconic that it hurts. But despite all that . . . it is a remarkable work and is a pleasure to behold: it fulfils its responsibility to the street, to the people who pass by it and never enter it, who use it, if that's the word, as a backdrop to their daily itineraries. I noticed a man a few metres from me. He was dressed in a touchingly desperate attempt at smartness. Tie, frayed shirt, threadbare but dapper suit, polished cracked shoes. He was grasping a suitcase whose covering of Naugahyde was so worn the cardboard showed though like the skin of a dog with mange. He was staring into a hairdressers called Colette Guy. He then walked a few paces along the street and stopped in front of the window of an adjoining house which is a barristers' chambers. He put down the suitcase. He reached into his pocket and clutched something. He leant forward towards close to the window unfurtively, not a voyeur. He lifted his hand to his face. He was holding a pair of nail scissors. Using the window as a mirror he trimmed his moustache, patiently, precisely, with deliberation. He tilted his head to scrutinise himself in a manner that made me wonder if he wasn't a former soldier down on his luck. There was no vanity in his gestures. It was as though he was adjusting his moustache to conform to a remembered ordinance of length and shape. Maybe he was going to a job interview. This tableau combined with the pod-shaped courtrooms which though they suggest a distillery are improbably allusions to the insobriety of judges, and with the cowls above them which recall penitents' hoods. And with the extraordinarily graceful undulating roof. And with the light-sensitive shutters painted the handsome navy blue which is the colour of the French establishment. For me, though I guess not for the man with the scissors, it was an affirmation of the freedom that is granted by streets. This was a complex interplay. An intimate private act performed in an animated public space before the eyes of justice and before my eyes – he can hardly have been unaware of my gaping. The decor included a sleek tram on Line A, two near-identical loden coats so engrossed in conversation that though they almost brushed against our man they didn't notice him, a wobbly board advertising a restaurant's *prix fixe* menu, an area bounded by the courts' entrance, by a section of ancient city walls, and by the magistrates' school's extension. This is a city of very grand, very stately, very boastful set-pieces. Here, in contrast, there is a kind of ragged harmony which is almost picturesque.

Given the unavoidable fashionability of narrative in every discipline from psychotherapy to ethnology, and given architecture's thraldom to the fashionable, it might be reckoned surprising that narrative forms little part of architecture. But that would be to forget the crude division between the linear and the static, the sequential and the three-dimensional. Places are read serially. And to a degree they are *created* serially: some cities can be interpreted almost dendrologically. The medieval walled core, the century-by-century expansion of rings beyond the walls yet still clinging to them, the burbs which tried to escape the city and coalesce with the country. We experience them sequentially as we move through them. We create, often without realising that we are doing so, narratives of our everyday topographies – these are personal to us and mnemonically potent. And so is narrative central to writing. The shaping of memory and imagined memory, of self or the self we longed to be, of self in relation to place as much as in relation to people. We can, on the other hand, obviously have no memories of what's new, of what is nothing but freshly manufactured product, of what is erected for the profit of the omnipotent construction industry – which would sooner build than restore, which lies to the government about how many new houses are needed in Britain, which craves land for without land it would be defunct.

Nostalgia is a basic human sentiment. It literally means merely the yearning for a long-lost place we once knew. Architecture despises nostalgia while failing to acknowledge that it is dependent on it. At least, for the past century it has despised it. It seems to ignorantly confuse nostalgia with the so-called nostalgia industry. It ought to go without saying that nostalgia has nothing to do with the nostalgia industry because that industry is selling us someone else's memories, some generalised yesterday, some collective rather than an intensely private past. The triumphs of revivalism would not have been possible without some sort of nostalgic longing: it's not, of course, that Viollet-le-Duc or Ruskin or Morris owned direct memories of the Middle Ages but they were afflicted by their memories of medieval structures. *Plus ça change* and all that – today's architects are afflicted by their memories of early modern movement structures. Indeed much of today's modernism is no such thing. It's not modernism, it's neo-modernism, synthetic modernism, an expression of nostalgia for modernism. We are witnessing the modern revival just as our late Georgian and Victorian forebears witnessed the Gothic revival. The present is constantly reinventing the past. Implicit in nostalgia is a second yearning, for the self we were when we inhabited that long-lost place.

Nostalgia is central to much writing and painting: Gray, Goldsmith, Friedrich, Chateaubriand, Millet, Egg, Dickens, Tennyson, Housman, Stevenson, Hardy, Joyce, de Chirico, Fitzgerald, Borges, and Auden – who was a modernist, of a sort. A sort that is not necessarily impressed by modernism in other fields: he wrote in *Letters From Iceland*:

> *Preserve me from the shape of things to be*
> *The high grade posters at the public meeting*
> *The influence of art on industry*
> *The cinemas with perfect taste in seating.*

A priori, this seems to be directed at that hectoring tendency which made an ideology or cult of modernism, collectivised it, standardised it, laid down rules (just as André Breton did for surrealism) and captiously claimed for it a gamut of moral properties.

But it also reveals the poet's fearful indignation at a world which would yield nothing to write *about*. A world bereft of the patina of age and of correlative objects and of the uncomfortable and the awkward, where the illicit and the sordid no longer flourish because officious jobsworths have swept them away. The overlooked can only survive so long as authority is lax – when authority goes looking for the overlooked the game is up. As it is today in the Lea Valley in east London. The entirely despicable, entirely pointless 2012 Olympics – a festival of energy squandering architectural bling worthy of a vain third world dictatorship, a jobbery gravy train, a payday for the construction industry, a covetable terrorist target – will occupy a site far more valuable as it was. It was probably the most extensive *terrain vague* of any European capital city: the English word *wasteland* is pejorative and lazy. Further it more or less states that the place has no merit – so why not cover it in expressions of vanity.

It has, regrettably, proved to be an almost caricatural illustration of the chasmic gulf that exists between the needs of writers and the aspirations of architects. A writer, at least this writer – and I am hardly alone – sees entropic beauty, roads to nowhere whose gravel aggregate is that of ad hoc Second World War fighter runways, decrepit Victorian oriental pumping stations, rats, asbestos sheets piled in what for obvious reasons cannot be called pyres, supermarket trolleys in toxic canals, rotting foxes, used condoms, pitta bread with green mould, ancient chevaux de frise, newish chevaux de frise, polythene bags caught on branches and billowing like windsocks, greasy carpet tiles, countless gauges of wire – sturdy strands it takes industrial kit to cut through, wire gates in metal frames, rolls of barbed wire like magnified hair curlers in an old time northern sitcom, chicken wire, rusting grids of reinforcing wire – flaking private/keep-out signs that have been ignored since the day they were erected, goose grass, artificial hillocks of smelt, collapsing Nissen huts, huts full stop, shacks built out of doors and car panels, skeins of torn tights in milky puddles, metal stakes with pointed tops, burnt-out cars, burnt-out houses, abandoned cars, abandoned chemical drums, abandoned cooking oil drums, abandoned washing machine drums, squashed feathers, tidal mud, an embanked former railway line, fences made of horizontal planks, fences made of vertical planks, a shoe, vestigial lanes lined with may bushes, a hawser, soggy burlap sacks, ground elder, a wheelless buggy, perished underlay, buddleia, a pavement blocked by a container, cracked plastic pipes, a ceramic rheostat, a car battery warehouse constellated with CCTV cameras, a couple of scraggy horses on a patch of mud, the Germolene-pink premises of a salmon smoker, sluice gates, swarf alps, a crumpled Portakabin, a concrete block the size of a van, bricked-up windows, travellers' caravans and washing lines, a ravine filled with worn car tyres, jackdaws, herons, jays, a petrol pump pitted and crisp as an overcooked biscuit, traffic cones, oxygen cylinders, a bridge made of railway sleepers across duckweed, an oasis of scrupulously tended allotments.

That's what I see: layers of urban archaeology. It's what painters such as Carel Weight and Edward Burra would have seen, what

George Shaw and Julian Perry still see. A site of richness and multiple textures which feeds curiosity. It is obviously decaying. But decay, as anyone who has watched meat rot knows, possesses a vitality of its own. Decay supports parasitism. There is a fungus I long to find which forms a mycorrhizal relationship exclusively with the flesh of a dead horse. Such vitality is infinitely preferable to sterility and stadia.

What an architect sees, blindly and banally, is not richness and severality. But, rather, something that is crudely classified as a brownfield site, that is tantamount to being classified as having no intrinsic worth. And because of the iniquitous generalisation that accrues from classification, it is seen as no different from every other brownfield site. A brownfield site is a non-place where derivative architecture can gloriously propagate itself with impunity. A brownfield site is a job opportunity, a place where the world can be physically improved. The architectural urge doesn't acknowledge the fact that it'll all turn to dust.

(2008)

THE STRAND

THE STRAND IS, PROPERLY, Strand – without the definite article. Too bad: it will be referred to improperly. It is one of London's best streets. It is one of London's worst streets. How can that be?

The Strand, remarkably, possesses four pubs which have not been subjected to makeovers, which retain their decorative integrity. This fluke of collective survival is undoubtedly ascribable to the street's immunity to elegance, taste and so on – a result of its being almost exclusively frequented by tourists. It is one of the rare remaining areas of London that has not been touched by the domestically centripetal impetus of the last few years. Its only residents are temporary: inside hotels or outside hotels in foetid alleys where heating pipes and the issue of air-conditioning vents raise the ambient temperature.

The impression of London gained from the Strand by those inside must be that of a city as peculiarly coarse as that which greeted visitors two centuries ago. Urine-soaked blankets, human excrement, vomit everywhere – but at least the sand thrown down on the last keeps the chewing gum off your shoes. There ought to be relief from the cameos of despair and degradation in the city's businesses that exist alongside them.

I walked from the Royal Courts of Justice, where the Strand becomes Fleet Street, to South Africa House on the corner of Trafalgar Square. And then to make sure I had missed nothing I walked back again: if, as wisdom has it, one needs only forty-eight hours to take the temperature of a city, two hours on a single street might seem excessive. What if such exposure jeopardises one's long-held impression of a street? I pass along it often enough to know that the Strand is a linear inventory of the commercially banal. It is 1,300 metres long and has three branches of Pret A Manger, two branches of Boots, a Costa, a Nero, a Coffee Republic, a Clinton Cards, a Paperchase, a Topshop, a Pizza Hut, a Vodafone, a Carphone Warehouse, a Superdrug, a Ryman, a Dixon, a Whittard, a Jessops and so on. I doubt that this is an English record, regrettably. It is, rather, the same old story.

Now and again there's a tokenistic one-off shop or stall that fell through the corporate net. Not, of course, that such ownership is any guarantee of quality. The hot-dog stall outside South Africa House stank of bad meat and indiscriminacy. A tent at the corner of Villiers Street pandered to the ignorantly suggestible's taste for replica football shirts. Stanley Gibbons and the Stamp Centre opposite are the last vestiges of the Strand's extensive philatelic trade which now shares premises with dealers in signed photos. Brando £750, Rickman £75, Michael Schumacher £295, Mae West £225. Differentials that suggest that there is something amiss with autograph collectors – but we knew that anyway. The Government of Gibraltar is a one-off of a different sort. Among the propagandist titles for sale are *The Governor's Cat*, *Strong as the Rock of Gibraltar* and *Gibraltar Calling: A History of Telephone Services in Gibraltar 1886–2001*.

By the time I arrived back at the Law Courts I should have been down in the mouth. I was, rather, delighted with a walk unspoilt. I had realised that, in this instance, shops make astonishingly little difference to the city's complexion. They hardly impinge on it – they're wee zits rather than boiling Etnas of acne. And they will, in any case, soon be gone the way of the Civil Service Stores and Lipton's and Timothy White's and Augustus Barnett and Lyons Corner House. (Astonishingly, in the small maze west of the Thistle Charing Cross, a Corner House Street remains, haunted doubtless by the shades of Nippies.) The buildings that shops inhabit for their moth-life will (mostly) endure.

There is no reason why the Strand cannot be renewed. Its eastern end is one of London's glories due to a series of gentle juxtapositions. These are obviously stylistic – two varieties of baroque, Gothic and beaux arts; they are less obviously spatial. The straits of Fleet Street widen as they mutate into the Strand to accommodate St Clement Danes, a church with a bulbous behind like a Missus by Donald McGill. It's a cheeky contrast to the upright pseudo-medievalism of the Law Courts. Axially it is just misplaced so that the subtlest play of oblique angles occurs between it and the buildings beyond it. The same effect is mirrored in St Mary le Strand's relationship to its Aldwych neighbours. These two steepled 300-year-old churches are positioned, the one in front of the other, like vessels moored in a sea of vehicles that will soon be scrap. London is not the capital city of a nation of shopkeepers – that distinction goes to Paris or Rome or Madrid. Nor is it the capital city of a nation of discriminate shoppers. There is no possibility of the sweet life here. What there is, is a massive and crude vitality in the chance collisions of the years. To appreciate this costs nothing. Let's all go down the Strand and 'ave a banana.

(2004)

ST PAUL'S

CLASSICISM IS ALWAYS with us. Like the poor? No, although poverty of the imagination is frequently its paramount characteristic. Rather, it is like clunkily rhyming verse, or doggedly linear narrative, or portraits whose sole virtue is that they are 'just like' their subject, or melodies that cannot fail the old grey whistle test. It is blessed with that most vital of modern attributes, accessibility. It is the built equivalent of eezee-lisnin': chummy, undemanding and entirely familiar because the new resembles – or at least alludes to – what preceded it.

Classicism necessarily seldom breaks the mould. It is bound to custom, practice and a cosmetic adherence to perennial tradition. It is never out of fashion: there is an infinite supply of footballers eager to commission a Mississippian mansion in Knutsford or a neo-Georgian villa in Chingford, the more fibreglass columns the tastier. Such excrescences are easy to deride, but that is no reason not to deride them.

There is, however, a different, elevated level of classicism, far from pilastered kitsch, which claims that it should be taken in earnest and which tends to raise more questions than it answers about the nature of our future cities and how we should accommodate their past. Of London's major set pieces St Paul's Cathedral is the most distinguished, though not, I suspect, the most loved: it is too stately, aloof and worldly to excite the sort of affection granted to such exercises in quaintness as Big Ben and Tower Bridge. It is a marvellously protean building: classified as baroque but too temperate, northern and humanely restrained to share that name with the inebriated swirling stones of Bavaria and Sicily; a temple that wears the vestments of Rome and the Counter-Reformation yet could never be taken for anything other than Protestant; a treasure house of sublime craftsmanship to amble through and a uniquely bold landmark from countless distant vantage points.

This last attribute is persistently tiresome – no tall building can be proposed without it being challenged (by English Heritage et al.) on the grounds that Londoners possess a divinely sanctioned right to an unimpeded view of St Paul's from everywhere. Which seems a bit rich given the immediate environs of the cathedral. Take, for instance, the view from Blackfriars Bridge. It is marred not by a high building but by a bad building, a clumsy 1930s phone exchange, just south of the cathedral – to which it coweringly tugs its forelock, to no avail. This bland mediocrity had two effects. It alerted planners to the threats that St Paul's faced. And it ensured that Lord Holford's post-war reconstruction of the entirely razed area north of the cathedral should not refer stylistically to the cathedral: it should be autonomous, unengaged with its mighty neighbour. As a craven apologist for 1960s architecture, I ought to be able to make a case for Holford's now defunct Paternoster Square – a joyless taste of Lodz in London but try as I might . . .

As a habitual sceptic about the practice of building 'in keeping' and the worth of classical 'homage', I should have misgivings about what has at last replaced it, but I don't. This is a special case. Quite what St Paul's really deserves is moot: anything short of Wren's entire master plan of which the cathedral was a component will always seem paltry, but dream on. What it has now got is at least inoffensive and reasonably congenial. To create a public space in the shadow of one of the continent's supreme buildings and to do so tactfully yet without deference is no mean achievement.

The new Paternoster Square is a place of urban recreation that is more likely to be sought out than hurried through, that will be enjoyed rather than despised. The most casually visible section of the multipartite development, Juxon House at the top of Ludgate Hill, is a provocation. It gleefully ignores the past half-century's bien pensant proscriptions of pastiche and facadism, proscriptions that were articles of faith. But if pastiche and facadism are executed with accomplishment, are they really so wicked?

Juxon House takes its cue not from St Paul's but from the vicinity, which is an inventory of earlier essays in classicism as an inspirational discipline rather than as a template for banality. In the streets near St Paul's every gradation of influence is manifest, from imitation to glancing nod. The Edwardian section of the Central Criminal Court pilfers from the cathedral; its mid-twentieth-century extension, a fittingly prison-like fortress, merely speaks the same language, gutturally. To the east of the cathedral is a gigantic, vaguely classical, office block like the dullest officers' mess; to its west a merrily rambunctious chunk of high Victorian boastfulness.

All classical life is here. And there is yet more beyond Wren's freshly re-erected (and freshly over-scrubbed) Temple Bar, which has spent the past century and a quarter loitering in a park near Junction 25 of the M25. There are columns with decapitations on them, pagan and bloody, as though the sculptor Emily Young was excising her memory of St John the Baptist and Holofernes. There is a large space enclosed by buildings which just about share an idiom. They variously recall: Alexander 'Greek' Thomson, who never built in London despite having drawn the finest of the competition entries for the Albert Memorial; 1980s po-mo (isn't it a bit soon for a revival?); and the measured elemental classicism of Mussolini's new towns Latina and Sabaudia, places that owe everything to the melancholy of de Chirico, who had in turn been inspired by Turin, the city of Wren's formidable younger contemporary Filippo Juvarra.

If that is one way of closing the circle, another is to erect a monument in the style of the Monument, yet make it from a single slab of whitbed, the purest, most homogenous stone from the Coombefield quarry at Southwell on Portland, source of the cathedral's material.

(2005)

LONDON BRIDGE

LONDON BRIDGE. 'Is falling down.' Once had houses on it. Twice demolished. Sold to Lake Havasu City, Arizona. Rebuilt in the early '70s. Is where a crowd flows 'under the brown fog of a winter dawn' in *The Waste Land*.

That was eighty years ago. We don't suffer London Peculiar dawns like that since the Clean Air acts. And we don't experience crowds in the way that our forebears in the nineteenth and early twentieth centuries did. The apocalyptic dystopias dreamed up by anti-urbanists were peopled by seething masses welded into the terrible organism of the mob, which is also a standard-issue subject of expressionism.

As late as the '40s, Eliot himself was stating that 'the mob will be no less of a mob if it is well fed, well clothed and well housed': he was prescient, even thin-lipped old Anglicans get it fairly right sometimes.

But not entirely right. Mobs tend not to be as mob-handed as they were. The involuntary flock, especially, is diminished by staggered working hours, disparate work places, private transport, the ascent of exurbia. And affluence (along with polychrome film and digital reproduction) has lessened uniformity, though fashion's centripetal pull will always act as a counter to excessive severality.

I observed London Bridge one recent Saturday from a site which was new to me. The third floor of the former Guardian Royal Exchange Assurance building. It was completed in the same year as *The Waste Land*. And that's all it has in common. Anyone who crosses the bridge towards the City will see it in front of them. But few will notice it, for it contrives to be large in extent and diminutive in effect, huge but hardly looming. Its style (a sort of approximate Caroline classicism with a cupola) and its size achieve no compact: the one is essentially domestic, the other is, evidently, commercial. But the view south towards Southwark is as delightfully fresh as that a decade or so ago of Waterloo Bridge from Christopher's in Covent Garden (a former brothel rather than an insurance office). It provided a novel angle on something that one had believed entirely familiar. The Guardian Royal Exchange Assurance building is now the House of Fraser, a department store on a scale that one might find in a moderately populous provincial town. Christopher's was (and is) to be frequented for reasons other than the view. Can the same be claimed for the House of Fraser? Yes: but not perhaps in the way that the kilted lairds who presumably own it might wish.

There is, for instance, the adjacent view from the southern end of Gracechurch Street down Fish Street Hill. There could be no greater contrast with that from the House of Fraser. Here is a London conjunction which might seem happenstantial but which is both deliberate and independent of high-quality design. In the foreground, the Monument; way beyond it at the bottom of the slope St Magnus the Martyr, hemmed in. Both of these are at least partially by Wren. Between them, a lamely detailed po-mo block of the mid-'80s which entirely justifies itself with a reticent corner tower that provides a vital punctuation and closes the space between church and column.

Thus one sees these three verticals as though through a long lens that is compacting them. Also to be gazed at: Caius Gabriel Cibber's magnificent allegorical relief on the Monument's pedestal – evidence, if any were needed, that the baroque is a sculptural and architectural idiom, not a painterly one. There's something to think about while not buying anything in the House of Fraser.

(2003)

CANARY WHARF

HERE IS THE SERGEANT in *The Third Policeman*: 'If I ever want to hide,' he remarked, 'I will always go upstairs in a tree. People have no gift for looking up, they seldom examine the lofty altitudes.' This is true. A truism even, when applied in a way that Flann O'Brien did not intend. It is a commonplace that streets in Britain seldom reveal their essence at ground level. This, as we know, is due to their disfigurement by the signs, fascias, logos and brandmarks of the mostly trashy shops which have homogenised the country, a tendency we have whiningly complained of for as long as anyone can recall, but which proceeds unabated: direct action of the sort taken by José Bové, the French farmer and activist who dismantled a McDonald's restaurant shortly before it was due to open in 1999, is perhaps the only answer. But who can be bothered?

And who bothers to look up to see what we have lost? Above the level of retail's despoilations there is manifest evidence of decorative variety, of the shades of different ages, of regional peculiarities, of builders' quirks. We are apprised of this richness, but we rarely act on its presence, we rarely tilt our head back to discover what's there. The extent of our negligent indifference to the lofty altitudes struck me the other day. I was not walking in some ancient borough whose high street has been casually stripped of its beauty (Warwick, Winchester, Worcester). I was at Canary Wharf. Never having lived or worked there (save for a day in Number One Canada Square's lift shaft), its appeal to me is as a sort of tourist destination. And like many such destinations it tempers its pleasures with pedagogy. Its first lesson is paradoxical – for this is surely a place where the untrammelled free market acted with extreme licence; where the master planners might have been the globally prestigious practice of Graft, Chance and Baksheesh; where the notion – let alone the actuality – of infrastructure was blithely disacknowledged. But out of the chaos of almost quarter of a century ago an exemplary kind of locus has evolved. Now, much of the early architecture is deemed trite: but then the architecture of a generation ago has always been deemed trite, it is the present's revenge on the immediate past, children's snub to parents.

But in this instance it is not the actual style of the buildings – Gotham City Bombastic – that is of moment. Canary Wharf is visually and sensorily satisfying because the integrity of its streets is nowhere compromised by retailers' vandalistic self-advertisements. A note of enforced restraint is struck. Covenants are apparently enforced as they are in certain of London's eighteenth- and nineteenth-century estates. There is evident exercise of what used to be called 'aesthetic control' until Michael Heseltine, then Secretary of State for the Environment, quashed it – because it inhibited the regeneration of such sites as Docklands.

Bored, luminous-jacketed security operatives with important clip-boards lurk everywhere, busily detecting bombers and reminding lay punters that this is not a wholly public domain, that it's as well to behave sensibly. Those punters might guess as much from the all-pervasive order, neatness and cleanliness. Across the river, Southwark Council allows pavement rubbish to accumulate – presumably an Old Labour homage to the 1970s. Here the streets are swept as assiduously as are Paris's thoroughfares.

Canary Wharf proclaims the briefness of the history attached to it. It doesn't dissemble its status as a very new quarter in a very old city – in that city, but not of it. It refreshingly lacks most of London's defining qualities. Thus: it is not piecemeal; it does not change round every corner; it does not fight shy of the monumental; it is served by the two parts of the pitiful transport network that actually work; it possesses numerous usable communal spaces; and, as I say, it is well policed. Of course, it might be anywhere.

This dislocation is precisely what makes it so satisfying: it might be anywhere.

One might then speak of its deracination, but that would be inapt. It has roots all over the world. Just about the only example it doesn't –

draw on is the megalopolis surrounding it. It is in contrast to old London's chaos and increasing charmlessness. It feels safe; at least, it does by day. Because it is contained by an udderish loop of river and several docks, it is reassuringly finite. When, as I often do, I saunter about it at weekends I am struck by the infrequency with which one hears English spoken. It seems to have become a home from home for affluent immigrants from Germany, France, Japan and Italy who expect cities to provide a doucer life than that which is the British norm. We must increasingly think of cities as multipartite hotels.

Individual attractions abound. Foster and Partners' entrances to the Jubilee Line station are indebted to their similar structures in Bilbao. There are abundant lumps of public sculpture intended to create the illusion of a city by de Chirico: they fail, of course, but nobly. A kitschy little garden of waterfalls and lights is enlivened by a nerdy-looking fellow with rucksack, anorak and vacuum flask. Mudlarks trawl among tyres and supermarket trolleys at low tide. Beyond them, on long-abandoned jetties, cormorants shirk their diving duties. As dusk approaches, the tops of the towers of infinite money adopt a festive glow that is more heartening than Regent Street's wretched lights. Just look up.

(2004)

THE NEW RIVER

ON OCCASIONAL SUMMER evenings in the 1930s the London-based members of the Order of Woodcraft Chivalry assembled near Haringey Stadium at the late-Georgian Northumberland House, a psychiatric hospital. It was here that one of their number, Norman Glaister, practised. He was a pragmatic Utopian, a pioneer of therapeutic communities, and was instrumental in the establishment of two self-help camps for young unemployed men called Grith Fyrd (which means, roughly, 'An Expedition for Peace'). One was in Hampshire, where the Order of Woodcraft Chivalry had been established as a pacific alternative to the Boy Scouts during the First World War. A number of the original log cabins still survived in the woods outside Fordingbridge the last time I looked, though – such is the way of heaven on earth – they are now surrounded by a caravan and leisure park called Sandy Balls. Woodfolk lit fires, powwowed, communed with the earth, flirted with pantheism, rambled, made music and folk-danced beside the Avon.

It was, perhaps, the promise of replication of that last activity which caused the reluctantly townee Woodfolk to meet at Northumberland House. Its grounds were bounded by the New River, on whose right bank, sure enough, they folk-danced as night fell. Some, no doubt, wore the green shirt of the related Kibbo Kift. The scene must have been charming and absurd. In Hampshire meadows the only witnesses to our terpsichoreans would have been bemused moo-cows. In London all the eyes of south Haringey must have been upon them. For, on the other side of the New River (which is almost four centuries old), there are modest Edwardian houses which overlook the two generations of post-war LCC estate – 'Festival' and Corbusian – which now stand on the site of Northumberland House. This is the least appropriate reach of the New River beside which the sensitive folk/morris/molly dancer can ply his aberrant hobby. It is atypical of this marvel of hydraulic engineering because it is exposed and public.

Further, this section is unusual in its failures to dissemble its fabrication and disguise the fact that it is a canal. For most of its original course from Great Amwell in Hertfordshire to near Sadler's Wells in Islington the New River, the first exercise in the supply of fresh water to London, is hidden – a secret river. It seldom reveals itself. And for such a baldly utilitarian enterprise it is astonishingly pretty, a meandering artifice which retains much of the bucolicism it possessed before its environs were built up with railways and factories, terraces and villas.

Its existence probably incited the speculative developers of those domestic types to buy land alongside it. So as well as a mere feeder of reservoirs it may be regarded as a series of spatially straitened but aesthetically exuberant parks, as an adornment, as an inspiration to those who built along its banks. The notion of severality of function is, regrettably, one that was gradually abandoned throughout the twentieth century because it did not accord with modernism's puritanical espousal of 'honesty'. The New River's decorative possibilities have long been appreciated and fostered in a way that those of most canals and conduits have not: the contrast with, for instance, the River Lea and its navigations to the east is marked. One might even suppose that it was the New River's tradition of engagingly cosmetic beautification which prompted in the 1950s the creation of a protracted, narrow public park in Canonbury which is stylistically at variance with the austere municipal practice of that decade. This park runs along the eastern side of a no longer functioning section of the river. On the other side are the private and partially hedged gardens of substantial villas built more than a century earlier towards the end of the classical hegemony.

The park's planting is abundant: wattles, bamboos, weeping willows, weeping ashes. There are thrilling false rocks created from concrete. These are commonplace in French nineteenth-century parks (Buttes-Chaumont and le Bois de Vincennes in Paris, le Jardin Lecoq in Clermont-Ferrand, le Rocher des Doms in Avignon, etc.). But with the notable exception of the Mappin Terraces in London Zoo they are rare in Britain save in grottoes. The northern segment of this park is further embellished by the finest display of duckweed (aka duck meat) I have seen in a lifetime's loitering round millponds and foetid water. The river, which is at most three metres wide, is an unvarying viridian path, as shiny as vitrolite. The illusion of solidity is total. It is liable to deceive dogs and small children – to the delight of the species of super-pike which wait below.

It is not with our eyes but with our imagination that we see what it was like when it was freshly dug. The New River's course has changed down the centuries. London has tried to swallow it. And has failed. The scapes it has caused to be created are evidently within London but not of it. The southernmost point of the river as water supply is today a place of the most enthralling oddness. One of the brimming reservoirs between Stoke Newington and Finsbury Park is now used for dinghy sailing. Boathouses have recently been constructed. Beyond the distant shore the gable-tops of dwarfed houses peep over the meniscus. Everyone who takes to the water is Eastern European. There are constant grunts and intermittent screams from a stripped classical former waterworks building: these are the sounds of karate.

Obese crows practise their football rattle imitation. And over all this looms the Castle, a high-Victorian folly which deserves to be sectioned: there can be no greater praise. Of course, it was not necessary to build a pumping station in the form of a Crusader fortress.

(2004)

CRYSTAL PALACE

THE IMPORTANTLY TITLED Theory of the Value of Ruins states that
in order to endure in human memory a civilisation must build great
monuments to itself. For it is only great monuments that possess the
capacity to become great ruins, and that it is by the grandeur of its
ruins that a civilisation will be remembered. Wrong.

And wrong not merely because this theory was proposed by Albert
Speer. It is safe to contend that the civilisation – hardly the word in
the circumstances – of which he was an engineer is not primarily
remembered by its ubiquitous monuments, whether they be ruinous
or intact. But is any civilisation? Speer's notion is very likely nothing
more than an expression of the hubristic present demanding the
future's fealty, an incitement to unborn generations to gape in awe. It
is an instance of hyperbolically distended wishfulness. Here, writ large
in broken stones, is a manifest of the 'place in history' preoccupation
so peculiar to the despotic, the vain, the arrogant, the absolutely
powerful. One thinks of the despicable Blair.

When the majority of humankind which has no desire to exert
authority over its fellows surveys ruins, it does so with blithe
insubordination to tyrannical will. It gazes on the chance collisions
of carving and nature, the fortuitous beauty of amputation, the glory
of decay, on decrepitude's pattern-making, stunted marvels, entropy's
sublimity. Small wonder, then, that one of the more common forms
of folly created during the golden age of such caprices was the sham
ruin, devastation's mimic. We should deduce from that fashion that
the appeal of the genuine ruin is not determined by its evocation of
the regime or society or culture that it once, supposedly, represented,
but by its shape, by its very appearance, its very essence. A building
which remains whole (no matter how restored, no matter how bogus
that wholeness may be) is susceptible to being read as a token of the

combination of circumstances that made it. Neglect and destruction strip a building of the meanings it bore when it was intact. Ruination, like death, is a great leveller. The ruin is a void, a vessel.

What is probably London's largest ruin gives its name to the area where it stood: Crystal Palace. The glass and the metal of which this supreme feat of Victorian engineering was built have disappeared. My father, a young man in digs nearby at Streatham Common, watched it burn on the night of 30 November 1936. But, more, he heard it burn. The groans, explosions, whistles, rustles and shrieks made it sound like a huge animal being sacrificed by pyromaniac ritualists. What remains today is not part of the structure but the protracted stone plinth on which it was raised above the terraced gardens and sloping park. This plinth is arcaded and intermittently punctuated by broad steps which lead to scrappy waste ground, to one of the area's three transmitter masts, to a bus park. It is shorn of purpose.

One obviously knows why it was built, yet its lack of context invests it with mystery and melancholy. I find it immensely and inchoately moving: a sort of memento mori, I guess, a sign of everything's finiteness and arbitrariness. It entirely fails, however, as an evocation of its era. There seems to exist nothing to trigger a leap of the temporal imagination to the Great Exhibition – that exuberant riot of colour and vulgarity which appalled such high-minded puritans as William Morris – nor to the crowds who flocked to the increasingly tawdry entertainments which were staged after it was removed to this site from Hyde Park.

They were transported here by two competing railways which occasioned the swift creation of a series of suburbs where there were, previously, not even hamlets. One of the railways, the High Level, disappeared in the 1950s: the site of its terminus was for years occupied by prefabs which gave way, eventually, to charmless brick hutches. So the main attractions have vanished. But their adjuncts remain and they are sure keys, they unlock an age. The suburban streets and avenues near by comprise a peerless high Victorian time capsule. And they are all the more telling for their eccentricities, their failure to conform to the retrospective taxonomies of architectural history.

With the exception of J. L. Pearson – who designed the magnificent, irately redbrick St John the Evangelist at the junction of Sylvan Hill and Auckland Road – no nationally reputed architect worked in the area. C. J. C. Pawley, the author of the arty St James's Court in Victoria, designed a dozen sheerly extravagant, secular gothic villas among the pines and rhododendrons of Auckland Road. The even more obscure Sextus Dyball, a speculative builder who was his own designer and quite insouciant of fashion, constellated the wooded heights with some of London's strangest fancies. Rockmount in Church Road, St Valery on Beulah Hill and a group on Belvedere Road possess the confident, not to say boorish, smugness of the *haut bourgeois* houses in Parisian suburbs such as Neuilly and Le Vesinet. Yet another non-architect designed the Swedenborgian church (now flats) in Waldegrave Road. In its form it is run of the mill. Its material, however, is bizarre. A kind of pink concrete which has proved to be climatically pervious. It is pitted like the interior of a Crunchie bar.

In the southernmost corner of Crystal Palace Park are thirty life-size models of prehistoric monsters, installed in 1854: the first dinosaur theme park, a pedagogic entertainment. I have been a loyal visitor to them since I was in my teens. I have marvelled at them, filmed them, persuaded visitors to this city that the trip to its deepest recesses is worthwhile. No more. Bromley council has subjected these hallucinatory prodigies to a most brutal restoration. I'd have rather seen them ruinous.

(2004)

BRIGHTON

WE USED TO BE TAUGHT that cities were founded on a particular
industry: steel, fish-smoking, boots, ale, learning, railways, pottery,
garrisons, market gardening, small arms, worship, shipbuilding, etc.
Helpfully educative illustrated maps would show trawlers alongside
Lowestoft, maltings at Burton, cotton mills in Manchester, refineries
on the Solent. Brighton was always omitted. The illustrations would
have been unsuitable for tender eyes since they would have depicted
a seedy private detective photographing a couple in flagrante delicto
or a raffish bar peopled by ginny former Gaiety Girls, remittance men
and bogus majors or a louche 'general dealer' charming an elderly
widow while stitching her up.

It is, of course, improbable that Brighton was ever actually more
wicked than any other English city. That, however, is beside the point. No
other English city has so rich and potent a mythology of transgression
attached to it. Such a mythology – or, according to taste, such a font of
clichés – may well be at odds with the actuality we witness, but it will
nonetheless still captivate us, still persuade us to seek corroborative
evidence. Brighton is a tainted ideal as much as a physical entity. The
ideal is most obviously associable with fictions of the mid-twentieth
century, with Graham Greene, Peter Cheyney, Patrick Hamilton,
Robert Hamer. The latter two were magisterially self-destructive
artists, puff'd and reckless libertines who trod the primrose path and
whose Brightons are foetid, corrupt, overblown, sinister. Brighton
might be a mere vessel into which the polymorphously perverse can
tip their orchidaceous poison. A regrettable fate? Maybe. But it's an
appropriately self-fulfilling one because, despite the protestations of
watch committees and rotarians and prim aldermen down the years,
this is precisely what Brighton was made for.

No other English city can boast that the particular industry on which it was founded was the provision of untempered sybaritism, illicit pleasure, princely licentiousness. All of which come at a price: the Prince Regent, never a quiet dresser, became an ample dresser – 'an old fat Mandarin'. His distended breeches, preserved in the museum, are an XXXXXXL monument to excess. And his pleasure dome, the Royal Pavilion, one of the prodigies of all English architecture, was for more than a century after its creation regarded with sniffy contempt. The measure of that contempt, both moral and aesthetic, is indicated by its lack of progeny. In Brighton it spawned the Sassoon Mausoleum in Kemp Town, now a bar, and a small house hidden away off Western Road. And in north London, on the edge of St John's Wood, there is a villa built for a Victorian magician which also took its cue from the Regent's marvellous toy: but that's showbiz.

The Pavilion's exterior, for all John Nash's felicities and caprices and coups de théâtre, forms a stylistic cul-de-sac. It belongs to the end of a tradition. It is the last and greatest of eighteenth-century follies even though it was not finished till the third decade of the nineteenth, when it was already out of fashion. Given the Regent's profligacy, it would have been derided no matter what it looked like. Its critics were not to know that George IV, as the Regent had become by the time the Pavilion was completed, would be the last British monarch to indulge in adventurous and enlightened architectural patronage. Taste was changing. England was on the point of suffering a sort of evangelical rearmament.

An expression of frivolity, even if executed in absolute earnest, could hardly be expected to excite sympathy even if its fantastical interior did anticipate the high Victorian appetite for accumulated clutter and energetic maximalism: more is more. The Regency's most celebrated confection is, quirkily, the least 'Regency' of buildings in a city which has a greater stock of such buildings than anywhere else. They suit the seaside. Indeed, they condition our expectation of the seaside. Cream stucco, bow fronts, fanciful capitals, delicate ironwork: the formula is simple, yet the responses it invites are several and contradictory.

How can Brighton be at once stately and blowsy, buttoned up and abandoned, douce and savage? It possesses all these qualities. And this is because here we blithely impose them as we would be disinclined to at, say, Cheltenham or Clifton, which are contemporary and architecturally kindred, but unburdened by the weight of reputation. As we would be disinclined to at Victorian Hove, whose collision with westernmost Brighton is that of a thin-lipped bishop with a buxom actress.

The shift from stucco to red brick is not simply a matter of fashion. It also signals, in the space of a few streets, the usurpation of that architecture which best represented the guiltless pursuit of happiness and sensual gratification with an architecture that stands for solemnity and moralising and priggishness – which stands, anyway, for those qualities in public: everybody knows that our Victorian forebears were, in private, laudanum-addicted child-beaters.

Now, fashions in building design are as ephemeral as fashions in thought or polity. What, astonishingly, has not proved ephemeral is the conviction that architecture must be plodding, earnest, 'serious', morally rooted. Brighton has suffered as much as anywhere from the puritanically wrong idea that buildings should 'work', that they should primarily be functional. Successive generations of Brightonian aesthetic midgets have spent more than a century and a half not learning from the Pavilion's geometric delinquency and from the Regency's propensity for legerdemain that building can and should be theatrical, exhibitionistic, illusionistic, lighthearted, camp. Build thus and the gap between the physical and the ideal is closed.

(2005)

TUNBRIDGE WELLS

TUNBRIDGE WELLS IS, of course, the Mecca of Middle England's Fundamentalist Tendency. Its villas are exclusively inhabited by ranting, irked bigots who have failed to understand that irony is the true weapon of the impotent. Instead they employ spleen. They fulminate against: humans, animals, homosexual marriage, heterosexual cohabitation, long hair, shaven hair, pedestrianisation, cars. These people's world-view was – it goes without saying – formed in the 1930s in a Jaipur officers' mess or a Murree rest house.

The Tunbridge Wells of enduring caricature is, then, a place of centenarians suffering vertiginous blood pressure with a sprinkling of a few chokingly dyspeptic upstart nonagenarians, all of them nostalgic for hanging and watch committees. To say that this Tunbridge Wells does not exist – never has existed – is to underestimate the compact between the public and the writers of 'comedy'.

Few things are less funny than 'comedy' (telly passim). But that lack doesn't stem its grim tide of right-on targets and received ideas and aunt sallies the size of a house. So the Tunbridge Wells of a thousand complacent knee-jerks has been forced into being by sheer repetition. It is to be found in the collective mind of the easily amused.

Its correspondence to the singular Wealden town that shares its name is slight. Singular, that is, in its physical appearance. The disposition of its buildings and the delightful peculiarity of certain of its set pieces are, I believe, sui generis. That cannot be said of its provision of the essential emporia of this century – nail-extension parlours, scented candle boutiques, Pilates gyms, software solution labs. Nor of the essential personae: a multiply pierced, hooded populace which asserts national rather than civic identity by pulling off the skilled task of simultaneously eating, drinking, lurching and yelling in the street. So far so depressingly normal. But the backdrop against which the estuarial oaths and expertly projected burps are to be heard is one of southern England's most undervalued pleasures. And, perhaps, most unexpected – for certain swathes of the town (an epithet whose validity is moot) may easily be mistaken for northern England and specifically for West Yorkshire where, as all comedians know, flat-capped whippets dine off tins of Thora Hird Mix. Where, also, dark-hued sandstones are as abundant as they are rare in the South. The opulent melancholy of nineteenth-century suburbs constructed from such materials is best conveyed in Atkinson Grimshaw's work (and in that of the painters who flatteringly forged him during his lifetime).

That painter – a one-trick pony, maybe, but what a trick! – would have felt on familiar territory here on the very border of Kent and East Sussex. Just outside Tunbridge Wells are the High Rocks, sandstone outcrops entirely atypical of southern England and so much an oddity that they have been turned into the centre of a sort of resort with a 300-metre-long car park and a gabled, bargeboarded, inevitably sandstone pub of circa 1850 to cater to punters, ramblers and climbers.

Until about twenty years thitherto, this stone, although evidently locally available, appears seldom to have been used at Tunbridge Wells. So the point of the place's most extensive and most inspired expansion is marked by geology as well as by style. That style belongs to the most overlooked period of the nineteenth century: its second quarter. Classicism's survival at another spa, Cheltenham, may be explained by a desire for continuation and homogeneity. At Tunbridge Wells – whose inchoately dispersed buildings before 1830 were variously wooden, tile-hung, brick – the impulse seems to have been to achieve grandiosity of a novel kind. Because Decimus Burton had worked as Nash's assistant at Regent's Park, the assumption is conventionally made that he was essaying a further exercise in rus in urbe. But on this occasion there was no city to which to attach rustic artifice.

Burton's canvas was precipitous open land sparsely dotted with agricultural cottages and the few buildings around the original wells which had been a watering place for two centuries. The template he established at Calverley Park, which he himself adhered to at nearby Camden Park, and which was followed a mile away at Hungershall

Park and Nevill Park, was one of great originality and whimsical accomplishment. And the full extent of its example has perhaps never been measured because its imitators have generally been reckoned – not without cause – to fall beyond the architectural pale.

These developments, far from re-creating the country in town, do just the opposite. They face rough fields which are not even planted like parkland. Yet they are unprecedentedly close by each other, each on a plot which is exaggeratedly diminutive for the size of the dwelling. The view to Nevill Park from Hungershall Park is bewilderingly strange. One gapes across an expanse going on half a mile – fields, stream, cattle – to a row of some dozen villas lined up beside each other on the crest of a hill. To say that they appear displaced is to nowhere near express the puzzlement they cause, stranded and urban, like a man in a City suit standing to attention on a midden's peak.

The nature of the successors to these prodigious roads is hinted at by their entrance. At Calverley Park there is a more or less monumental arch. At Nevill Park there is a lodge. Both devices proclaim the separateness of that which lies beyond them. Continue through Nevill Park, or, for that matter turn down any one of a dozen rhododendron, pine and birch curtained roads in Tunbridge Wells, and you arrive at that ne plus ultra of Home Counties life, the private estate of thumping great houses of no architectural distinction, but with abundant space for two SUVs, one Porsche, one Boast.

(2004)

PORTLAND

THE FORMULATION 'the soft South' is lazy. It is also wrong. Its persistence suggests that we are as likely to meekly accept a received idea as to trust to our eyes and our senses. We see what we are conditioned to see rather than what is before us or what is framed in the car window when we are heading to those sites which we are enjoined to see, sites mediated by purveyors of postcards. We are liable to be blind to the bits in between. And in the South, certain of those bits are quite as hard, quite as bleak, as anything up North.

This is not, incidentally, a southerner's boast born of a closet striving to underdress more thoroughly than climatically impervious Novocastrians: that would be the vainest of endeavours. It is, rather, to suggest that this particular North–South divide is as fragile, inconclusive and qualified as the rest. There's a list of places as far south as south of the Thames which feel – and sometimes look – as though they have migrated from north of the Humber: Chatham, Gillingham, Sheerness, Dungeness, Shoreham, Portsmouth, Southampton, Bridgwater, Plymouth, Redruth, Camborne. What they possess in common, apart from their illusory dislocation, are inter alia: docks, cranes, tars, locks, heavy industrial plant, gaunt maritime buildings, a licence to distil gin (So'ton and Plymouth only), ozonic breeze, prisons, estuarine reek, gulls built by Supermarine, piercing light. What they lack in common are such relieving features as contorted waterside trees and sandy beaches.

That place-name list omits the most startling of southern hard cases. Portland is a bulky chunk of geological, social, topographical and demographic weirdness. It is the obverse of a beauty spot. 'Beauty' in that construction implies the picturesque. Portland is gloriously bereft of this quality. It is awesome. There's nothing pretty about it. Mainland Dorset itself is not all softness, not all sweetness. To the east,

Purbeck's austere grandeur verges on bleakness and is quite apart from the douce meadows of the Stour and the Frome, from the high-hedged patchwork of Geoffrey Household's Marchwood Vale. To the west: the Chesil Beach, known till lately by a northern English appellation, the Chesil Bank. The longest pebble beach in Europe, a phenomenal place, so steep that when seen from the sea it might be a great vertical wall of polished but unknapped flints. And when we gape down from the top of that wall at people beside the pouncing breakers our ocular frame shuts out the sky, amends planes and swooningly configures the sea higher than the land so that we suffer a sort of vertigo. And an intimation of deafness: the pebbles scrape against each other in eternal attrition, groaning, bellowing, like a giantess giving birth. It hurts like a loop of musique concrete that you can't suppress. It is the Chesil Bank's narrow link with it that renders the Isle of Portland an isthmus. Just. No isthmus has ever made a more convincing show of being an island. The causeway to Portland from Weymouth's palmy, pantiled suburbs crosses the lagoon called the Fleet where oyster wars are forever raging.

Thereafter the causeway is protected against the prevailing sou'wester by the mighty pebblescape. To the east is the distant chalk of Osmington where I ate whale for the one and only time in my life and, closer by, the sullen Royal Naval installations. Like so much martial kit, it is a challenge to nature. There is a lot of camouflage that proclaims that it is camouflage – perhaps not the point of camouflage. There is a lot of bigness that challenges the brute scale of nature, but fails. When the causeway ends and it touches the oolitic rock there is Chesil, aka Chesilton. Then the road snakes up the cliff to the artless plateau – an unselfconscious work of serendipitous art. Up and on the road goes. Past the car parks where you can 'read' the *Sport* and practise not moving your mouth to *The Sun* while your cars block the vista beyond to Lyme and Start Point. And then to Portland Heights. This is when I begin to feel transported.

Portland Heights is a piece of middle-American kitsch, a motel of the first generation of British motels, but in a place where no one is going anywhere. Portland is a cul-de-sac. Maybe that's why I am so drawn to it. I come here often. And the times I come here do not elide. The first time I saw Portland I was eleven. But by the time my father's car had chugged to the plateau, the afternoon was pewter-skied. I feel one of my paternal obligations is to take my children to this place where stones poke out of slippery ground like carious teeth from gingivitic gums. My youngest daughter, Coral, looked at me when I parked in a quarry and said: 'Where have you brought me Dad?' Well, as I explained, I had brought her to the epicentre of stone. And I had written a novella set here. And we could climb the rocks at Portland Bill. And I could show her how the streets on Portland are oddly wide. Wrong for the size of the houses which are, of course, all of Portland stone, so where the land ends and the walls begin is undefined.

Then we set out around Portland. It turned out to be the day of Stephen Curtis's funeral. Curtis was an immensely wealthy adviser to Russian oligarchs who had died in an unexplained helicopter crash while flying between Bournemouth and his sham castle on Portland. It seemed that the entire population was lining the route – or had been enjoined to line the route. It felt like an exhibition of obsequious serfdom. There were black stretch limos and men mountains in black coats. Here was a novel and unexpected layer to Portland's weirdness: for an afternoon it had become a province of Russia. It occurred to me that Coral Meades might always think of it thus. So I took her to St George, Reforne – a church madly situated in the middle of quarries. This luminous oddity is worthy of Vanbrugh, who never built a church. But he did build in Dorset, just as he built in his beloved north-east. Here was a man who recognised the congruence.

(2005)

VANBRUGH IN DORSET

ARCHITECTURE IS NO LESS PRONE to fashion's dictates than are clothes, motors, hairstyles, political programmes and systems of thought. The entire history of the world can be as much ascribed to fashion as to want or necessity: war is a fashion, totalitarianism is a fashion, federalism is a fashion. And so was the baroque a fashion.

It was the Roman propagandist mode, the architectural idiom of the Counter-Reformation. It was thus bound not to do well in a Protestant land. Here's an early exemplar of the confusion between a kind of architecture and a kind of theology or ideology. The brief-lived English baroque was knocked on the head by Palladianism for not purely aesthetic reasons. But, of course, Palladianism did/does appeal, because it is formulaic, unerringly safe, colourless and so attractive to a sensibility which fosters watercolourists, a sensibility which has an idiotic faith in good taste.

Sir John Vanbrugh was indifferent to the precepts of good taste. He was imperiously coarse as a playwright, a technically numb Orton. He was exhilaratingly atheistic. And opportunistically proud: the church commissioners sussed the cut of his godless jib and never commissioned him, though if you look at the lower levels of Hawksmoor's churches you will see Vanbrugh. When Vanbrugh and Hawksmoor worked together they produced great art – Blenheim, Castle Howard, the City churches. When Vanbrugh worked alone, later on, he achieved sublimity: Seaton Delaval, just north of Newcastle, is an exemplar of the architecture of fear.

Eastbury, on the edge of Cranborne Chase, in Dorset, the third biggest of Vanbrugh's buildings, was out of fashion by the time it was finished – by which time the Palladian Roger Morris, who did the bridge at Wilton House, had been brought in to effect a partial makeover. He was brought in by Bubb Doddington, whose uncle had commissioned the house, or palace. But when Bubb's estate passed to Lord Temple of Stowe, he blew up Eastbury, reduced it to the very quarry above ground that Swift and Pope had accused Blenheim of being. What remains of Eastbury is a stable yard and a service wing. But what a stable yard! What a service wing!

There are few great artists whose example is not taken up one way or another. Eastbury's remnants look 'Victorian': specifically they presage railway buildings and the portals of prisons, which is only apt, for Vanbrugh spent five years in the Bastille and, so to speak, wrote about incarceration in every building. His belated influence is also manifest in nineteenth-century martial structures and fortifications – Palmerston's follies along Portsdown Hill or, more spectacularly, The Verne on Portland.

The estate adjoining Eastbury is called Chettle. The house is con-temporary with Eastbury. Thomas Archer was not of Vanbrugh's stature but he is important if only as the single architect of the English baroque who 'borrowed' (mostly from Borromini) rather than devised a native idiom.

The effect that these two houses had on east Dorset and south Wiltshire was quite something. The case of Blandford Forum is the most celebrated. A fire there in 1731 destroyed most of the town, with the exception of a house that is a masterpiece of the gauged brickwork of 1660. Blandford was rebuilt by the Bastard brothers, who had worked as masons at Chettle and who can't have failed to know Eastbury, which is only four or five miles away. The result is Borromini with straw in his ears, a homogeneous small town which manages to be stately without pompousness.

You can travel for miles throughout most of England and never set eyes on anything which has been touched by the baroque taste. But here Chettle, and mainly Eastbury, went on being copied long after the rest of the nation, unbeknown to lost Dorset, had leapt on to Palladianism's muted bandwagon.

The two most ambitious essays in the Vanbrughian mode are not the work of rustic masons but of a grandee called John Pitt, who as an MP and Surveyor General of Woods and Forests must have been aware of what was going on in the rest of the land. Pitt's own house at Encombe is horizontally emphatic and, for all its size, is dwarfed by the Purbeck hills around it. Much closer to Eastbury, he reworked a sometime hunting box of the 1640s. The entrance front is nothing special. But on the garden side he went in for a strange primitivism involving massive Tuscan columns. Pitt did not, however, have the last word. In, presumably, the second or third decade of the nineteenth century the house was rendered with stucco, lending it a cosmetically 'Regency' appearance. It's probably the only Vanbrughian building in existence which could be accused of prettiness and it makes one realise how reliant the greatest of all English architects was on the quiddity of stone.

(1994)

UNFINISHED: WOODCHESTER

AUGUSTUS JOHN LIVED the last forty years of his long and disgraceful life at a house called Fryern Court, just outside Fordingbridge in the Avon valley, which is the effective boundary of the New Forest and Cranborne Chase. I went there two or three times as a child. It wasn't an experience I much enjoyed – indeed I didn't enjoy it at all. The King of Bohemia terrified me. Which was, admittedly, pretty easy to achieve. Inchoate terrors dogged me. And here was a source of terror which was physical, old, irascible, glaring, bearded, boozy. The pile of empties outside the back door was epic. It towered over me. A vast vitreous pyramid. A monument to intoxication, to abandon, to the pursuit of an altered state – a pursuit which mystifies children. Well, certainly mystified this child. John wrote in his autobiography that 'We have become the sort of people our parents warned us against.' Had I known that then I'd have been in complete agreement: I was evidently not only a terrified child but a prissy one too.

It also mystified me that this old man should be regarded with such awe. Long before I was born John and my father had managed to lose a drinking club its licence in circumstances which were never fully disclosed to me. My father was rather bashful about whatever it was that had prompted the local beaks to act thus. Indeed he retrospectively regarded such a successful épatement of the bourgeoisie and its propriety as some sort of source of shame: at the time it had probably been a battle honour. The third miscreant in this provincial adventure was a painter called Peter Lucas who in the years before the war was one of John's courtiers. That is the only epithet to apply to a surprisingly sizable group of people who were attracted to the Hampshire, Dorset, Wiltshire border by his presence. It was never a colony in the way that St Ives became or that Hastings is today. Peter Lucas fascinated me in a way that John never did.

That is probably because I never met him. *Certainly* because I never met him. Thus I never had the opportunity to discover whether or not he reeked of cheap wine, tobacco and general adultness: terrified, prissy, and olfactorily fastidious. The child is *not* father to the man. But despite never having met him nor even having seen a photograph of him Lucas was all present in my childhood. Early in the war he was called up into the navy. His vessel was the *Hood*. The night before it left Plymouth he got drunk, again, fell down a flight of steps and broke his leg. He was still in hospital when the *Hood* went down with no survivors. This lucky man spent his recuperation at my mother's house: my father was in India. He passed his time painting. When he wasn't painting he was conducting an adulterous affair with Bella, Lady Normanton, chatelaine of Somerley, a few miles from Fryern Court, and a former ballet dancer – a risqué trade in those days. When his leg was mended he ran off with her. And that was the last my parents ever saw of him. Some years later they found in a shed a painting of his.

Lucas's omnipresence in my life was due to that painting, which hung in my father's study. Today it hangs in my house. He was a consummate technician. He possessed a gift for the sort of eerie hyper-realism that one finds in the work of Meredith Frampton and Christian Schad. Like them, I suspect, he had studied and sought to emulate the surfaces of van der Weyden and van Eyck. Which makes his thraldom to an energetic dauber like John all the more puzzling. This painting, which I stare at every day of my life, is hallucinatorily potent. It's a freeze-frame from a dream. It depicts a man turning from the painter as though to shield his face, with his right arm stretched out in a gesture which is theatrical, operatic. On the ground beside him a woman stares at the painter/spectator. She is clothed in a full-skirted white dress whose shoulder strap she is adjusting. The textures of the man's apparently velvet clothes and of the woman's crinoline are achieved with extraordinary verisimilitude. The tension between the two figures is graphic – and ambiguous. Have they been caught in the preliminary stages of flagrante delicto? Or is he rejecting her advance? What is going on?

Unfinished: Woodchester

Beyond them are a precisely realised mackerel sky and a group of baroque buildings – there's a Netherlandish brick gable which might derive from one of Wren's City churches, there's a limestone palazzo with pilasters. And then there is a near void. For the south-western corner of the canvas is merely sketched in pencil: a fountain, an arcaded facade, a section through some sort of house, a vanishing point which is all too well named.

The composition is incomplete.

The artist had more pressing concerns to attend to. Thus my earliest exposure to painting was to the unfinished. The partial was what I knew. It was the norm. Lucas's canvas was destined to be forever a work in progress. He didn't return to work on the bald corner. It turns out that he ran rather rackety south-coast clubs, taught at Folkestone Art School, painted somewhat academic seascapes. By the time I got round to tracking him down he was dead. I owe to him, I guess, not just half a century's aesthetic delight in this untitled, unknown, unfinished work but, beyond that, a more generalised appetite for that which is fragmentary. Such work – in no matter what medium – holds multiple appeals.

When I was a small child, I conflated the painting's sketched section with painting by numbers – which was fresh from America and as *dernier cri* as square watches and cars with fins. I conflated it with colouring books and, more, with de facto colouring books. Every line drawing in Wild West annuals was there to be rudely attacked with my Lakeland crayons. A line drawing was, after all, incomplete: it was an invitation, a blueprint. In those days when photographs were invariably black and white the gaudy polychrome of my cowboys in eau-de-nil trousers, mauve shirts, azure waistcoats and carmine stetsons was a consummation I zealously prosecuted with, doubtless, my no less carmine tongue leeching out of my mouth in rapt concentration. Filling in: that's what it was, though I dignified it as something grander. I was working like a proper painter who also filled in shapes that were defined by lines. One way and another we shed such delusions.

Some scruple or some intimation of parental reprisal mercifully prevented me, however tempted I might have been, from having a bash at colouring in the actual painting. We eventually learn to appreciate that the unfinished work is not an invitation to join in, to take on the project and complete it.

We learn that it is the very unfinishedness of the unfinished which defines it. We come to see that this is the essence of the work and that this quality transcends everything else: subject, movement, style, school (if applicable), technical devices and, especially, the maker's intentions. We learn to respect this thing called authorial integrity. Which is a fairly recent invention. A medieval cathedral – even a cathedral such as Amiens or Salisbury, both of which were originally constructed in a single continuous campaign – took more years to achieve than those of a single lifespan. They were the labours of generations. The most obvious effect of this was a bereavement of stylistic unity: that, anyway, is how we see it. The most obvious effect to those who lived during the period of their making was that of never witnessing completion.

From a person's initial sentience to eventual senescence the building was, so to speak, in utero, it was always a future building, a building site, unachieved promise. The experience was not of an entity but of what was yet to be called *process*. We can, perhaps, just about experience this today at Gaudí's Sagrada Familia in Barcelona which is not even a cathedral but an Expiatory Temple: you'd be forgiven for thinking that it is the modern age's worship of the false god of speed that is being expiated. Speed of transport, speed of telecommunication, speed of capital's movement – and speed of construction. From my home I've watched Lord Foster's Swiss Re Building, the Gherkin, go up at a phenomenal rate. The excitement is in the ascent. Not in the product. The Sagrada Familia is, conversely, currently in its twelfth decade of construction. Few of us will live to see its now inevitable resolution. Which is perhaps as well. For so long as it remains fragmentary it foments anticipation. And anticipation causes us to overlook the fact that much of this huge building is an exercise in sacred kitsch. There is some correspondence here with Stevenson's apothegm that it is better to travel hopefully than to arrive. Everyone, I suspect, is familiar with the sensation of turning from a searcher into a finder. It is a sensation that is dogged by anticlimax and disappointment. Nothing lives up to expectation.

Unfinished: Woodchester

A kind of romanticism attaches to the fragment which will always remain partial. Or *should* always remain partial. It simultaneously excites hope – and luxuriant melancholia, for there is a reason why wholeness has been thwarted. Why what might have been never will be. Mortality is most usually culpable. The garret poet – to amend Cyril Connolly – has succumbed to syphilis for the sake of an adverb before he can add the last stanza. A broken heart may be culpable: the lovelorn lord quits work on the bergerie that he is constructing for his paramour when she jilts him.

The Mystery of Edwin Drood and *The Last Tycoon* and Kafka's novels are oddly satisfying because they disappoint: there's some sort of paradox at work. We never know for sure what would have happened. They leave a gap for the exercise of imagination beyond that which normally forms the compact between reader and writer.

However, subsequent writers' essays in completing these fictions – and the word completing should be encrusted with multiple quote marks – are at best games, more usually they are simply impertinent forgeries. They miss the point that prose carries a writer's DNA. The same applies to composers who add movements to unfinished symphonies and – most glaringly – to architects who attempt to reproduce the original idiom of a building to make it whole. Or who attempt the idiom of a past era for a new work in its entirety.

There is no other area of creative endeavour in which humankind strives so keenly – and so vainly – against the tide of time. This isn't to claim that the results are necessarily aesthetically nul – though they mostly are – but rather to state that they deceive no one. Or should deceive no one. Take neo-Georgian design. The overwhelming majority of examples of this (mostly) insipid mode proclaim the date of their construction as surely as if they had written it – in Latin numerals, of course – in a lozenge on their facade. The earliest outings, largely the work of amateurs, are unmistakably Victorian. Officers' mess neo-Georgian by the War Office architect W. Ross is quintessentially 1930s. Quinlan Terry's villas are exterior decoration of the age of Cecil Parkinson.

Each age – witness costume drama – sees the Georgian age

differently, not least through the refraction of previous eras' copyism. The same goes for the Gothic revival which began as a stagey jest – Hawksmoor's Westminster Abbey towers; Vanbrugh's toy castles and fortified gates at Eastbury and, probably, Chatham; Strawberry Hill and its frou-frouish successors. Et cetera. But even when, from the 1830s, the Gothic was prosecuted in moral earnest and espoused archaeological correctness, it seldom belied its date. But this doesn't inhibit a sort of institutional wishfulness that revived Gothic should be taken as the real thing.

Albert Speer posited what he pompously entitled 'The Theory of the Value of Ruins'. It's not much of a theory. It merely claims that a civilisation will, after aeons, be judged by the grandeur of its ruins: Greece; Rome; the Thousand Year Reich – about which the best thing that can be said is that it was the Twelve Year Reich. Speer's own Reich Kongresshalle at Nuremberg is now a roofless parking lot for impounded and abandoned vehicles. This is improbably what that indulged war criminal was dreaming of as he evolved his theory. But, nonetheless, *despite* its grandiloquent banality and *because of* the very squalor of Nuremberg, Speer's theory is partially true, or truistic. It at least acknowledges a ruin's power . . . though those of us who do not seek to rule the world may prefer a ruin's power to be a memento mori, to be a symbol of decay, to be a token of the return to dust. The problem is that stone and brick are stubborn materials.

This is a process that we seek to arrest. Humans have always attempted to stem their own ageing process: Medea used hoar-frost, horned owl's flesh, water-snake skin, stag liver. Subsequent optimists used mandrake, atropine, powdered orchid and, famously, monkey glands. Today we've got Botox and vitamins: an entire industry devoted to propagating the lie of immortality.

Somehow – over the last two and a half centuries – a form of transgenic contamination occurred and we started to preserve the inanimate. Not only ancient buildings which were intact but ruins too. The appetite for preserving the latter may have derived from the picturesque cult of the ruin which cut out the middleman by creating ready-made ruins: Worcestershire abounds in them, just as Kent abounds in George Devey's equally sham houses which were designed to look as though they had been added on to down the centuries.

But let's stick with Worcestershire. West of the Severn is one of England's most lovely landscapes, all hop yards and pyramidal oasts and precipitous hills . . . I sound as if I'm doing its marketing. I'm not. The greatest house for miles around is Witley Court. It was begun in the early seventeenth century: Jacobean. It was enlarged, in the newly prevailing classical style, at the end of that century. It was further added to in the early nineteenth century, again with no regard for what had gone before. And then it was vastly made over in the late 1850s in a columnar idiom which was by then hopelessly out of date. But so what? It is fantastical by dint of its size alone. In the 1930s part of it was burnt out. It was abandoned. Rain, wind and vandals completed the process.

I first saw it as a child, when it was happily neglected. I used, subsequently, to beseech my father to drive me there from my grandmother's home in Evesham, the other side of the county. It displaced Eggesford House in north Devon and Wardour House in west Wiltshire which had thitherto tied for first place in my personal pantheon of ruins. Witley Court remained a lichenous, hazardous marvel till the middle of the last decade when English Heritage got its hands on it and decided that what west Worcestershire needed was a stone-cleaned ruin, a brand-new ruin, a restored ruin. The ready-made ruins of Sanderson Miller and the picturesque are monuments to rationalism beside the Witley Court of today, which stands as an anachronistic aberration betokening temporal hubris and the insolent conviction that the present knows better than the past.

Where does such restoration stop? The absurd reduction – and like all such reductions it is eventually not so absurd – is that Witley Court should be entirely rebuilt and refurnished just as it was. Just as it was when? When it was built? When it was burnt?

Woodchester Park should be grateful for its comparative indigence. To preserve the status quo is one thing. To conserve – as we know from confit of duck, preserved bird, or from confiture, jam, preserved fruit, is a step in another direction: it is a transformational step. But to *recreate* is an exponential and regrettable *leap*. The boundaries of these processes must not be trespassed upon. Woodchester is defined, as I say, by its unfinishedness.

And that is, after all, a delightful irony. If – through the medium of Benjamin Bucknall – it is Viollet-le-Duc's only building in England, then its ruinous state is a reproach to what he did at Carcassonne and at Pierrefonds. Serves him right. But serves those places well, for this is what those walls and that château might have been like had he stuck with theory's ideals rather than with . . .

(2003)

[SCRIPT #1]
JERRY BUILDING: UNHOLY RELICS OF NAZI GERMANY (1994)

**PARODY VW COMMERCIAL. LUMP OF CONCRETE DROPS FROM HEIGHT BEHIND JM —
SWASTIKA INCISION/TITLE**

Nazism did not die in the ruins of Berlin in 1945, nor on the
gallows at Nuremberg in 1946. It merely buried its uniform,
slipped into mufti, and sauntered into the post-war world.

**PRORA, RÜGEN: EXTERIORS FROM HELICOPTER, INTERIORS: ENDLESS CORRIDOR
OF OPEN DOORS. JM WALKING TOWARDS CAMERA FROM EVER-INCREASING
DISTANCE — ZOOM OUT. HE SLAMS DOORS AS HE PASSES THEM**

Greetings from Prora, which is an architectural manifestation of
Nazism's example to the future. There was no such thing as a Nazi
style of building, but there was such a thing as a Nazi size of
building, and it was vast; XXL and then some.

PRORA HELICOPTER SHOTS — LIKE AERIAL TRACK

Prora was Europe's first industrialised holiday resort, begun in
1936 and unfinished at the outbreak of war, so never inhabited
by the 20,000 Aryan gods and breeder goddesses for whom it was
intended.

JM OUTSIDE BUILDING SPRAWLED ON GROUND AS THOUGH RELAXING . . .

Its legacy is unmistakable. This is the prototype of the slummy
hutches that were erected all over Europe twenty-five years later
by the package tour industry. It was National Socialism that
showed the entire continent how to trash its coastline.

Specifically, it was the party's 'leisure' organisation,
Strength Through Joy.

**JM TO BEACH WITH BEACH BALL. RUNNING DOWN WIDE STAIR FLIGHT TOWARDS
BEACH**

Strength Through Joy was the brainchild of the national
cheerleader Robert Ley, who named a cruise ship after himself and
who hanged himself with a towel in his cell at Nuremberg. His big
maxim was: 'There are no private individuals any more.'

**ARCHIVE: COMMUNAL GAMES. PHYSICAL JERKS, CALISTHENICS, SWIMMING,
SAUNAS. HAPPY HEARTY FACES — PLEASURE, NOT A CHORE**

Ley's notion of leisure — for anyone other than himself — was
compulsory communal games, compulsory communal walks, compulsory
lectures in Germanic culture, compulsory communal gymnastics.

All this suited the tragically obedient German people with
their taste for all that is compulsory and all that is communal.

It was a golden age for the joiner-in, for the sycophant and
the sneak. In many regards the Third Reich was like a hearty
outdoorsy cadet club, a nationwide scout troop.

**JM DUMPS BEACH BALL. JM ON ROOF OF PRORA. WHAT APPEARS INITIALLY TO
BE A CRANE SHOT EXPANDS. HELICOPTER MOVES AWAY FROM JM AND ALONG THE
EXTENT OF BUILDING**

Prora is a two-mile-long barrack. It illustrates Nazism's
militarisation of civilian life.

SCRIPT I

In our retrospective abhorrence we persuade ourselves that the Third Reich existed in a void, that it was culturally autonomous. In fact Prora came out of international modernism and presages that movement's decline into soulless repetition. There was no unanimity on modernism among the top Nazis — Goebbels was sympathetic, Goering believed that small rooms weakened the race.

EINSTEIN TOWER, POTSDAM

There was however a uniquely German strain of architecture to which the entire regime was unequivocally opposed, and which it killed off.

The rampant individualism of expressionism was abhorrent to a party obsessed by aesthetic control. Erich Mendelsohn's amazing observatory at Potsdam was built for Albert Einstein who was subsequently to be denounced by the Nazis as a practitioner of Jewish science.

This is some sort of marvel. It's a sculptural oddity. It was more or less completely without precedent, and has never been matched save in the drawings of Frank Hampson for Dan Dare. This is what the future might have looked like had a tyranny not intervened.

BOHM'S ST ENGELBERT CHURCH, COLOGNE. EMPHASISE ABSTRACTION, SYMMETRY

Mendelsohn, a Jew, got out of Germany. Dominikus Böhm, the Catholic ecclesiastical expressionist, did not. He attempted to come to some sort of accommodation with the new regime, not necessarily out of sympathy with it, but out of expediency — he had to make a living.

BOHM'S ST WOLFGANG, CHURCH, REGENSBURG

He compromised himself to little avail. The audacity of his pre-1933 work is much diluted after that date.

In a totalitarian state there is, by definition, room for only a single faith: one people, one Reich, one god, one church.

JOHANNES AND WALTER KRUGER'S JOHANNESKIRCHE, FROHNAU, BERLIN. SPEAKING DUSTBIN/WHEELIE BIN FOR *MEIN KAMPF*

And that church was to be the National Socialist church, which intended to replace the altar with crossed swords and a copy of *Mein Kampf*, the worst-seller which became a best-seller after Hitler came to power, and made him a millionaire within a year.

The cross was to be replaced by the swastika. The German word for swastika is *Hakenkruis* — literally hooked cross — so every time you saw it or said it you were aware that the ubiquitous symbol of the Third Reich was a denial of the Christian cross, a perversion of it.

ST KUNIGUND CHURCH, NUREMBERG

The wholesale implementation of the National Socialist Church was never effected, but various liturgical amendments were made. For instance the marriage orders were altered towards the end of the war to allow pregnant unmarried women who had lost the father of their child in battle to obviate the stigma of bearing an illegitimate child by marrying the helmet of the dead hero.

Jerry Building

BOOKHOLZBERG (20 KM W OF BREMEN) AMPHITHEATRE AND COTTAGES > ARCHIVE: NEO-MEDIEVAL PAGEANT, MUNICH

Nazism turned nationalism into a sort of religion — thus welding together the two most belligerently pernicious tendencies of humankind. It peddled the idea of a new Germany that would reach the glorious heights of some mythic Germany of the past.

Nazism dug deep into Nordic legend and Teutonic folklore and Aryan heritage.

BOOKHOLZBERG THATCHED COTTAGES

These cottages were erected simply as the backdrop for an annual Nazi pageant representing the overthrow by doughty peasants of a despotic Archbishop of Bremen. The audience was invited to draw parallels between the spontaneous insurrection and the great National Socialist revolution.

GROLLAND, BREMEN, NORWEGIAN SIEDLUNG

This was also the décor for history in the making, for the stage management of the magnificent adventure of National Socialism.

BERLIN-FALKENSEE, HALF-TIMBERED VOLKISCH HOUSES

The word *Volkisch* signifies something more than folksy; it signifies that which grows from a particular patrimonial sod. It has connotations of tribe, breed and racial exclusivity. Volkisch building is inextricably bound in with Nazism's doctrinaire irrationalism, and its contempt for the idea of human progress. These reincarnated buildings are profoundly irrational. There was no reason in the 1930s to have built according to the precepts and plans of several hundred years before. There was no reason to have imposed the past on the present.

ARCHIVE: MONTAGE IDEALISED, BAVARIAN PEASANTS DANCING

It was even proposed that the nobly toiling neo-peasants should be obliged to wear heritage costumes to complement their heritage dwellings. Such costumes could be obtained by mail order from the Party's Homesteading magazines.

ALT REHSE (20 KM SW NEUBRANDENBURG), COTTAGES. 'GREEN'

Anti-urbanism is at best crankish, at worst a springboard to horror. Like all reactionary hot air — back to basics, traditional values, that sort of tosh — it should be fit only to be mocked.

The chief ideologue of the lethal cracker barrel philosophy called blood and soil was Richard Walther Darré, born in Argentina, educated in England at King's College School, Wimbledon, a member in the early '20s of a succession of back-to-the-land brotherhoods, and subsequently Reichminister of Agriculture and the SS's Chief of Race and Settlement. Darré's background was by no means atypical for a top Nazi — Hess, Rosenberg, and, of course, Hitler all came from outside Germany.

Germany was a magnet for malign cranks. It was a vessel into which they could pour their poison. In which they could turn wrong into right; the Holocaust's engineers did not believe they were doing wrong. They were acolytes carrying out a religious duty. If ever we need a demonstration of why we should never respect

SCRIPT I

ideology or religious dogma, Nazi Germany provides it.

The Volkisch movement politicised thatch and beams just as the Nazis' social Darwinism politicised biology and genetics. A twee thatched cottage of 1936 in England is a memento of a sweet, perhaps rather naive suburban dream. A twee thatched cottage of 1936 in Germany is a memento of Year 3 of the New Order, and, at Alt Rehse, of a monstrous inversion. This is where the 1,100 SS doctors who worked in the concentration camps were trained.

WOLFGANG KOPP, MAYOR OF ALT REHSE. MONOLOGUE IN FRONT OF COTTAGE

According to what I have been able to discover, Alt Rehse was the only 'Führer' school for the German medical profession from 1935. When I looked at lists of lectures at the school it was obvious what was being taught here: euthanasia, elimination of unworthy life forms, eugenics, and from the very beginning the 'Final Solution to the Jewish Problem' — these were the subjects that they were studying.

The idyllic setting of this village, the entertaining that took place, and the rural landscape camouflaged what was really going on right from the very beginning. Essentially, a cover story was put about that this village was a sports centre for the German Olympic rowing team.

ALT REHSE, POND

The indoctrinatory process which turned the healing art upside-down, and which turned doctors into anti-doctors, lasted forty-six weeks, during which time the barbaric idealists were comforted, indeed encouraged, by this hallucination of the old Germany, an old Germany which might be emulated by maiming, and by murdering.

SS SIEDLUNG, RAVENSBRÜCK. RUINOUS HOUSES IN WOOD

At Ravensbrück, the Death's Head Corps, the SS, drove slaves and burnt bodies by day before retiring to their *gemütlich* houses to play cheerfully with their families, as though they had just done a normal day's work — which, of course, they had.

Nothing that the supposedly decadent cities threw up has ever compared in evil to what happened in the small towns and boondocks of Germany.

ARCHIVE: FRUITCAKE CULTS/PRIMITIVE PARADES — ANIMAL MASKS, ANTLERS, ETC.

In the early years of this century there grew up a weird collection of cults and brotherhoods which combined primitivism, woodcraft, animism, nationalism and anti-Semitism. The Order of the New Templars, for instance, or the Germanen Order.

EIFEL NATIONAL PARK, ORDENSBURG VOGELSANG (10 KM N SCHLEIDEN) STATUARY AND SOLSTICE FIGURE

In the wake of the morally catastrophic defeat in World War One, they fused together in the National Socialist Party, which was certainly Nationalist but hardly Socialist, and hardly a political party. It was, rather, a neo-pagan sect, whose rites were to include a reinvented form of sun worship. This was built for summer solstice ceremonies.

Jerry Building

ARCHIVE: SKIING BY CANDLELIGHT/TORCHLIGHT. NOCTURNAL CULTISH RITES > NUREMBERG RALLY

When Rudolf Hess flew to Scotland, Ian Fleming, then an intelligence officer and yet to become the creator of a series of Hitlerian villains, suggested that the only man to interview Hess was Aleister Crowley, the black magician and sometime initiate of the Order of the Golden Dawn, whose salute the Nazis had borrowed. Fleming was thus among the first to acknowledge that the Nazi government was an occult organisation.

You have to imagine an entire nation taken over by Jim Jones or David Koresh or Luc Jouret. Imagine the United States of America in which everyone is a member of the Ku Klux Klan.

PADERBORN, WEWELSBURG CASTLE

Because it is sited at the conjunction of several of the fictions called ley lines, Heinrich Himmler believed that the Renaissance castle of Wewelsburg was the centre of the planet. In 1934 he commandeered the castle and rebuilt it.

This man, who caused such untold suffering, of course established his own concentration camp on the edge of the village in order to supply labour for these projects. The work was deemed too noble for Jews, so the camp was composed mainly of Jehovah's Witnesses. It was Jehovah's Witnesses, then, who were privileged to die personally for Himmler. It was Jehovah's Witnesses who gave their life in the glorification of the man who enslaved them.

CASTLE CRYPT

This subterranean room was both chapel and crypt. Urns containing the ashes of the SS's highest initiates were to have been placed here. And this is where those initiates met, and celebrated rites to fortify themselves in their exterminatory ministry. It's the most terrible place.

It's disquieting, sullying even, to stand where the genocidal zealot stood, empowering himself, multiplying himself through auditory trickery, and mystical bullshit.

VERDEN (30 KM SE BREMEN), SACHSENMORD & SACHSENHAIN. AVENUE: STANDING STONES, SHOOT LATE AFTERNOON

There's evidence of this neo-paganism in backwoods all over Germany. This is not a prehistoric site. Heinrich Himmler had 4,500 stones brought here to this former battlefield near Verden. They commemorate the Germanic chiefs slaughtered by Charlemagne.

In 1936 Himmler conducted a solstice ceremony here. It celebrated the birth of the sun child who was born from the ashes of Christ. It was attended by 10,000 members of his Jesuitical order: the SS. He encouraged them to drop their baptismal names and to adopt pre-Christian Nordic ones instead.

WEWELSBURG VILLAGE, BAKER'S SHOP, EXTERIOR CARVINGS, RUNES JAHR 3, ETC. OTTENS HOF

Himmler was a patron of exercises in racial pseudo-history and cosmic pseudo-science. He took particular pleasure in receiving arse-licking missives from concentration camp doctors detailing the experiments carried out at his suggestion.

SCRIPT I

In this *Gasthof* he entertained the scholars of the New Order — the parodic academics who reported to him in his neo-feudal fiefdom. This place renders agrarian kitsch sinister and corn-dollies a prop of genocide.

OTTENS HOF INTERIOR

These are some of the projects undertaken by the ersatz scientists, the Deutsches Ahnenerbe, the SS's ancestral research department. They show how far into irreason and credulous infantilism Germany had slipped.

1. They sought to prove that Jesus Christ was an Aryan — this an idea first posited by Richard Wagner, the patron composer of anti-Semitism.

2. An archeological dig in Tibet was intended to uncover the fossilised remains of a race of giants, from whom Aryans were supposedly descended.

3. A scientist, who went on to work at NASA under Wernher von Braun, complained that he'd lost two years of his career working with radar to demonstrate that the Earth is the inner core of another hollow planet.

4. Himmler himself believed in reincarnation — for Aryans only — in numerology, astrology, folk remedies, and homoeopathic pest control.

At the heart of Nazism there were hippies in uniform. Even then they were referred to as armed bohemians.

WÜRTZBURG, RESIDENCE

Architecturally, too, the Nazis looked for ancestors. Hitler was an architectural dabbler whose initial enthusiasm had been for the baroque.

ALTES MUSEUM, BERLIN
ARCHIVE: MILITARY PARADE AND FOLK DANCING WITH BOUZOUKI MUSIC

But under the influence of Alfred Rosenberg, another failed architect, he developed an obsession with ancient Greece, and with the Greek revival of the early nineteenth century. He declared that the Germans were Nordic Greeks, and that Greeks . . . were Aryans.

He had motifs from the Parthenon on his cutlery in order 'to symbolise his deep inner union with Greek antiquity'. This devotee of symbolic teaspoons was also in thrall to Imperial Rome — because the Romans had been, yes, Aryan and Nordic.

BERLIN BUILDINGS BY SCHINKEL (IN ADDITION TO ALTES MUSEUM)
SPEETH: WOMEN'S PRISON WÜRZBURG, SCHINKEL: KONIGSPLATZ, MUNICH

Furthermore, neoclassicism was also associable with the only period when German architecture had been pre-eminent in Europe. That was the period of Schinkel . . . of Speeth . . . of Von Klenze.

REGENSBURG WALHALLA

Neoclassicism is susceptible to giantism — and sheer size is something that Hitler valued over the niceties of style. It is also free of the hints of vulgarity that inform the baroque, and Hitler possessed the watercolourist's deference to middle-brow good taste.

Jerry Building

Von Klenze's Walhalla was an important precursor, not only neoclassical, not only grave, not only large, but a shrine to German heroes.

WALHALLA INTERIOR, TICKET BOOTH

Heroes such as King Heinrich . . . known on account of his love of birds as Heinrich the Fowler. This man was one of several of whom Heinrich the Foulest, Himmler, believed himself to be a reincarnation.

Promoting the Lebensborn project for the breeding of pure Germanic stock, Himmler said: 'When conception takes place in a Nordic cemetery, the child will inherit the spirit of the dead heroes who are buried there.' Lists of suitable sex cemeteries were published in the *Schwarze Korps*, the SS's periodical. The German dead, it seems, are always with us, and Nazism was nothing if not a death cult. It glorified death, from which it promised return.

HERRSCHING (40 KM SW MUNICH), TECHNICAL SCHOOL

It was fascinated by death. Its sanctioned murderers wore skulls, its high priests killed millions, then killed themselves. Death was truly its speciality, and its iconography is persistently morbid. Its neoclassical buildings are often funereal; they're like mausolea for the living. The living in this case were trainee tax inspectors.

In whatever country it was found, this kind of neoclassicism was a dim, middle-of-the-road style. And so it remains save in Germany where it is contaminated by association. It is perhaps not impertinent to amend Hannah Arendt's memorable phrase and talk of the banality of evil buildings.

BERLIN-MAIFELD SPORTSFORUM, & MUNICH, DEUTSCHES TECHNIK MUSEUM

And every building possessed a symbolic as well as a utile purpose. There was nothing complicated about this. Every building symbolised the state — National Socialism was persistently vain. It existed in order to celebrate itself and to exhort it subjects to celebrate it. The German people were force-fed a diet of swastikas, eagles, vastness. . .

VOGELSANG STATUARY & BERLIN-MAIFELD SPORTSFORUM

Over and again they were faced with exhortations in the shape of bloated statuary. It represented the perfected race that Germans were on the way to becoming, a race of imbecilic, genetically impoverished body-builders. How were they going to breed?

Nazism was, however, as imaginatively wanting as it was morally deficient. Emulation could only be conceived of in terms of mimicry. Hitler was forever planning for eternity, but eternity was going to look like 2,000 years ago.

NUREMBERG, KONGRESSHALLE, LUITPOLD ARENA, PARTY RALLY GROUNDS. ARCHIVE: RALLY

Albert Speer, the architect of the party grounds at Nuremberg and of much else besides, devised what he self-importantly pronounced to be 'the theory of the value of ruins' — which was little more than the commonplace observation that great civilisations reveal themselves to the future through the remnants of their buildings and that these remnants must thus be impressive — and very big. And to achieve an impressive and very big ruin you have to start with an impressive and very big building. What a theory!

There can be no question that Speer built big — and he certainly impressed the hundreds of impressionables who attended the week-long September rallies in a state of impressed delirium.

Like some priestly Busby Berkeley, he stage-managed mass mesmerism, mass enchantment, and the central rite of the party, in which Hitler brought the blood flag of the Nazi martyrs into contact with all the other flags, in a gesture that has been compared with that of a cattle breeder inserting a bull's penis into a cow's vagina.

CHILD'S PEDALO ON LAKE

Speer, though, has many devotees — the Speer-carriers, the keepers of the toxic flame. Chief among them is one Leon Krier, a man who genuinely believes that Speer is the greatest architect of the twentieth century. This is a piffling opinion and would be of absolutely no moment were it not for the fact that Krier is one of the chief architectural advisers to the Prince of Wales. It's useful to know where those dim populist opinions actually come from.

KONGRESSHALLE CAR POUND

Of course, Speer and his master never conceived that their great civilisation would be so swiftly reduced to ruins. Nor can they have anticipated that those ruins would be treated with anything other than veneration, like those of Imperial Rome. They dreamed of immortality. The Kongresshalle of the Thousand Year Reich is now a car pound. Not much of a posthumous insult, but it's a start.

BERCHTESGADEN. VIEWS FROM BUS UP TO EAGLE'S NEST — BUS COMMENTARY

Ladies and gentlemen, your attention please. Welcome aboard the bus on our trip to the Eagle's Nest.

Hansi and Lotte and Magda and Fritz never got to see Hitler's Eagle's Nest, the Kelsteinhaus. This was as much a secret from the majority of the people as was Eva Braun.

The road up to the Eagle's Nest is probably one of the most beautiful mountain roads in Germany, and a great feat of engineering. It was built in as little as three years, from 1936 to 1938.

The tour bus commentary by Lord Haw Haw commends the speed of the road's construction but neglects to mention the word 'slave'. The guidebook neglects similarly to mention the word genocide, but bangs on about the alleged crimes of the French regiment that liberated the place. But enough of that.

TUNNEL IN MOUNTAIN

Almost half a century after the end of Nazism, the official end, there is no unanimity about how to deal with the sites that were most holy to it. While the party grounds at Nuremberg rot gracelessly, Hitler's mountain eyrie has become a place of pilgrimage — of Adolfolatory.

Jerry Building

INTERIOR LIFT > EXTERIOR LONG LENS VIEWS OF DISTANT MOUNTAINS PILED UP ON EACH OTHER

And from here on the highest peak in Germany he could survey his domain, lose himself in visions or whatever. It's like the lair of a creature that is more than human, a thunderous Nibelung, a Teuton god of war. It's also like the tree house of a mad child who believes that he can rule the world.

BLACKNESS. TUNNELS BENEATH PLATTERHOF HOTEL. STAIRS, CORRIDORS, TORCHLIGHT

The contrary form of infantile protostructure is the burrow, the warren, the uterine comforter. The Reichs leadership was obsessed by them, and built them long before war made them actually necessary. These dark labyrinths are associable with secret societies, and with the desire of humans to take on animal form.

In war soldiers and civilians alike had no choice but to behave like beasts. They were forever forgetting to observe the chasm that separates animals from humankind. They doubtless took inspiration from Himmler's assertion that Germans were the only people in the world who know how to treat animals properly, and from their beloved Führer's dictum that 'the Aryan is closer to the animal than he is to the Jew'.

HAMBURG, HEILIGENGEISTFELD FLAKTURM. INTERIOR/STAIRS > EXTERIOR

Eventually the nation that conducted itself like a disgusting zoo was obliged to scurry and burrow and hide in the dark. This fortress in Hamburg housed up to 50,000 people bombed out in the wake of the Allied air raids. It did its job. Direct hits dislodged anti-aircraft guns, but the three-metre-thick walls and its four-metre-thick roof held firm.

The Third Reich's few original buildings, those which are aesthetically rather than historically momentous, belonged to the architecture of belligerence and desperate defence.

GUERNSEY, PLEINMONT OBSERVATION TOWER

Even more sculpturally plastic forms were employed in the occupied western territories, notably on Guernsey. Of course, these structures are utile, but they're also patently and perhaps unwittingly representational. Their imagery is unmistakable; they recall masks and visors and helmets and chainmail fists.

These concrete gun emplacements and watchtowers form a link between the expressionism that the Nazis feared and so destroyed, and the concrete brutalism of the '60s and '70s. London no doubt owes its South Bank to the Third Reich.

Because of the enormity of its crimes, and because of the extent of those crimes, and because of the moral squalor which infected it, the Third Reich is regarded as some sort of aberration, twelve years when a supposedly civilised nation went mad. It soothes us to think of it thus; it certainly soothes subsequent generations of Germans, this exculpatory myth of collective madness.

DACHAU

Visit Dachau, the 1,200-year-old artists' centre with its castle and surrounding park offering a splendid view over the country.

The concentration camp at Dachau was opened three weeks after the Nazis took power. It was the first of 1,037 such camps.

Almost every town in Germany had a concentration camp. The idea that the ordinary Hansi in the street didn't know what was going on is as much a lie as the great myth that every German saved at least one Jew. You have to ask: Why did they save them if they didn't know what was going on?

ARBEIT MACHT FREI GATE

In addition to concentration camps there were, by the end of the war, 7,000 forced labour camps. The Nazi regime was founded on slavery. Most of the buildings shown in this film were built by slaves.

JM HOLDS UP PORTRAIT OF SPEER, THEN IMAGE OF SPEER WITH CAMP PRISONERS

Here is the master builder, Albert Speer. And here he is consulting with members of his workforce.

Nazism's architecture is inseparable from the means of its construction: 32,000 people died in this place — starved, beaten, shot, experimented upon.

It was, of course, not by its architecture, not by Speer's bloated temples, not by its creepy thatched cottages, that the twelve-year Reich ensured that it would indeed last in human memory for the other 988 years.

RAILWAY TRACK VANISHING INTO DISTANT WOODS

A German railway track entering a wood will for ever mean mass death.

END

2

URBS AND BURBS

The Quiet Pleasure of Getting Lost

ANTI-URBANISM'S ANCESTRY

WE ALL KNOW the 1890s: Yellow Book, green carnation, telegraph boys, concupiscent putti, gilded caryatids, Corvo, velvet, epigrams, astrakhan, orchidaceous arcana, Uranians, occluded cults, abhorrence of nature, Octave Mirbeau, art nouveau's sinuous perversity, Rome's lure, artifice for artifice's sake. We all know that that singular decade was lived in a hothouse furnished with brothelish ottomans and that the air which our forebears breathed was never – God forfend – fresh: it was always heavy with incense and drowse-making cologne.

We know too that this litany of potent received attributes left no legacy outside itself. It had no epigoni. It led nowhere. It was, if you like, a cultural cul-de-sac. At least it was in this country. Why was this? It does not follow that 'decadence' should decay and die, any more than that a self-proclaimed two-hour dry-cleaner should have your clothes ready in that time; decadence was merely a name attached to a broad gamut of endeavours and states of mind.

These 1890s were the 1890s of the city – which was an apparatus of corruption, a post-lapsarian sink mired by capital, vanity, vice, temporal wickedness, the quashing of virtue, by sophistry and the un-English property of cleverness. Now this is, of course, not peculiar to London of the 1890s. This notion is as old as cities themselves, as old as the cities of the plain, as old as Babylon and Rome. No city exists – it doesn't matter how vital and energetic it is – without supporting the myth of the country girl who arrives in it to go into service and finds herself entrapped in sexual slavery; the London of Cleveland Street made that perennial nightmare come true for country boys too.

But no nation in modern times has reacted so strongly to these urban inevitabilities as England did. Here's Lord Rosebery in 1891:

Anti-urbanism's Ancestry

There is no thought of pride associated in my mind with the idea of London. I am always haunted by the awfulness of London: by the appalling fact of these millions cast down on the banks of this noble stream working each in their own groove and cell. Sixty years ago a great Englishman Cobbett called it a wen. If it was a wen then what is it now? A tumour, an elephantiasis – sucking into its gorged system half the life and blood and the bone of the rural districts.

This bilious howl merits scrutiny. Nature – the stream, the Thames – is noble, like the howler himself. The millions cast down beside it patently are not. Whatever nobility of soul they might have possessed when they lived on the land has been stripped from them by the cancerous city. London is both a disease and an organism that is demonised and zoomorphised as a creature which draws into its frightful maw all that is best in England, that is the fine pure stock which is to be found in the country. So who was this milord with the gall to co-opt Cobbett, a pamphleteer not known for his admiration of the aristocracy? Who was this antagonist of what was then the 'greatest city on earth'? Lord Rosebery, when he expressed his vituperative opinions, was speaking in his capacity as, of all things, Chairman of the London County Council. With friends like these . . .

Rosebery, who was subsequently to become a short-lived prime minister, had behind him as precursors a formidable tradition – not just Cobbett but William Cowper, Dickens, Engels, Ruskin, Charles Kingsley. No matter that Dickens's genius was in large measure conditional upon and succoured by the very wretchedness he sought to expunge, no matter that he wrote about social and hygienic circumstances that had all but ceased to exist – he was an artist who happened to wear the clothes of the social reformer, philanthropy, on his sleeve. But he was so formidable an artist that even as London was cleaned up and even as it suffered environmental improvements, his fictive version of it seemed to imprison the minds of those who scrutinised it, who wrote about it with something approaching apocalyptic despair.

Still, during the years that this pessimistic and artistically engendered vision of London held sway, the majority of the prescriptions thrown at the city's ills accepted that it was improvable, they accepted the inevitability of the metropolis, they acknowledged its might while simultaneously decrying its moral infirmity.

But there were exceptions. Or rather, there was a prescription which was essayed over and over again in marginally differing guises – this prescription was the land colony which had its roots in Winstanley's Diggers during the Commonwealth. But the lure of communality, self-sufficiency, sandals and vegetarianism did not take hold in a big way until industrialisation and the spread of the enclosures gave it a reactive boost in the early years of Victoria's reign, when suddenly England was constellated with seekers of Arcadia going back to the land with poor tools, ignorance, faith in a series of more or less charismatic leaders and indifferent prospects of survival in alien conditions.

Transcendentalists, Chartists, Shakers, New Levellers – they all tried it, and they all sank up to their knees in mud. This strain of anti-urban utopianism was merely a trickle, a winterbourne, throughout most of the century. And, fervidly as they might have wished it, it is improbable that any of the participants in these climactically inappropriate experiments would have dared to predict that their example would combine with another to form the environmental mainstream of twentieth-century England.

The other example is that of the aristocracy which, it goes without saying, had always maintained rural seats away from London and the cities. But it is the indigent utopians who showed the way. It is their subcultures which fomented the centrifugal impulses which have marred English urbanism for the last century. Lord Rosebery is quoted, approvingly, in the introduction to what can be seen, retrospectively, to have been one of the most damagingly influential tracts of its or any subsequent age. *Tomorrow: A Peaceful Path to Real Reform* was published a century ago next year (1898). It was reprinted four years later as *Garden Cities of To-morrow*. And none of us has escaped its baleful effects.

Its author, Ebenezer Howard, gave great forelock to Rosebery and

Ruskin – but then he would: he was neither aristocrat nor polemicist. He came from the utopian side of the regrettable alliance. As a young man he had travelled to Nebraska to found a land colony – which, of course, failed. His trick, and the source of his insipidly written book's massive appeal, was to rid utopianism of religiose, earthly paradise garb and to clothe it in mundane pragmatism, in practical, potting-shed terms designed to find favour with a nation that prizes the banalities of common sense over all other systems of thought. He didn't write a poem, he drew up a blueprint; he shunned the abstract in favour of the concrete; his sensibility was not that of a visionary but of a dogged planner; he was intellectually bereft and his prose is numbingly dull. At the dawn of Waugh's century of the common man he was just the ticket.

The hostility of the Fabian Society, which considered *To-morrow* derivative and unpalatable, was toothless – regrettably, but hardly astonishingly. The Webbs further believed that:

> *We have to make the best of our existing cities; proposals to build new ones are as about as useful as arrangements for protection against visits from Mr Wells's Martians.*

But Mr Wells himself, forever on the qui vive for new excitements, was impatient with such cautious and rational counsel:

> *The city will diffuse until it has taken considerable areas and many characteristics (the greenness, the fresh air) of what is now country . . . the country will take to itself many of the qualities of the city. The old antithesis will cease, the boundary lines will disappear.*

Right so far. But he was wrong to add that:

> *There will be horticulture and agriculture within the urban regions and urbanity without them.*

This precisely is what didn't happen. The boundary lines disappeared, certainly – but instead of the mutual exchange of properties that Wells foresaw, the new alignment proclaimed itself in a sort of locus which we are now forced by its sheer commonplaceness to take for granted but which was then new. Newer, indeed, than Howard, who was no aesthete and who was indifferent to architectural style, could have foreseen. There is a qualitative difference between the suburbs which took their cue from Letchworth and those which had preceded them.

Suburb is actually a misnomer for the twentieth-century English places that are so called. Sub-urb makes explicit the idea of dependence upon the city. Architectural history of all colours is united in its fondness for the atypical; it grants benign anomalies an importance and influence that they don't really possess. So we hear a disproportionate amount about such microcosms as Nash's Blaise Hamlet, about his Park Villages, about Darbyshire's Holly Village, about Bournville and Port Sunlight – all of which are entirely against the grain of nineteenth-century suburbia. Nineteenth-century suburbia was architecturally centripetal. It sought to join itself to the city. It used urban styles, urban planning. It acknowledged its link to the core of the city. Whether it was grand like Bayswater or humble like Bow, it looked inwards.

It is our bad luck and our cities' misfortune that the Rosebery/Howard/Wells's recipe for solving cities' problems by ignoring them and starting afresh should have coincided with the coming of age of a generation of architects whose flock-like impetus was folksy, rustic, merry Englandish. It is our further misfortune that this generation was perhaps more collectively gifted than any other which has existed in this country. Through the example of Letchworth, through the example of Hampstead Garden Suburb and New Earswick and Old Oak Common, we were persuaded that the notion of a better quality of life was indissolubly linked to neo-vernacular and pseudo-rural domestic architecture.

This had everything to do with fortuity and architectural fashion, nothing to do with some specifically utopian conception of fitness for purpose – witness the Chartist colonies of the late 1840s: their buildings were characteristic of that period. The remains of Snig's End by the Severn near Tewkesbury are brutally urban, almost utilitarian – they make no concession to where they're situated, they don't mimic in their materials or style the indigenous buildings of the Vale.

Utopia UK in its 1900 version could hardly have been more different. Now the equation of ersatz bucolicism and the common good was set, if not in stone, then in hung tiles and cosmetic beams and luddite leaded lights. I know that it's easy to sneer at this stuff, but just because it's easy is no reason not to do it. But sneering should not preclude simultaneous celebration, and awed appreciation, of the sometimes hallucinatory weirdness of, say, Queens Gardens in Ealing or the infinite green pantiles of the house built of shallow German bricks at Pinner Hill for the former champagne salesman Ribbentrop or the pagoda-ish Park Langley Garage at Eden Park in Beckenham. Voysey was copied on an industrial scale. Lutyens designed 300 houses – and 400 of them are in metropolitan Surrey. Trickle-down meant dumbing down.

Letchworth was conceived as autonomous, as self-contained, but failed in those ambitions; it pretty soon turned into a dormitory town, albeit a rather unusual one with no pubs and a strong contingent of refugees from various communes – folk who rolled naked in the morning dew, commuted to London, donned rational dress like togas when they returned. Actually this proved to be irrational: the roads were not metalled and the wet clay played absolute havoc with those garments' hems. As a social experiment it was a non-starter. But as an architectural and planning model it was peerless. And there can be no question that it was a turning point – all that followed it is infected with a douceur, a lightness, a cheap optimism which the Victorians simply had not bothered with.

These new places are frozen in flight. They are fugitives from the idea and actuality of the city, and they are disguised in the garb of where they want to be; they're wearing country threads; they are saying that they belong to the Chilterns or the Weald or the Surrey heaths. These places deny the city even though they are dependent upon it for their wealth. But the imagery of delusion is more powerful than the actuality of quotidian routine; these places owe their essence to a sort of topological faith, they engender a proud sense of belonging which is almost sectarian or tribal – no, I don't live in Purley, I live in Woodcote. They gainsay their dependence on the city so cannot properly be reckoned suburb.

Outer London is not in the European tradition; it is, rather, a precursor of Los Angeles, a conglomeration of conjoined districts which for all their incidental felicities are woefully homogeneous, dismally conformist – compare this architectural meanness with that of Brussels's suburbs, or Antwerp's, where a furious invention is manifest. This may have something to do with the fact that Belgian architects habitually sue each other for plagiarism. But it has much more to do with the fact that Belgium, like most of Europe, is costive with land. England is positively prodigal. England possesses an insatiable appetite for greenfield sites. Loss of such sites has been one price we have paid for this national bent towards rusticity – like tourists we destroy what we come to admire; unlike tourists we don't go away again.

A far greater price, however, has been the effect of this ceaseless diaspora on the cities themselves. England is unique in this continent in having the construction inner city as a synonym for deprivation, social mayhem, poverty, nonexistent infrastructure, wholesale neglect. There's a degree to which this is no longer applicable to inner London, which is culturally and demographically separate from the rest of England. It is so various, so beguilingly diverse, so perpetually mutating, so vital, so much the de facto capital of Europe that it conforms only to its own rules. One of which is that it has collectively reacted over the past twenty-five or thirty years to the outward movement of the previous seventy years. What is disparaged as gentrification, the process by which one mauve front door led to an entire street of them and thence to the transformation of one sink borough after another, this process has been wholly beneficial because it has meant that those with the wherewithal to have some say in the city are minded to do so because they actually inhabit it – it is no longer a workplace they merely squat in by day.

Now transformation has been achieved without the assistance of political intercession, save perhaps of Geoffrey Rippon's spot-listing of most of Covent Garden. Contrast that with the macro-bodging of the LDDC, whose first chairman lived two hours away in Wiltshire, a county whose small towns are afflicted with the same syndrome as the formerly great cities of the Midlands and North. There the diaspora continues. The first thing a Mancunian who makes any money does is move to Cheshire, to Knutsford or Hale or Alderley Edge. Brummies migrate to Henley-in-Arden and Hampton and Stratford. It is almost as though to live in town is an admission of worldly failure. Yet the very people who follow this conventionalised route rhapsodise about the French and Italian towns they visit on holiday and often retire to – there are currently 40,000 British expatriates in *le grand sud-ouest*; one doesn't hear about French people coming to live in Taunton and Dorchester.

These French towns, of course, enjoy an autonomy which their English analogues are denied but much more importantly they are not nocturnal deserts, they don't switch off at 1800 hours. I can't remember whether we're still stakeholders this week but I know that we're liable to develop a stake in a place only if that place has some hold of community or sentiment over us. It doesn't matter how much money is thrown at cities, it doesn't matter how many nostrums are offered by such bodies as urban development corporations and regional development agencies, until there is a grand cultural shift away from the perceived desirability of the country, until citizens with clout and flair return to live, so to speak, above the shop – until that happens, well, Liverpool is going to go on being Liverpool.

The vicious circle of ex-urban migration, which fosters inner-city dereliction which encourages further migration and so on – this motor can be stopped. Not by preposterous gestures like damming the Tees to make a recreational lake for Middlesbrough or by the tokenism of public art, but by preaching like an invert Rosebery. The country is sapping the strength of the cities and is disappearing as a consequence; the ineffable estates of executive houses are not internal cancers or wens, they are, rather, external buboes and boils. And it is the country which is now the free-for-all toxic playground that cities once were – agrichemicals, organophosphates, polluted rivers, filthy fertilisers, diseased cattle, ailing sheep, poison fowl. Add to that the aesthetic bereavement of agricultural buildings, the parochial xenophobia, the chippy animosity of the natives, the inbreeding, the two-headedness, the penchant for domestic crimes, the sententious drivel that is dignified as folk wisdom, the quasi-prairie landscape, the breeze-block piggeries, the olfactory assault of slurry and silage and oilseed rape, the fields bathed in the blood of sheep which have been shot and butchered on the spot before being shoved in hired artics for transit to bent butchers. It is not for the sake of preserving all this that we should resist plans to build four and a half million homes on fields during the next decade or so. It is for the sake of the people who are supposed to be in need of homes.

A century-old remedy which has had deleterious and divisive social effects as well as environmental ones is not the solution to the end of the nuclear family and the consequent need for domestic space for the divorced, the children fleeing home, the whole army of live-alones. Nor is it the solution for the admittedly sometimes overlapping army, the burgeoning army, that works from home. These two groups along with the elderly are the paramount demographic prospects of the near future of tomorrow. They require cities, not garden cities, not the debased imitations of garden cities which have been the old century's flawed norm. Today's peaceful path to real reform is not to follow utopians and aristocrats and the aspirant aristocrats of the bourgeoisie out of cities but to think inwards and upwards. This is not merely a matter of reclamation but of conquering air-space, of extending vertically – which demands, of course, a drastic change in our collective mindset: we are the most terrestrially inclined of European nations; regrettably the epithet has been elsewhere appropriated, but we, the English, are truly les grandes horizontales.

How can this at last change? The catalysts are not happy ones – but then nor were those which initially prompted the urban exodus which, as we've seen, became a causeless self-powered force. The catalysts for the live-alones, for the wired, webbed, the globally electronic work-alones, for the elderly and the halt are loneliness, cabin fever, the lack of recreational amenities, the absence of human intercourse. It is to counter these elemental ills that we need cities. Not as places of work or fabrication – which has, anyway, been removed – but as dense social concentrations, interpersonal exchange centres, protracted fun palaces, sites of pleasure and erotic adventure and of the very quality which is etymologically bound to cities – civility.

But what, I can hear you ask, about the greenery, what about the consoling trees, the agents of rustic delusion? Well, in greater London there are 270 streets called Oak something or other; there are 210 Elms. There are 130 Willows. And for those of a terminally English disposition, there are 114 Birches.

(1997)

ZARAGOZA

SOME OF US collect medieval churches or salmon rivers or navigable creeks or potholes or railways or 'great' restaurants. The sites I collect are cities. If I have such a thing as a hobby it is urbanistic self-perdition. It began in London, quite happenstantially: I believed at the age of seventeen in the cartographic literality of Harry Beck's diagrammatic Tube plan. Not the first wrong idea of my life, but among the more influential. Because Beck placed Westbourne Park to the north-west (rather than east-north-east) of Holland Park, I got lost, and I discovered that I liked getting lost. But the better I got to know London, the fewer opportunities it occasioned.

My problem is that I come paramartially equipped with an on-board compass. So, improbable as it may seem, I take myself to places which I don't know so that I can walk the grid and end up trekking through, say, the burnt-out cars of the South Bronx. Some hobby.

A potent incidental pleasure of wandering aimlessly is the realisation that artists who have made cities their own are often rendered more naturalistic than we might previously allow. Borges's labyrinth was simply (simply?) Buenos Aires's grid. Brussels's suburbs are so quirky that one is tempted to think of Magritte as an almost reportorial painter. Rome is still peopled by Fellini's extras.

I'm currently in Zaragoza. I have so far failed to get lost. But there's time enough yet. Idolatry is not my forte. As soon as I appointed anyone an idol when I was a boy, he died. That put stop to the habit. But I retain, let us say, an amused admiration for Luis Buñuel. Zaragoza was his home city. This is where he was, till his late adolescence, educated in a Jesuit school. One might think that his entire oeuvre was nothing more than an inspired revenge on the Society of Jesus, a heretical raspberry. Even a cursory scrutiny of the city suggests that there was much more to react against than pedagogy. The place is still an epicentre of Catholic superstition and Mariolatrous kitsch: what can it have been like in the teens of the last century?

I soon ceased counting the number of shops dealing in plaster representations of Jesus in swaddling clothes, in plastic penitents wearing the team colours of their brotherhood, in lifestyle-size Arks of the Covenant, in low-relief depositions, in saccharine Calvaries, in Mary, Mary, Mary. The Virgin came down from Heaven on a pillar a while back. The vast basilica Nuestra Señora del Pilar is devoted to her cult. Excepting its scale and its age – most of it was built 300 years ago – it is of a piece with the industrially produced twenty-first-century icons in the shops.

Yet we delude ourselves that the retablature and the baroque Lady Chapel are works of art while the plastic haloes – you pay more for those which light up – are tat. The first lesson of Zaragoza – a shriller lesson than that of Santiago de Compostela or Rome – is that kitsch is not some by-product of Catholicism but at its very core.

Buñuel, the divine blasphemer, the sacred apostate, turned ignorant kitsch into knowing camp, the way aesthetes will. Franco's government, sedulously supported by Rome, forbade him from working in his native country and proscribed his work. I suspect it rather overestimated his films' capacity to corrupt and underestimated his dependence on the very superstitions he mocked. The satirist is, after all, parasitical upon the body he attacks. The blasphemer requires a healthy clerisy. Political anti-clericism, of which Zaragoza owns a bloody history, is equally dependent, for it defines itself through its dialectical opposition.

Apart from a multiscreen cinema named after him there is no memorial to Buñuel.

(2003)

BUENOS AIRES

THIRTEEN MISERABLE HOURS out of Gatwick you disembark with a military band tuning up in your sinuses and a mouth like a lorry driver's crotch and find you've been deposited in Madrid, or Milan, or in a Gaudí-less Barcelona, or an Alp-less Turin. This is Buenos Aires – and the sheer lack of culture shock is itself shocking. You fly half-way round the world only to arrive in a synthesis of the great cities of southern Europe. Are you really on another continent? Or has British Airways merely pulled another dirty trick – has the pilot been describing countless loops of the Old World before putting down in, say, Valencia?

The only sign that you haven't been conned is arboreal. Buenos Aires's trees are wonderful, and, thankfully, they comprise a sylvicultural gamut which is alien to Europe. This is South America, then – but it doesn't want to be South America. It wants to keep South America at bay. The characterisation of a Buenos Airean, a porteño, as an Italian who speaks Spanish, behaves like a Frenchman and wishes he were English is way out of date. The French and English parts of the mix have long since gone AWOL. The current hegemony is WISP – white Italian–Spanish. Nonetheless, it is to some idealised pan-Europe that this city gazes across the diarrhoea-brown water of the Plate estuary, which is so wide it might be, almost is, the ocean. (What colour do this city's schoolchildren, who wear white overalls in the manner of French children fifty years ago, use to represent the sea? Do lovers tell each other that 'your eyes are as brown as the sea'?) This is a city which is, then, founded in nostalgia in its purest, most literal sense – in an aching yearning for a forsaken place across the sea, a yearning which expresses itself so variously, so ubiquitously, that it defies the reason and the desire that this city should still be here. It should have freed itself and made its way back to the Old World. But, of course, it never has, any more than the three sisters will get to Moscow. It is in a state of perpetual anticipation, about to travel hopefully. 'Buenos Aires,' someone said to me, 'is a hotel. You can spend an entire life just passing through.'

It is a commonplace, to me anyway, that once you have looked at Brussels's suburbs and at Charleroi's inky canals, Magritte appears in a different light – he is not a surrealist but a social realist faithfully recording the quotidian life around him. Buenos Aires and Borges are inseparable. That, of course, is self-evident in stories such as *The South*, stories which his exact contemporary and admirer, Vladimir Nabokov, castigated as 'all *lunfardo* and local colour'. (*Lunfardo* is the street slang of the poorer barrios.) These stories are all passion, honour, revenge with a knife – and many of them indeed are set in the formerly run-down area of San Telmo, which was popularly known as El Sur (The South), but which is now no more threatening than any other antiques quarter. The traders display the same ingratiating cunning and ill-dissembled desperation as their peers at Clignancourt, or the Rastro, or up Portobello. And their wares are the same too: there is nothing to say that this is the New World, much that convinces you that you are in one of the markets in the last sentence. It's all repro Louis, ormolu and ebony, fake Aubusson, Scottish baronialism, dull landscapes in overwrought frames.

A subsequent generation of South American writers, the ones who made flowers talk and trees dance, tended to write off Borges as a sort of Eurocentric Uncle Tom. They pointed to his debts to Swift, to Stevenson and Chesterton as evidence of his inability, despite his genius, to make a literature which was peculiarly Argentinian. But this criticism neglects to acknowledge that the very qualities they abhor in his work are quintessentially Argentinian. It is akin to the Victorians who fretted about having no architectural style of their own, and failed to see that reproducing past styles was their style.

Besides which, even the most casual acquaintance with the city where he was born and where he lived nearly all his life will tell you that his recurrent themes – the maze, infinity, repetition – are metaphors for its relentless topography; they are deferred paeans to the rectilinear labyrinth, to streets which stretch so far they make their own horizon at a vanishing point, beyond which you know they continue to another horizon and another vanishing point. The grid goes on and on. It is magnificent in its dogged monotony, claustrophobic, incarcerating. The tyranny of the straight line and the right angle is invariable.

At the micro level, anything goes. At the macro level, there is a well-mannered adherence to the building line, to the integrity of the grid. This was ever the case. And so too was architectural sobriety. The nineteenth century has bequeathed to today's porteños magnificent parks heaving with allegorical statuary and blooming jacarandas. There is little architectural excess. And styles which almost depended on excess – psychotic Gothic, art nouveau – are notable by their absence. There is a collective fear of solecism, of being found out as a hick.

This conservative timidity lasted well into the twentieth century. Buildings of the mid-'20s are based on European models of thirty years previously. All this changed, with exceptional alacrity, in the early '30s. Buenos Aires got Corbusian modernism – and has never lost it. It is probably the greatest modernist city in the world. It is inventive, fresh, architecturally monocultural – but within that culture there is infinite variety.

The buildings of the grid are not rectilinear themselves. They are curved, stepped, vertiginously pitched, gravity defying – and curiously chaste. Extraneous ornament and applied decoration are seldom seen. The wildness and polychromatic vigour which characterises so much of the continent's building is off-limits. And, of course, 'international' modernism was a European mode. It makes you think that apartment blocks and banks and offices should be wrapped with banners bearing the most common of Buenos Airean signs – *'importada de Europa'*. It occurs in the windows of shops selling clothes, drinks, power tools, watches, foodstuffs, jewellery, white goods. It is even to be found on the opaque bands wrapped around the hard-core porn mags which the multitudinous news kiosks display. The message is clear: photography of penetrative adventures is another thing they do better in the Old World.

Which is, perhaps, a surprising admission given the city's overt appetite for mercantile sex, an appetite which is evidenced by the 1,200 small ads for personal sexual services that the *Clarín* (not imported from Europe) carries every day, along with three further pages of chat-lines, love hotels, houses of assignation, speciality shows, etc. By their classifieds shall we know them. The kiosks, incidentally and appropriately, display these wares at groin level, at eye level to ten-year-olds – which spares the kiddies the embarrassment of having to ask for a hand up to the top shelf.

Before a country churchwoman, Libby Purves, takes up her pen to rage against this iniquity, it should be pointed out that, far from promoting sexual violence, the toleration of new forms of the most ancient industry makes Buenos Aires the most pacific of megalopolises. Women can, and do, walk alone in the early hours, without fear of molestation. They can sit alone in cafés without hassle and in sure safety – their potential aggressors are not victims of Protestant repression. Nor, for that matter, of Catholic repression. Uruguay, across the Plate, is officially secular, Argentina is de facto.

The apparent lack of churches is initially heartening. But it is only, regrettably, an illusion. It's just that churches are, quite properly, subsumed into the urbanistic regime which prevents buildings from standing alone. So their presence is advertised by old crones on the buses called colectivos crossing themselves as the competitive, chromed vehicles pass them in their headlong privatised rush.

The real religions of Buenos Aires are astrology, tarot, numerology, and spiritualism for the workers. And psychiatry for the bourgeoisie – there are more shrinks and therapists per head of population than anywhere else in the world. But beyond these solaces is the one which really counts – football.

Maradona retired while I was there. The newspapers and the TV were taken up with nothing else. They acknowledge what this country's politically obsessed media are incapable of recognising – that politics is not really that important, that there are other things in life. Like football.

I went to La Boca the day of the elections, the day after Boca Juniors had played River Plate and two fans were murdered. I went there, not because I wanted to pay some sort of homage to a player who strikes me as not being even the greatest of Argentinians (that would be Mario Kempes), let alone the greatest in the world. Nor, forbid the idea, that I wanted to see the touristic, folkloric, arty street called Caminito, which is a Bayswater Road market of loudly coloured, horribly mismatched buildings. I went there, to Garibaldi Street, because this may have been where Eichmann lived, having arrived in Buenos Aires courtesy of the Vatican and Odessa in 1950. May have been – there is no other Garibaldi in 'federal' Buenos Aires. There may be dozens in the further agglomeration.

I wanted to imagine the Israelis who watched him day after day from a car as he returned from the Mercedes (it would be Mercedes) plant where he worked, and captured him, and behaved with a propriety and rectitude which is hard to imagine. This Garibaldi is a street of self-built houses, mostly but not exclusively of corrugated iron – and Eichmann did build his own house. But even if this is not the right Garibaldi, it is the apt one – for down its centre runs a single rail track, a reminder of better days.

La Boca is in the south, it's even further south than The South. Buenos Aires gets richer and richer the further north it grows. The British railway engineers built their Surreys, their Simlas at San Isidro and Los Olivos – free-standing Victorian Gothic churches, Norman Shavian villas, and, unastonishingly, railway stations airlifted from Box Hill or Westerham. A very particular Englishness was recreated.

A few stops further north on the coastal train track you arrive at the terminal, called Delta. Henceforth, it's boats. Beside the rivers of the Tigre Delta there are shacks on stilts, drowned shacks, shacks which have made way for dream houses. It's very like the Thames once was – all cutesy hideaways. But unthought of, evidently, by German war criminals.

And unused too by Borges. The very wateriness of it and the boathouses of the Buenos Aires Rowing Club have no place in the heartbreaking ballads of the city which begin among the boundless scents of eucalypti at the Villa Triste Le Roy.

(1998)

HANSA

I WAS BEING driven at dusk through suburbs of the northern German city of Bremen. The city where Beethoven's First Symphony was first performed. So the cab driver elected to play hideous rock music. His mullet, apparently dressed with schmaltz, lolled lifelessly on a nicely shiny antimacassar. The aural and ocular offences prompted me to gape more intently than I might otherwise have done at the passing scene.

There was something naggingly puzzling here. It occurred to me that I could be in any former West German burgh of a certain size and affluence. Bremen's suburbs are park-like: avenues of grand trees, abundant verdure, sinuous rides, wide paths. And steep-roofed villas of the latest nineteenth and earliest twentieth centuries whose bourgeois opulence becomes apparent as their rooms are illumined. This taxi ride did not quite prompt a sensation of déjà vu: I had to dig for it, exhume a distant memory.

One day, ten and a half years ago, at the same hour, I had been driven by a volubly characterful cabbie though Munich's suburbs. These had seemed uncannily similar to Bremen's. Munich is, however, almost 400 miles south of Bremen, in Bavaria. The two cities surely belong to different cultures? Yet here was proof in tiles, render and gables of a pan-Germanic commonality – which is most certainly not to be found in the cities' ancient core, in their mercifully ancient cooking, in their history, in their denominational complexion. Nor is that commonality present in Germany's contemporary self-estimate. Günter Grass's celebrated denunciation of reunification as a threat to the stability of this continent perhaps underestimated the strength of the federal impulse, of regionalism's grasp.

It had certainly not taken into account the interesting localism of the Munich driver. He pointed at a group of keen smokers loitering in nylon shirts and cardboard shoes. Ossis, he told us, easterners, job-seeking, job-stealing. Why don't they go back where they came from (the former GDR)? If they don't go back, they should be gassed. He suffixed this with a guffaw to indicate that he spoke in jest.

Now, the pan-Germanic idiom evidently came in the wake of the establishment of the German, or Prussian, Empire. It was short-lived. It did not survive the First World War, after which building styles once again became localised. Yet it had lasted long enough to have been one of the instruments that fomented a notion of national, rather than regional, identity and which implanted that ideal in the minds of the generation the extent of whose abominations was fully revealed in the spring of 1945. The Third Reich's aesthetic preoccupations constitute a sort of pathology: its best-known, most grandiose, projects are its least typical.

Its stylistic preoccupation was, oddly, far from overtly nationalistic. It was to fetishise the particular and the parochial, and to exaggerate the diversity that had been self-consciously pursued in the thirteen years between the Treaty of Versailles and the seizure of power. Not all the diversity, however. Only those elements of it that were supposedly rooted in an officially sanctioned (and largely bogus) German past. This was a peculiarly hermetic Germany: the very word *deutsch* originally signified a narrowly indigenous people and a language uninfected by Latinisms, borrowings, cosmopolitanism.

Bremen did not belong to that Germany. Like Wismar, Hamburg, Stralsund, Greifswald, etc., it was – and nominally remains – a Hanse-stadt. This was not the Germany that the Third Reich invoked. It was mercantile and international. It became political and martial. It expanded and died. The Hanseatic League was probably God's first try at the EU. Whether the EU will bequeath such a legacy of marvellous towns and buildings is moot. Actually, it isn't. It won't.

The Hanseatic style was an essay in what was not then called branding. Like, say, Roman building long before it and modernism long after it, it paid little heed to the vernacular practices of the places on which it was willingly imposed. These places wore the recurrent devices of this idiom as a badge of belonging. Houses, guild houses, churches, convents, schools, town halls and town gates all shared a kindred elaborate brickwork and a verticality which is so exaggerated that we suffer the illusion that we are seeing them in a fairground's distorting mirror. This verticality culminated in energetic gables, spires, crenellations, tourelles. So the roofscapes and skylines of the Hanseatic towns are fantastical: Stralsund rivals Venice. But it is hardly known in Britain due to our southern bias and incuriosity about the Baltic.

The source of the Hanseatic style was Lübeck. This great port's motifs came to be adopted by cities as distant as Bruges and Riga. But it is in Lübeck's neighbours, Hamburg and Bremen, that the tradition of idiosyncratic brickwork was most enthusiastically revisited in the 1920s and early '30s. Revisited, not reproduced.

The nascent modern movement, all smooth surfaces and orthogonal geometry, made little impression in these northern cities whose designers possessed a taste for angularity and counter-intuitive restlessness – which had nothing to do with function and everything to do with dramatic gestures tilting at the horizontal relentlessness of the flatlands. This was expressionism. Bremen has no slopes. Its most characteristic buildings compensate for nature's want. Boettcherstrasse, a long alley of the 1920s, runs south from the market place and expressionism runs wild: there are bizarre homages to the mercantile past, art nouveau-ish flourishes, farouche gargoyles, crazily patterned bricks. It is lit in such a way that those who wander through it at night cast terrible shadows and find themselves in the world of Dr Caligari. We owe this prodigious folly to the inventor of decaffeinated coffee.

(2005)

DURHAM

ADMITTING TO NEVER having been to Venice is, I have learnt, tantamount to confessing to never having seen a production of *Hamlet*, to thinking Kipling was a baker, to believing that Beethoven is a cuddly St Bernard in a 'feel-good' (i.e. emetic) film. This admission prompts incredulity, bafflement, pity, condescension.

A few days ago I visited the City of Durham to participate in a debate: it was the first time in my adult life that I had been there. An omission that, when revealed to my hosts and fellow speakers, was met with little more than mild surprise. No rolling of eyes, grimaces or exaggerated frowns of wonderment. Why not? If Venice is reckoned to be a mandatory component of a topographical syllabus then surely Durham should also be included. It is without question among the greatest of sites in England. And here, of course, is the core of the problem. The English are in thrall to elsewhere. We seek the exotic. We are convinced that the grass is greener where the grass is actually sun-bleached. Distance enchants us. Our resistance to internal travel other than for work is no doubt determined by habit and custom. But equally climate, the ill-named 'hospitality' industry, the cost and reeking decrepitude of much public transport . . . these militate against whatever will we possess to acquaint ourselves with own country.

It's our loss. Marvelling at this steep, spectacular city, I could not help but tell myself what a clot I am for not being the owner of a camper van who is happy to subsist on a diet of crisps. Forget fussing about decent food and a comfortable billet. How could I have denied myself the unalloyed treat of a site which is so satisfying, so multiply satisfying?

The very word 'site' is rarely pertinent in discussion of English towns. Sure, every town possesses, of necessity, a site, but it not usually a dramatically determining element. England is not a country of hill towns. The lie of the land does not favour them, and the unlikelihood of invasion has rendered them martially unnecessary. Bridgnorth, Ludlow, Lewes, Launceston, Richmond (the one beside the Swale): the list is brief. Durham's grandeur eclipses the rest. It is far more exotic than everyday palms and familiar white sand. An accumulation of unmistakably English buildings – cathedral, castle, houses, colleges – combines with a prodigious geology to create an entity which is thoroughly un-English, which might belong to a forgotten corner of northern Europe.

The site is a sort of isthmus, surrounded on three sides by the Wear. The river is populated by countless skullers being bellowed at by their bicycling coaches on the right bank: 'Nice rhythm lads, keep the rhythm lads!' The path along the left bank is entered by St Oswald's churchyard, a scene from a maudlin Victorian narrative painting. The ever-changing view of the cathedral recalls a different sort of Victorian art. It is emphatically vertical, as though mediated by the inventions of the romantic architectural fantasist Axel Haig. Here, it seems, is one of his sets brought to life in stone and wood.

Everything points up. Everything is conditioned by the cliff's being almost sheer. A tiny Greek-revival gazebo sits in front of the cliff as though it is a three-dimensional object against a stage-flat. To the cathedral's immediate south houses grow out of the top of the cliff on arcaded buttresses which form such a weathered elision with the stone that it is impossible to determine where nature stops and building work begins. The cathedral itself is, unastonishingly, entirely out of scale with all that surrounds it. Mature trees are diminished like shoots. An exquisite Georgian road bridge and a mill beside a weir might be models.

The effect of the endless glorious details of sacred and secular structures, of high art and untutored vernacular, is fleetingly that of pure happiness. A market trader had helpfully leant his stock of mirrors against the side of Framwellgate Bridge so the cathedral was multiply, cubistically reflected. I followed a sign along an alley to Brocks Cottage – it had not previously occurred to me that badgers behaved thus – and then up a ramp to the top of the cliff. Here, first, is an attempted essay in aptness: a building of the 1960s that takes its cue from castles in general, if not from Durham Castle in particular. Appended to its wall is a plaque announcing that it was the winner of a Civic Trust Award. I yearn to find a building of the past forty years that has failed to win an award. Opposite, on a nicely varied house with facings in the unusual material of red sandstone, another plaque commemorates John Meade Falkner, author of *Moonfleet* and – better, creepier – *The Lost Stradivarius*. He was very likely the only English novelist who has made his living selling armaments, which he did for Armstrong Whitworth.

Then Palace Green. At one end is the castle, layer upon layer of fortification, augmented by layer on layer of Gothic frivolity. So here is a real castle that has been partially dressed as a sham castle – and is all the more civilised for it. At the other end of the green is the north front of the cathedral, the entrance front, for the west end is, eccentrically, taken up with the Lady Chapel, the Galilee, a heart-stopping work of the gravest beauty with crazily zigzagged arches; that decorative device is found throughout this incomparable church. So, too, are bulky columns and inspired art: the Galilee has a northern, presumably Flemish, triptych, the remnants of a wall painting and a cross that might be a folded easel. The south transept is dominated by a wonderfully bizarre clock, more than forty feet tall, the doors of which are by a follower of that most limpid, cold painter of interiors, Pieter Saenredam, who anticipated the neue sachlichkeit by three centuries. As for the font cover, this is a chunk of temporal exuberance, the Restoration's clamorous rejoinder to the austere miserableness of the Commonwealth, and, like everything around it, a joy.

(2005)

BRISTOL

WHEN I WAS APPROACHED to address this (RIBA) conference I was asked whether I knew Bristol and if so what, if anything, I thought of it. I was asked with a casual jocularity that failed to mask a certain wariness, a certain concern that I might hold some animus against, for instance, a dodgy mullion in Cotham or an imperfect pediment in Bedminster and that the expression of that animus might offend the RIBA's president – a Bristolian; it is well known that Bristolians are thinner-skinned than the rest of us.

Happily I was able to put my interlocutor at his ease – and to do so with a clear conscience.

I've known Bristol all my life. After Southampton, where my mother's family lived and which is only twenty miles from Salisbury where *we* lived, Bristol was the second city of my childhood and teens: it has impinged on me in a very different way from Southampton, which was morbidly affecting, dark, rain-slicked, tough, scary. Bristol was – astonishingly – always sunny, a bright douce treat three or four times a year. I still have photographs taken here when I was three and a half. I share the frame with, variously, a suspension bridge which today I can only get across by crawling, a lion, a polar bear, a zebra. I guess that all save the bridge have long since been turned into pies.

Of all matter food – and drink – is the only matter that we urge not to remain outside us, that we give an internal home to: perhaps this is why our memories are so connected to it. I ate my first deep-fried egg at the Marine Café on the Triangle, I had my first Chinese meal directly across the street, I first tasted pizza in a basement near St John's Gateway. I got to know St Nicholas Market by bunking off from some awful play by Dürrenmatt at the Old Vic and being led by a school friend to a bar called The Rummer – an early enterprise of the Berni Brothers since you ask.

Bristol was also sartorially important. That dogtooth pattern shirt from Daniel Neale when I was twelve. The herringbone suit made to measure by Mr Graham Manning of Christmas Steps, a place which so delighted me that I wanted the fittings to last forever. I can recall a shop widow on College Green and a pair of chisel points which my mother refused to buy me on the grounds that my father would consider them too common – which was precisely the point. I can recall the name of the purple-heart dealer who plied his trade in a coffee bar on Park Street. I can't recall the name of the young woman my mother had taught at primary school who had migrated to the big city and who had fallen – as the Victorians would have said – into prostitution. She was found strangled in Leigh Woods – a murder which somehow contaminated a group of houses in that direction immediately on the other side of the suspension bridge: they are high-Victorian picturesque, some of them chalet-like, some of them cosmetically oriental as though intended for a hill station like Murree or Ootacamund. They had been jolly. Now they had become – and remain – slightly sinister.

The Devil's Cathedral was never jolly. Indeed when I was tiny I would look at it through parted fingers as though it were the set of a spine-chiller, which, of course, it was. My spine-chiller. Opposite it is the steep Arno's Vale cemetery where one of Peter Nichols's great telly films, *Hearts and Flowers*, was set. Further into the centre along the same road there's a terrace encrusted with grime, and behind the grime, climatically impervious industrial terracotta. And behind the terrace the best-named place in Britain – Totterdown, where late Victorian houses do indeed teeter madly on impossible slopes. No one would go there on account of its architecture . . . but the appeal of place is quite apart from that of architecture.

This then is what might be called Bristol One – or, rather, a tiny fragment of Bristol One. My personal Bristol. A serendipitously accumulated Bristol of chance views, fortuitous encounters and thankfully stubborn memories.

Bristol Two is a Bristol that is widely shared. That many of us know. Some of its facets coincide with those of Bristol One: Clifton Bridge, evidently, Arno's Court's former stables aka the Devil's Cathedral – so called by Horace Walpole because they are built of shiny black slag which recalls the volcanic stone of Catania in Sicily and Clermont-Ferrand in the Puy de Dôme: cities which are, incidentally, architecturally defined not by a particular stylistic idiom or even by the quality of their design but by their building material.

Bristol is a city which possesses an abundance of structures which are one way or another undoubtedly of the highest stature. We can all reel them off. St Mary Redcliffe, the cottages at Blaise Hamlet, the tobacco warehouses, Kingsweston, the Wills Tower, Edward Everard's premises – and three outrageous essays in Victorian Venetian high comedy: the Granary on Welsh Back, the old Lloyd's Bank in Corn Street, the sometime museum close by the Wills Tower which is now a restaurant of sorts. It goes without saying that they were not intended to be comedic – but they are by dint of their solemn self-regard. There is a sort of beauty in pathological pomposity. That they are jokes against themselves does not for a moment lessen them.

As well as these individual structures there are, of course, such set pieces, such magnificent set pieces, as Clifton, Queen Square, King Street.

Bristol Two is the Bristol of icons. The icons which we tick off: if we haven't seen them, we haven't seen Bristol. We are all of us prey to itineraries of what is already mediated by prose and photographs and television, the prepared scapes, the ready-made vistas, the off-the-peg sublimity, the store-bought sites of emetic picturesqueness. It was always thus: hence the majority of Grand Tourists of a quarter of a millennium ago took in the same places and were forever running into each other. We see what we have been taught to see. We visit places we have been enjoined to visit. We line up our eyes along the axis of an anonymous photographer's lens. We try to duplicate the very position that a perspectivist long ago adopted for his easel. We behave in this regard with apparently conditioned gregariousness.

It is impossible, as they say in football, to legislate for Bristol One. It has, so to speak, chosen us, imprisoned us. I never elected to have an art deco Smith's Crisps factory tattooed on my brain in perpetuity . . .

Bristol Two is different.

The purpose of monuments extends far beyond their utility: they memorialise their maker, their patron, a commerce, an ideal.

Their function is no more modest than that of a statue which an autocrat raises to himself. Clifton Bridge can hardly have been claimed to have been necessary from the point of view of transport or communications; the Wills Tower, one of the tobacco dynasty's many gifts to this city, is as much an advertisement for the product's benefits as an exercise in pedagogic philanthropy. I guess that hardline prohibitionists might consider it wickedly dangerous, a stone ode to passive smoking.

As a general rule: the greater a regime or a culture's fondness for monuments, the greater the deprivations and repressions its subjects will suffer. At a sartorial level this may be expressed thus: a society's civilisation is in inverse proportion to the percentage of its population that is in uniform.

Now, if I suture together Bristol One – the sentimental mnemonic triggers – and Bristol Two – the quasi-official, popularly sanctioned sites, their aggregate comes to a tiny, and atypical, fraction of a single per cent of this city's buildings.

Which leaves a vast mass that is not exactly unseen – we are aware of its presence – but which is unscrutinised, which passes passively, which is not animated by our gaze, which is unengaged by our curiosity, which is taken for granted.

This is the silent majority. The buildings which occupy the spaces between the buildings which have made their mark, which have impinged on us singly and collectively. It is this vast mass which forms the backdrop to our life, which is the wallpaper of the roofless galleries we call cities and towns, it is this vast mass which is the fabric of urbanism. I'll reword that: which is the fabric of what should be, what might be urbanism.

The character of a place is to be found in the ordinary – the apparently ordinary. The beans in the cassoulet, not the confit and the sausage. Or, for those who of you who are WI members, the cake itself rather than the cherries and the filling.

Bristol

We have a self-deluding habit of making the atypical stand for the typical. We convince ourselves that English villages are routinely like Lacock or Castle Combe rather than like the accumulations of bungalows, pseudo-bucolic executive estates, crescents of ill-maintained council houses, abundant rusting water butts, reeking silage towers, breeze-block piggeries and crack-dealing yobs with ASBOs, which are actually the norm. We similarly convince ourselves that a city is its highlights.

Twenty-five years ago that most craftsmanlike of architects Peter Aldington suggested that what was needed was not more grand gestures but, as he felicitously put it, *a better standard of ordinariness*. That, I believe, still remains the imperative. Bristol demonstrates the primacy of ordinariness because it possesses an unusually high standard of ordinariness. The circumstances which have occasioned this are complicated.

The peculiarity of the topography is an important factor. The combination of precipitous hills, chasmic valleys, rivers and navigations and consequently, the delightful lack of a grid, seems happily inimical to grand gestures or, anyway, to extensive grand gestures. It is a terrain and a labyrinthine pattern which has encouraged disciplined one-offs.

The result is, if you approve it, an exhilarating patchwork which is resolutely non-consensual. Heterogeneity is the norm. Beneficent collisions and wacky juxtapositions are ubiquitous. Differences of scale, material, style: differences which are exacerbated by buildings being seen in vertical as well as horizontal relationship to each other.

If you don't approve it you call it piecemeal, scrappy.

It is certainly the urbanistic obverse of the great port with which it for so long traded. Bordeaux's classical unities are not on the agenda here. Individuality and eccentricity *are* on it. Though eccentricity is an inapt word since it presumes a rule to diverge from, a centre to be the exception to. Here there is no rule.

Bristol's genius resides in its benign anarchy. That does not, of course, apply to Clifton – to which I'll return in a moment.

The history of English urbanism – though not of Scottish or European urbanism – is equally the history of *sub*urbanism. Or the history of a longing for suburbanism. And of a resistance to urbanism. Mistrust of urbanism and of cities themselves – the two are not altogether congruent – is very deeply embedded in the English psyche.

If a culture is essentially a compound of custom, practice, tics, habits, traditions and collectively received ideas then we have an anti-urbanistic culture to fight against for the very survival of this country as a place worth living in.

It was, astonishingly, exactly half a century ago in the June 1955 edition of *Architectural Review* that the twenty-four-year-old Ian Nairn published his great polemic *Outrage*, a work of scorn and ire and prescriptive inspiration. Save that it inspired no one. A fat lot of good it has done. Nairn railed against . . . well, he railed against pretty much everything that was bad about the unchecked spread of low-quality built environment, an environment that has continued to get worse since. And will very likely go on getting worse – burb will breed burb will breed burb. The sub has long been redundant, these burbs have no urb to be sub to. Hence his coinage of subtopia. He felt that we have been beaten into submission. Gore Vidal's dictum that irony is the weapon of the impotent might be extended to cover polemic. It was probably Nairn's despair that drove him to devise the Nairn Diet.

I met him twice, towards the end of his life I need hardly add. One lunch I spent with him he ate a packet of crisps and drank fourteen pints of beer, the other time he was thinking about his figure so ate nothing and drank a mere eleven pints. He died a few months later of cirrhosis. Cruelly, in Ealing, the so-called queen of the suburbs and the sort of place he abhorred.

Back to Clifton.

It is certainly a suburb. A most unusual suburb if only because it doesn't make my heart sink. There's more to it than that. Like most of Bristol it is not easily legible: its layout is not orthogonal but haphazardly tied to the contours of its slopes. It invites us to lose ourselves.

But unlike most of Bristol it is not a quick-change artiste, it does not turn into something new with every corner. A broadly classical idiom was maintained long after it had slipped out of fashion elsewhere. It achieved its integrity by resisting fashion, by its architects and builders deploying self-effacement for the common good. Some of its terraces were at least half a century out of date. They were anachronisms. But we don't notice that today.

This is what Betjeman called the classical survival – we should be thankful for it. Like Cheltenham and like Brighton – which are as architecturally blessed but lack the sheerly dramatic site – Clifton owes its singularity to a property which causes many of today's architects a certain embarrassment. It is all *in keeping*.

There is nothing wrong with being in keeping. There is nothing wrong – commercially, artistically, let alone morally – with pastiche. There is evidently nothing wrong with pastiche if it's of modernism since that is what the current British mainstream comprises: neo-modernism, synthetic modernism – those are handles by the way, not denigrations. Pastiche is like anything else. To amend Duke Ellington: there's good pastiche and there's bad pastiche. The pasticheur is a sort of chameleon. The problems arise when the chameleon is a loser chameleon who has no sense of shape or is colour-blind: being colour-blind is a disability which real chameleons quite properly fear – for it can result in one of those life and death camouflage tragedies. So homogeneity is one of Clifton's great attributes.

Its second great attribute is that it is centripetal. Its architectural aspirations and its forms are urban. It is not trying to flee the city despite the lure of the downs and gorge and the strangler's woods and Somerset's lush fields. It is in no sense save the debased sense of estate agency a village.

It is, rather, aesthetically attached to the city. Its streets and squares are city streets and city squares. A bombastic construction such as Royal York Crescent may not be technically great architecture but it is great *city*scape. It makes no allowance towards its being perched on a cliff staring towards the distant Mendips. Clifton belongs to a long late Georgian, early Victorian moment when suburbanism did not imply anti-urbanism. That moment came to an end. Though there was nothing at all inevitable about its coming to an end.

At the time that Clifton was being completed the young Frederick Engels had just published, in Germany, *The Condition of the Working*

Class in England – a book based almost entirely on his observations, some acute, some callow, of industrial Manchester. For Engels – as for Hitler after him – the city's only virtue was as a concentration of the disaffected, the mob, among whom revolt might be fomented: once the revolution had triumphed cities would be abandoned, not least, one suspects, because they might serve as concentrations of counter-revolutionary fervour.

Many of his less guileful contemporaries saw no virtue whatsoever in cities. Indeed as Britain's might and wealth increased and the population of its cities' made exponential leaps so did the tide of anti-urbanism gather strength.

To deplore the very notion of the city and to insist that the only alternative to the city was to reshape mankind so that the city would be rendered redundant and a new form of organism be created from zero . . . this was so much a part of the bien pensant Victorian curriculum that it was not reckoned dotty or utopian or whimsical.

Far from it: this was a conventional, institutional shibboleth – as commonplace as anti-elitism is today, and as vacuous. They all signed up for the big fret: Thomas Carlyle, Matthew Arnold, the sexually squeamish John Ruskin, the ineffable William Morris who thought expensive wallpaper was going to change the world, Arthur Mackmurdo who abandoned architecture to write tracts urging humans to learn from the social structure of the beehive, such future fellow travellers as H. G. Wells and G. B. Shaw – who would announce during the worst of the famines caused by Stalin's collectivisation that the USSR was the world's best-fed country. These were the officer-class gullible. There were countless troops too. But no one had any idea what the alternative to the city would *look* like, not even the well-intentioned old fool Ebenezer Howard whose disastrous book was written – with fists dipped in lard – just after Engels tardily appeared in English translation spouting nostrums in response to the conditions of half a century before.

Howard's *Tomorrow: A Peaceful Path to Real Reform* which from its second edition became *Garden Cities of To-morrow* is an anti-landlord tract which is obstinately mute about the *appearance* of such settlements, it is insouciant about their architecture. It's as though he considers that appearance is a frivolity in comparison with a new economic foundation and shelter. What were the physical attributes of Nowhere?

Most small-scale nineteenth-century utopian essays had done little more than reflect mainstream taste and constrained budgets: Robert Owen's New Lanark was composed of what might have been workhouses; the Chartist colonies were industrial back streets displaced in meadows and orchards; Saltaire was a primly Italianate model village which was as cheery as a prison. These places did not sing a sweet lullaby of a bright, Start-rite new dawn. As aubades go they were positively prosaic.

Now it was Howard's good fortune and twentieth-century Britain's bad fortune that his derivative tract was published at a time when a new domestic architecture of brilliantly contrived escapism was at its apogee. Never was Never Never Land more persuasively realised than by the rurally fixated, childlike luddite of the arts and crafts. Never was twee, cutesy, unthinking, saccharine, smiley, eager-to-please, easy-on-the-eye winsomeness carried off by a greater genius than the young Lutyens. This may be because genius is seldom disposed to express itself in flights of dreamy treacle.

Parker and Unwin, the architects of Letchworth and Hampstead Garden Suburb, may not have been as blessed as Lutyens but they were good all right. Pragmatic utopianism had found its perfect architectural expression. Shaw called Letchworth the Heaven near Hitchin. The future mass murderer Lenin went to inspect it. The English garden suburb became the world's cynosure.

And the world, it has to be said, treated the recipe more judiciously than England did. France and Germany, Russia and Belgium realised what the English begetters never admitted to themselves: namely that this was a *fantasy* of bucolicism, this was *playing* at rusticity, this was a *game*. And so it had rules. A discipline to abide by. A set of constraints.

Sokol in northern Moscow and Elisabethville in western Paris and Le Logis and Le Floréal on the edge of Brussels are places that revel in their bogusness. They are joyously infected by an extreme *artifice*. There is no pretence of being part of any countryside other than that of their own devising. These are suburbs which still belong to the city. There is no will to secede.

In England, however, the garden suburb was like an actor subsumed by a role. We are all familiar with that embarrassing identification. A garden suburb is, of course, inanimate, insentient. I mean, evidently, promoters of the garden suburb – they believed in the cheap fiction they had woven.

They were beguiled by their own industrially produced folk-imagery: they forgot they were living in and building illustrations from a fairy tale. They believed that they were creating villages. And the millions of people who bought into that fantasy believed it too. Exodus from cities became an English norm. The aspiration to move to the country was eased by the sudden provision of so much new country, what might be passed off as country . . .

This diaspora was the most demographically significant tendency of Britain's twentieth century. The coarse irony was that it was a reaction to a gamut of problems that no longer existed. By the beginning of the twentieth century cities enjoyed drainage, sanitation, gas and in some cases electric light. Conditions propitious to the diseases which had ravaged cities had been eradicated. Diphtheria, cholera, typhus, scarlet fever, yellow fever, mauve fever, turquoise fever had disappeared.

So: the migration from cities was undertaken by fugitives not from present insalubriousness but from past fear, by refugees from an urban myth of malevolent miasma – which was more potent than the evidence witnessed by their eyes, their nose, their skin.

This was a migration whose motor was fuelled, then, not by reason but by faith. A faith which is all too evidently still espoused today.

And faith is always a problem: try telling a credulous communicant, and there is no other sort of communicant, that the emetic red stuff offered by a bridegroom of Christ is not actually Christ's blood but something made from rehydrated wine-style powder in Crediton, Devon . . .

Try telling the subtopian, land-hungry Englishman that his

wretched home in Dismal Close off Dreary Avenue is not a castle but a cancerous cell that is destroying what remains of his country.

For Cobbett it was the city that was the wen, the cancer. Today we know much more about that disease, we know how it delights in multiple sites which are not necessarily connected to each other, we know how promiscuously playful it is in where it elects to strike. It is worth noting that of western European countries it is only in England that the epithet *inner city* is synonymous with crime, destitution, social deprivation.

Why? Because it is only in England that the people who run cities and who generate the wealth of cities, live in exurban dormitories and so have no *personal* stake in those cities: communitarianism does not function at second hand. Sure, this is changing. Slowly. But it is changing through a new faith – in the balms of regeneration. And believe me it is a faith.

We, the faithful, know that Michael Heseltine entirely transformed Liverpool's economy and spirit with the miracle of the garden festival. We, the faithful, know that the Lowry and the Imperial War Museum of the North have so changed Manchester that today there is no gun crime in Moss Side and that what look like arsoned warehouses in Ancoats are chimeras. We, the faithful, know that Frank Gehry has demonstrated that if you get rid of right angles you put a smile on a terrorist's face. We, the faithful, know that alfresco restaurants and important perfumed-candle shops and bravely fuchsia Alcantara dock sides by Citterio spell a better future.

If I had less faith I'd whisper an unmentionable phrase.

Dirigism. Environmental dirigism.

But then I'd whisper it in vain.

For the only civil liberty that this too subservient country values is the liberty to hope for a triple SUV garage with benefit of water-feature garden.

(2005)

BIRMINGHAM

IT WAS AFTER I'd been in Birmingham for twelve consecutive days that it occurred to me that I had never spent so long in any British city outside London – Salisbury, a country town which is a 'city' because it has a cathedral, doesn't really count. The other thing that occurred to me was how much I liked it. The film I was making was already written and was, I hope, reasonably affectionate. But it was written, as usual, on the basis of prior knowledge, a lifetime's curiosity, (favourable) prejudice, memories, imaginative recreation and a couple of flying visits – too much reality gets in the way; my films are not about places but about ideas about places; naturalism is the bane of telly, etc., etc.

I was liking the place despite staying in an extraordinary hotel which suffered quasi-wartime shortages. The one draught beer was frequently unavailable. When I asked for some matches I was told that they were 'on order'. They were still 'on order' four days later. But Birmingham was, nonetheless, consistently amiable, absolutely bereft of the chippy aggression which characterises so much of the north, bereft too of the uptight *froideur* which is the southern provincial norm. It is, also, an almost excessively sylvan place. Its southern suburbs are lavishly green, and they are picture-book anthologies of all the domestic architectural styles of Birmingham's long innings as the manufactory of the world. And when it comes to public buildings Birmingham showed Britain the way: it was a virtual city-state which invented the notion of civic pride and provided its people with schools and libraries and technical schools and a university college – all of them in the loudest and reddest of red bricks.

And this, of course, is where its problems start. Redbrick was an expression of patronising belittlement long before this class-bound country adopted Nancy Mitford's frivolous division of U and non-U as a sacred text. And, more literally, industrial red brick incited the loathing of several generations of bien pensant aesthetes such as the voluble Alec Clifton Taylor: it was deemed 'inferior' to handmade brick or quarried stone. Birmingham suffered those generations' related distaste for the places where things were made, though not, of course, for the things themselves – guns, clocks, toys, cars, especially cars. And just when that generational prejudice was waning Birmingham pulled another trick out of its hat. It remade itself as the city which not only manufactured cars and their components but which showed its love for internal combustion by building a road system of a complexity and scale unmatched in Britain. It was, after all, a city with a short past and the motto 'Forward': it was almost programmed to renew itself with pragmatic vigour, high energy and low aesthetics. (Jeff Lynne's ELO could not have come from anywhere else: the greatest pasticheur in the history of pop music is all craft and no art – that's the beauty of his stuff.)

The massive traffic schemes of the '60s were not without precedent. Birmingham had been demonstrating its car friendliness for decades by then. In Moseley there are several villas of 1900–05 which were built with integrated motor houses as they were then called – the French *garage*, locally 'garridge', is first recorded in English use in 1902 but was not commonplace for another ten years. Birmingham was the first city in Britain to authorise one-way streets. And in the late '20s and throughout the '30s it constructed broad arterial roads out into the surrounding country. Beside them are the largest, swankiest roadhouses imaginable, built with a brute confidence by Mitchells and Butlers before the association between enthusiastic consumption of their product and an impairment of motor senses had been statutorially acknowledged and acted upon.

The most singular manifestation of Motown's autophilia-going-on-autoeroticism was also eventually proscribed by law: as a child in the '50s I would, every few weeks, go with my father to his family home in Evesham and then spend much of the rest of the day eating crisps and drinking pop in the car park of whatever pub it was he was boozing in with my uncles. These pubs were in the Worcestershire countryside – which is Brum's playground. Now cars were then much more expensive in real terms than they are today. And the men who worked on assembly lines and in the factories which produced brake pads and gear boxes and tyres and indicators did not necessarily have the wherewithal to own the products to which they contributed. Hence 'The Special'. There were specials everywhere, from Bretforton to Cleeve Prior to Bidford to far-off Malvern and Witley. Specials chugged up the Abberley Hills and up Clee Hill. Brummies were, maybe still are, bodgers, resourceful improvisers. They built their specials from the remnants of write-offs, from material that was begged and recycled, from components used in contexts for which they weren't intended, from bits which, I dare say, had fallen off the back of a factory conveyor belt. These vehicles had a fantastical ad hoc air to them. They were Heath Robinson's drawings come to creaking, roaring, backfiring life. They were rarely beautified, they were often unpainted. They were the mobile cousins of the shacks made from old railway carriages and corrugated iron which enhance the Severn Valley between Bewdley and Bridgnorth.

Actually, they were not always that mobile. Hence my familiarity with them. Their owners would use pub car parks as unofficial pits, carrying out running repairs with fierce concentration. They were so properly proud of their machines – which were, of course, a delight to a child the way a tree house is. It was the MoT test that killed off specials – although the spirit still lives on in the motorised trikes ridden by Hell's Angels.

Outside Kidderminster, near Harvington, is the largest Reliant Robin dealership I've seen. Dozens of these vehicles face the road, and it's an even chance that the aggregate of wheels is an odd number. Many of the customers for these joke-butts are not little old fellows in mufflers and flat hats but fearsome wannabe Wotans in leathers and Castrol-stained jeans. They strip down the three-wheelers, add handlebars and chrome – and away they go. Never let a Hell's Angel know that you know the source of these last specials; it'd be like accusing him of having a gun in his pocket that fires only caps.

The effect of constructing successive rings of de facto motorway through and round the centre of the city were manifold, contrary, paradoxical. It must be recalled (and it's something that in our retrospective wisdom we tend to overlook) that, pace Betjeman, in 1960 Victorian architecture was popularly regarded with the same contempt that 1960s architecture attracts today. And it has to be acknowledged that Birmingham was merely doing what other cities wanted to do but failed to do through lack of resolve and audacity – I'm talking evidently of British cities. It is because of the example of Birmingham and, to a necessarily lesser extent, of Glasgow, Newcastle, Sheffield and so on that the exaggerated and debilitating respect for the old simply because it is old has become so pervasive. Birmingham made access to its centre easy – and exciting: driving into the city along the Aston Expressway from the engineering marvel of the Gravelly Hill Interchange, aka Spaghetti Junction, still, after all these years, feels like driving into Godard's *Alphaville*; there is the same terror, the same wonder, the same awe. But at the same time it made egress from the centre easy. So commuting by car became all too simple, and that in turn fomented the spread of the burbs; and as the roads grew more clogged businesses moved further out – the 'Birmingham' Business Park is twelve miles out of town. The NEC is even further.

The centre of Birmingham, if anyone bothered to look, is much better than its reputation insists. But, of course, until recently none did bother: this is a city that has been perennially overlooked. It has never been in fashion the way that Liverpool, Glasgow and Manchester have; it has always had an aptitude for substance over cosmetic style. But if, to amend Buffon, style is the city itself – i.e. a potent quintessence – then it has great style. It is very sure of itself, but because its collective humour is self-deprecating, unboastful and peculiarly ironic it's not particularly good at promoting itself. And I guess the accent doesn't help. I love the accent. I take unalloyed pleasure in listening to it – not, I must add, a patronising pleasure but a straightforward delight in hearing English as it sounded before the Great Vowel Shift. This is Shakespeare's English (local boy): just try, say, Polonius's 'Neither a borrower nor a lender be . . .' in Brummie – the inflections guide the verse whose rhythms are lost on most contemporary English actors. Note to Adrian Noble – let's have a Brummie Hamlet. Note to the rest of Britain – go see Birmingham.

The rest of the world already goes. The non-anglophone are not put off by the accent. Nor are they put off by modern architecture, much of which is first rate: J. A. Roberts's work at the Bull Ring, the Rotunda and Smallbrook Queensway is something to celebrate; and so is the Alpha Tower, which is, I assume, by the most wrongly reviled architect of the half century, Richard Seifert – it certainly bears his trademark sculptural piloti and fenestration without reveals; the New Street Station signal box is a tour de force of plastically wrought concrete and has, rightly, been listed.

There are also several multi-storey car parks of genuine splendour – yes, I can hear the apoplectic guffaws, but architecture is about aptness and aesthetic certainties, not about a hierarchy of uses. Churches can be (often are) duff, while toilets (more useful than churches) can be landmarks. The begetter of England's best-loved toilet, Piers Gough, has contributed the pick of the city's po-mo buildings, a glass bar in the mostly rather leaden Brindleyplace development. Po-mo, in this context, may be taken to mean post-Motown as well as postmodern. The very latest Birmingham, the city of conventions and plastic name tags, is a creation of European money and possesses European ideas about pedestrianisation and public transport. So as well as a load of off-the-peg architecture which has neither the capricious invention of Gough's work nor the self-confidence of the '60s work, there are acres of square and road which are banned to cars. And given that the business tourists who frequent these spaces are all lost and looking for, say, Inter::Lard – The 24th National Chip Fat Convention, the place has taken on the air of an airport lounge without a roof.

The redevelopment of the '60s was probably not comprehensive enough. You can't build Brasilia on the site of an extant city. And the trouble with Motown was that shards of Horse-and-Carriagetown kept pushing through. There were bits of land that were stranded. There were buildings that were maroons. And the same has happened again. Brum's progressive spirit is impatient and infected by a cultural memory of making-do. So Foottown is constantly impinged upon by Motown. Different idioms pile up against each other in all cities but in Birmingham they are different idioms which all required a clean slate, which all depended on starting from zero.

Still, it has to be accepted – it is the brusque juxtapositions of the irreconcilable and the risibly antipathetic which now define central Birmingham, those and the gaps, the wastelands, the no-man's-lands which roads inevitably create. And these ragged bits are not without uses – they are used as car parks, as places where garages can display noxious, toxic S-reg bangers for sale (£275 ono), as the sites of drive-through McDonald's, and drive-in balti shacks, as graveyards for motors that have been joy-ridden to death . . . Whatever Foottown wants us to believe, this is the one place in Britain with the honesty to admit that the car is king.

(1998)

LONDON TRANSPORT

IF GOD HAD meant us to walk he wouldn't have allowed us to invent the wheel. About eighteen months ago I was giving Richard and Ruth Rogers a lift home from a dinner party: Dartmouth Park to Chelsea. After my third or fourth ruseful road-scam/rat-run/time-muncher the noble architect asked me, not so rhetorically that I could squirm out of answering: 'Do you drive a lot in London?'

The truthful reply would have been: 'I am probably a reincarnated cabbie. I've got the knowledge. I know every cut-through from Parliament Hill to Hilly Fields, from Gospel Oak to Honor Oak. NW5 to SW3 is a piece of cake. I drive in London because it's the most efficient way to get around. Want to know who I think should be hanged?'

What I heard myself saying was: 'Well . . . no, not a lot so to sp . . . just, uh, a bit – you know . . .where I live now it's imposs . . .'

Rumbled?

You bet. To admit to anyone in a certain stratum of bien pensant London that you drive is bad enough. To admit it to the Urban Renaissance Man, the most public proponent of all things green and modern and trafficless, is rather like boasting to the Lung Vice Marshall of ASH that you see off two packs of Capstan per day. Driving is bad: it's bad for you; it's bad for everyone else too. Non-drivers are at risk from passive driving.

I'm not sure how the Rogerses had intended to get home. But even had they (improbably) been thinking of calling a taxi they'd almost certainly have been out of luck. And if one had shown up – if – it would have been a mobile sink estate, an ashtray on wheels, a B-reg smoker held together with camera tape, lacy with rust and driven by a sleep-deprived psychopath who's never even heard of Child's Hill or Hatcham and doesn't bother to own the A–Z because he can't read maps. It's not just London's public transport which stinks.

This is a city which also has a major taxi problem. And that problem is actually exacerbated by creeping pedestrianisation. Buenos Aires should be the model for London in this regard: just about every other car is a (licensed) taxi; they are on the streets twenty-four hours a day; they are cheap and thus available to all but the very poor; they are subsidised; they are run for the good of the city.

London taxis are characteristically English – they are run for the convenience of taxi barons. So they're expensive. And they're never there. The West End at 2300 hours is virtually bereft of black cabs. But it is swarming with African taxi pimps: Soho, Covent Garden and Charing Cross are plagued by these guys whose importunacy is a threatening irritant, whose cars may or may not be insured to carry passengers for gain, whose knowledge is minimal, whose prices are astronomical.

One reason why there are no taxis at that hour (apart from the fact that cabbies are obliged to live in Gants Hill or Hatch End and so have a long drive home) is that much of the West End suffers from Westminster Council's proscriptions of traffic which have been prosecuted with autocratic crassness, with cosmetic 'consultations', and with an opportunistic embrace of vehicular correctness. If the intention was to turn Soho into an inaccessible labyrinth, then the council has succeeded all too well. It doesn't seem to matter to Westminster that Soho's shopkeepers and bar owners and restaurateurs are experiencing a diminution in trade as a result of this half-baked policy. I thought of ringing Wally Cretin-Berk at Westminster's traffic soviet but then thought again: one would only get the usual sententious obfuscation and platitudinous cant.

What one wouldn't get is the admission that the obstruction of cars (and taxis) is now a de facto policy pursued by numerous metropolitan local authorities. The aim is not to alleviate congestion or to make cities run better, but rather to punish the car driver.

Take for example Oaf Prescott's ludicrous wheeze of a bus lane from Heathrow: has it not occurred to him that the provision of this facility is rendered unnecessary by the atypical Heathrow–Paddington link – atypical because it is a rare, perhaps unique, instance of a new British public transport initiative that actually works.

But that wretched bus lane is evidently a one-off. Take the thousands of other bus lanes which provide space for empty vehicles travelling in convoy (a habit that has ceased to be a joke). Take the practice of slowing traffic with the belt-and-braces device of supplementing roundabouts with traffic lights – a patent manifestation of an over-authoritarian state interfering wherever it can to save us from ourselves. Take the squads of traffic wardens who apply only the letter of the law.

The policy of exponential pedestrianisation is not one born of any sort of pragmatism or practicality. If its propagandists and advocates had ever bothered to look at Leicester Square how could they have resolved to do the same to Trafalgar Square?

Leicester Square is a pit lined by reeking fast-food outlets and inhabited almost exclusively by gormless tourists and loitering representatives of the various trades that exploit gormless tourists. This is what happens to places which are stripped of traffic. They become magnets for low-level squalor.

Leicester Square is already avoided by most Londoners – does Norman Foster (who appears to have grabbed the Trafalgar Square chunk of the cake from his old friend and rival) really intend that a further swathe of epicentral London should be put off-limits and freed up to create a sort of trash sump for grockles? Who are cities meant to be for?

And, besides, why should it be architects, of all people, who are appointed to determine strategies for London's road use? Why not civil engineers or actuaries or traffic wonks? The notion that cars must be stopped even if public transport is inadequate is more than mere current orthodoxy or conventional wisdom. Vehicular correctness is, rather, a sort of inchoate ideology; it is a sort of semi-articulated faith. Which is to say that it's not susceptible to correction by reason or by getting its devotees to trust in the evidence of their own eyes.

Vehicular correctness is a soft-left, liberal-consensual mix of high-minded, herbivorous good intentions and markedly illiberal petty-minded authoritarianism – one of the government's unlovely specialisations that seems to have spread through local authorities. The roots of this ideology are buried in a midden of eco-friendliness, greener-than-thou planet-saving self-righteousness and old-fashioned arcadianism.

The problem persistently overlooked by champions of car-free London is that this city cannot be expected to behave like the imaginary European city which is the cynosure of all architects' eyes – no cars, public spaces, civility, good local shops, outdoor cafes, intravenous cappuccino, etc.

This is the wrong model. London is thinly populated. It sprawls. It had horse-traffic problems before the invention of the internal combustion engine. Its density is between half and two thirds that of Paris or Rome (where, incidentally, traffic is forever exhorted to go faster, to keep moving).

The Master Atlas of Greater London, edition 6B (part revision) 1992, contains the area within the M25, an area of going on 1,000 square miles, in 188 pages – of which perhaps eight or ten can be considered cartographic representations of an unequivocally urban topography. The other 95 per cent is suburban, exurban, edge, margin, delusory countryside – it is fantastically varied and it is often exciting. It's not the heart of London but it is where London's heart is.

London – how often does one have to say it? – is essentially centrifugal. And it is not composed of villages with village shops for urban villagers: sure, there are one or two such areas where one can live as one might in a *quartier* of Paris with every quotidian need catered for in a single street – Primrose Hill is like that, and Westbourne Grove, perhaps. And very few others.

However these places, inhabited almost exclusively by newspaper columnists, are held up as typical of London when they're absolutely uncharacteristic. London is characteristically loose-weave, diffuse, disorganised and too big to walk: the separation of uses means that only a small fragment of the populace lives near shops – if there are any shops left. Supermarkets are often the only choice, and they demand cars. If the government wants us to give up our cars it should not put its knees behind its ears for such retailers as Wal-Mart.

The lack of holistic, all-embracing urban policy is, of course, not surprising. Richard Roger's Urban Task Force report, *Towards An*

Urban Renaissance, may be persistently wrong-headed about cars, but many of its proposals are unexceptionable and apt.

None the less, its persistent insistence on how much Jerusalem is going to cost is unlikely to endear it to Culpability Brown who has already denied Oaf the funds to put wheels or brakes on rail rolling stock in his certainty that tax cuts will win more votes than a diminution in the number of trains crashing or failing to run on time.

Vehicular correctness is woefully indiscriminate. All cars, from St Toni Scrounger's limo – whose motorcycle escort (New Labour, same old outriders) almost ran me down the afternoon of the evening he had dinner with his idol Clinton beside Tower Bridge – to unroadworthy bangers posing as taxis are treated the same. They are The Car. Well, of course, they aren't.

A simple and cheap means of starting to reduce the number of cars on the roads is a vastly more stringent MoT test – a test which would certainly disqualify, perhaps impound and possibly melt down the patently unroadworthy vehicles which probably constitute fifteen to twenty per cent of the cars driven in this country. I am sure that Mr Steve Norris, a former car dealer, will see the point of this one. Of course, the knock-on of all these confiscations would be a commensurate increase in stolen cars – but no system is perfect. Is it?

Vehicular correctness doesn't recognise that eternal truth. It seems to believe that a perfect traffic system can be achieved: it probably believes, with Pelagius and the architectural determinists of mid-century, that humankind can be improved, that with subjection to the right combination of mini-roundabout clusters, sleeping policemen and bollards, we will all be better people. It is Utopian. It is goody-goody. It longs for year zero. It is curiously unworldly.

And it goes without saying that it signally fails to acknowledge that a city modelled on some Dutch provincial town full of cycle lanes and happy, smiling, pathologically tolerant, socially responsible, new model citizens is not half so alluring as a city of entropy and chaos which you move through in a private space that adds to the degradation.

(1999)

JOE BUILDING: THE STALIN MEMORIAL LECTURE (2005)

LECTURE THEATRE

Good evening. And a very warm welcome to the first Joseph Stalin Memorial Lecture.

It is unthinkable that anyone in Lithuania, having made a fortune from mushrooms, would set up a theme park devoted to Hitler. Yet this Baltic state's top man in canning, drying and trading boletes, chanterelles and morels has founded Gruto Parkas, also known as Stalin World . . .

STALIN WORLD. DETAILS OF STATUARY, GUN TOWERS, ROLLING STOCK

What is it?

A shrine to a despot?

A memorial to his victims?

A provocation?

A witting joke?

An unwitting joke?

A test to stretch the limits of taste?

A memento mori?

A gesture of politically incorrect defiance?

It is probably all of these.

Although it makes no allusion to many Lithuanians having collaborated with Nazi Germany, it is also that least Stalinist of endeavours — an attempt to tell at least a partial truth about the past rather than to rewrite it.

BEFORE AND AFTER PHOTOS. STALIN WITH TROTSKY. PUSHED OUT OF FRAME BY
ALTERED PHOTOGRAPH IN WHICH TROTSKY HAS DISAPPEARED

For Stalin the past was whatever he and his battalions of airbrush-operatives wanted to make it. It was as much a lie as the present and the future. The erased image presaged the erased life.

GRUTAS PARKAS/STALIN WORLD. STATUARY. STRESS INCONGRUITY
OF WOODLAND GARDEN AND EXHIBITS

Stalin's genocidal zeal was different from Hitler's — in its aims, in the sheer randomness with which its victims were selected. And the extent of its enormities was greater.

Hitlerians are regarded as morally infirm. They know it, and they tend to keep their squalid infatuation to themselves. They're furtive, like paedophiles in jackboots. They are beyond the pale.

Why then are Stalinists not beyond the pale? Why does Stalin's name not prompt the same revulsion that Hitler's does? We are selectively fastidious, selectively demonising.

It is, apparently, no solecism to express a sly admiration for Uncle Joe. Queen Elizabeth the Queen Mother's riotous annual turn at the Sandringham staff's Christmas party dressed as the mass murderer's mass murderer never began to incur the level of censure that the mob visited on her great-grandson Harry when the poor mutt donned a Nazi armband.

SCRIPT 2

STEPPES. VIEW TO INFINITY. SWAYING GRASS, VAST SKY. THREATENING CLOUDSCAPE. CLOSE-UP AND DISTANT SHOTS MIXED, SO WE KEEP LOSING SENSE OF SCALE

Certain of the effects of tyrannies are quantifiable — the number of deaths achieved by pogroms, genocidal programmes, famines, wars, forced labour, imprisonment. This is what the creation of a better world evidently involves. It involves easy things: turning policemen into willing murderers, turning the army against civilians, turning crime into the rule of law, turning fields into killing fields, turning a country's soil into an ossuary. It is much harder for a state to use its power gently than to abuse it.

During Stalin's regime countless millions of people died: shot, poisoned, tortured, starved, frozen. Many died, incredibly, to meet the death quotas determined by the secret police, the NKVD, as a sort of sacrificial offering to the bloodthirsty tyrant, for no reason other than to satisfy the ogre in the Kremlin who, in the words of Saint-Exupéry, 'murdered according to Marx'.

Joseph Stalin exercised something close to absolute power for quarter of a century.

This is the poet Osip Mandelstam who died in captivity in 1938, having recited these now famous, then suppressed, lines mocking Stalin. He recited them to less than a dozen people, one of whom turned out to be an informer.

> *Every day someone is executed.*
> *To our barrel-chested Georgian*
> *It's no more than picking berries.*

LENIN LIBRARY. READING ROOM. RANKS OF DESKS. TRACKING BETWEEN THEM. EVENTUALLY JM SMALL IN SHOT AT END OF ROW OF DESKS

It was Vladimir Nabokov's contention that there was no such thing as Soviet literature. The compromised circumstances of literature's creation militated against it being anything more than propaganda to deceive the naive and to flatter tyrants. Tyrants are notably susceptible to flattery.

Now, Nabokov's dicta were imperious and his loathing of the Soviet Union was visceral — witness his inordinate fondness for a tie which James Mason gave him: it bore the legend 'Fuck Communism'.

C/U JM'S TIE WITH 'FUCK COMMUNISM' IN CYRILLIC. INTERIOR. READING ROOM

But he was the greatest writer of his time and he remains almost thirty years after his death a tonic antidote to the battalions of fellow travelling apologists who deluded themselves that the stench of evil was sweet, who turned a blind eye to Stalin's crimes, and who excused them when they could no longer be denied.

The roll-call of these useful western idiots is long and varied. For instance: H. G. Wells; George Bernard Shaw, who declared during one of the great famines that the USSR was the best-fed country in the world; and the Queen Mother's chum James Callaghan — who observed after a trip to the repressive state of Czechoslovakia as late as 1960 that 'Socialism works'. Whatever that means.

SOCIALIST REALIST PORTRAITS OF WILSON, CALLAGHAN, FOOT, KINNOCK, SMITH, BLAIR — UNIFORMS, MEDALS, ETC., BUT NOT STALIN MOUSTACHE

Joe Building

It means that politicians in supposedly democratic states covet the licence of free-range tyrants.

So — can it be asserted with Nabokovian certainty that there is no such thing as Soviet architecture? There was certainly Soviet building. Just as there was Soviet facial hair.

VDNKH METRO STATION, MOSCOW

Architecture differs from literature. No matter what the regime, architecture depends upon a client's patronage. A writer merely requires pen and paper. An architect requires a commission.

Whether that commission is private or public, architecture is the result of a compact between creator and client. Architects have seldom suffered writers' persecution. But equally they have seldom enjoyed writers' freedom. And when the client is the most prescriptive and authoritarian of states, even that freedom is largely curtailed.

The compact is grossly unbalanced.

Every despot considers himself an architect — the architect of a society, the architect of its buildings. Buildings were to be the expressions of the new world. Buildings were a means to forge a new form of mankind. Buildings were a tool of the social engineer. Who would, of course, prevail because humans were potentially clean slates for the state to draw its crude lines on. Humans who did not become clean slates might soon become ex-humans.

ENTRANCE TO KOMSOMOLSKAYA METRO. JM IN FRONT OF TILED FRESCO OF CONSTRUCTION WORKERS

The act of building is a boast about the happy future which the state is providing for its people.

The act of building is a manifestation of perennial change, of constant revolution, of endless progress . . .

It is above all else the tangible, physical expression of promise.

VSKHV PAVILION OF RABBIT BREEDING. JM BESIDE STATUE OF BREEDER

Jam tomorrow . . . To travel hopefully is better than to arrive.

The Soviet Union was as much founded in the work of the nineteenth-century poetic realists Lewis Carroll and Robert Louis Stevenson as in that of the nineteenth-century prosaic fantasists Karl Marx and Frederick Engels.

Anticipation: that's the thing.

Anticipation of the ideal socialist state which is just over the horizon — we're almost there but not quite. The Russian word for an arterial road stretching into the distance is, tellingly, a *prospect*.

FRIEZE: WOMEN WITH RABBITS

Meanwhile, let us breed white rabbits. Utopia will be arrived at — eventually. Because the dream is more potent than the actuality, the actuality must be postponed — indefinitely.

MOSCOW STATE UNIVERSITY. INTERIOR. TRACK ALONG CENTRAL AXIS/CORRIDOR

Thus the most grandiose of Stalin's projects were never realised. They were triumphally announced. Competition upon competition was staged. Ever more ambitious drawings were published. The conditions of the brief would be extended to incite architects to push themselves further, higher. Yet the endless bureaucratic prevarication and the labyrinthine commissioning processes suggest that these projects were somehow never intended to be realised. Consummation was unachievable. They were as much a lie as Heaven on Earth.

PAINTING OF STALIN WITH ARCHITECTURAL PLANS AND PERSPECTIVES
The peaks of Joe Building were, ultimately, myths of the future indefinite. Myths wrought not by architects but by consummately gifted anonymous perspectivists — whose materials were paper and ink rather than stone and steel.

MOVING MONTAGE. PERSPECTIVISTS' VERSIONS OF FUTURE, UNREALISED PROJECTS: MOSCOW AEROFLOT BUILDING; LENINGRAD NAVAL ACADEMY; MOSCOW COMMISSARIAT OF HEAVY INDUSTRY
The fact that they were fictive does not detract from their reality. An object or person need not be actual to be real: look at God, the Devil, Anna Karenina, Eugene Onegin.
The highest peak, the palace of the Soviets, was, at 1,300 feet, to have exceeded every other building in the world.

MOVING MONTAGE. PALACE OF SOVIETS
It was the living god Stalin's homage to the dead god Lenin, whose statue on top was to have risen over 300 feet.
The cult of the dead god was not merely one of veneration. In Stalinist iconography Stalin is modestly portrayed: simply another comrade, seldom in centre frame. Nonetheless there is no doubt that he is Lenin's earthly representative. Lenin is the father, Stalin is the son. Thus Stalin, although self-invested with absolute power, enjoyed an exculpatory device: filial flaws could be ascribed to Lenin's example, just as cock-ups could be ascribed to saboteurs acting on the orders of Trotsky, the Judas — who from his exile described Stalinism as 'the syphilis of the workers' movement'.
As a representation of the dictatorship of the proletariat such architecture is laughable.
As a representation of the dictatorship of the dictators it could not be bettered.

VDNKHV PAVILIONS/STATUES/ENTRANCE ARCH DETAIL
Trotsky said that: 'The years 1928—31 [i.e. when Stalin had more or less gained absolute power] represented the bureaucracy's attempt to adapt itself to the proletariat.'
Put more bluntly, Stalin's cultural — and architectural — programme was the search for an idiom which appeals to idiots and illiterates, an idiom which is impervious to interpretation, which can be taken only one way.
This is what totalitarianism means. A despot's ideal is the building which imprisons the mind without having to imprison the body. Such a building offers no possibility of an alternative to its unequivocal message which, like a church's, is an exhortation to credulous obedience and dependence.

Joe Building

Stalinism was monopolistic just as the church had been at the height of its power. It was the conduit of all knowledge to an ignorant, unlettered populace — the source of all learning. It was a means of programming, and controlling, the people.

BLANK WALL OF AN UNIDENTIFIED BUILDING WHICH MIGHT POSSIBLY BE A SEMINARY: MAYBE THE HINT OF AN ARCH WHICH MIGHT BELONG TO A SACRED BUILDING. DONSKOI MONASTERY

In 1894, at the age of fourteen, the future Joseph Stalin — he was still Dzhugashvili, and would not adopt that name till he was over thirty — moved from his home town of Gori to enter the orthodox seminary in the nearby Georgian capital of Tbilisi.

He claimed to despise the church. But he would improbably have become the messianic tyrant he did without having suffered the seminary's cruelly rigorous regime of secrecy, searches, informers.

ARCHIVE OF CHURCHES BEING DEMOLISHED

Stalin must have been religious. He may have *proclaimed* himself an atheist. He believed so much in God that he murdered God in order that he could seize God's power, and become almighty — astonishingly he succeeded: God was the greatest of his many victims.

CHURCH EXTERIOR, ST JOHN THE WARRIOR

Pascal wrote: 'Men never commit evil so totally, so gleefully, as when they do so out of religious faith.'

INTERIOR, ST MARON CHURCH, RICHLY DECORATED ICONOSTASIS. JM IN FRONT OF ICONOSTASIS

God, so far as I can gather, is reached through *faith* — a low-grade form of learnt imagination.

But imagination nonetheless.

CUT BACK TO ICONOSTASIS. ALL OF THE ICONS ARE NOW MATTED TO REPRESENT STALIN

To summon up The Living God no imagination is required. Imagination is rendered irrelevant.

JM EXITING CHURCH

While Stalin was training to achieve his most unusual ambition, Tbilisi was booming. And when cities boom they build. The adolescent Stalin, brought up in a shabby small town, saw a new city arise around him.

ART NOUVEAU. SHOT SO THAT WE DON'T SEE SETTING/SURROUNDS. HOUSE AT SIMONETSKY STREET SHOT IN SEVERAL UNRELATED FRAMES

Its architecture made an impression on him. It determined his taste. A taste that would subsequently mutate into his empire's taste — because it was his will.

Tbilisi enthusiastically adopted art nouveau — which was not new. It was, rather, the last gasp of nineteenth-century revivalism.

GORKY HOUSE. JM ON SEAT AT BOTTOM OF STAIRS. POSTCARDS TO ILLUSTRATE BAROQUE AND ROCOCO. EXTERIOR, HOUSE ON CISTOPRUDNYJ WITH ANIMAL FRIEZES

Art nouveau gives us an indication of Stalin's early

thinking about the uses of architecture and its potential for politicisation. He observed how the opulent style of domestic and commercial buildings commissioned by Tbilisi's newly rich merchants was swiftly imitated by the poor.

ART NOUVEAU RAIL STATION. OUTSIDE DRINKERS
The poor aspire to be rich.
And if they can't be rich they will ape the trappings of the rich.
Art nouveau was a supreme form of maximalism. It was tabloid architecture. Importunate, loud, vulgar, easily understood, familiar, comforting, figurative.

EXTERIORS, RUSAKOV CLUB. BEGINS WITH CRASHING AUDITORY CUT TO LOUDLY DISSONANT SERIAL MUSIC. LOWER THE MUSIC, BUT LET IT CONTINUE
The architecture of the early Soviet period was bereft of these qualities. Its abstraction and structural candour were revolutionary — *architecturally* revolutionary, that is.
The equation of revolutionary art and revolutionary politics is naive in the extreme.
Constructivism was Russia's particular version of modernism. Which was routinely derided by its many detractors as international modernism, an epithet which deliberately echoed international capital — a derogatory code for Jewish money.

LYUBYANKA METRO ENTRANCE. SERIAL MUSIC
Modernism would gradually *become* internationalised — not least as a result of the emigrations enforced by the very tyrants who feared that internationalism might jeopardise the illusory nationalistic purity they sought to impose. Tyrants are big on purity.

WHIP PAN THROUGH 180° TO REVEAL JM IN FRONT OF LYUBYANKA
This is the Lyubyanka where purity was taught.

SCREAMS. KRASNIYE VAROTA: METRO ENTRANCE; PERFORATED WALL AT LENIN LIBRARY
What most strains of modernism *did* have in common was a resistance to the literally representational. They also had foundations in painting and sculpture which celebrated the machine to the degree of wishing to appear machine-made.

SHABALOVKA TOWER. BUTCH C/U OF ENGINEERING DETAILS: GIRDERS, BOLTS, ETC. — MADE TO LOOK BEAUTIFUL
More than any strain of western European modernism, Constructivism was preoccupied by engineering's potential . . .

SHABALOVKA TOWER
. . . to create an architecture which had never previously existed, an architecture which was *genuinely* new. It was realised that engineering could be used in the invention of a counter-intuitive sort of structure . . .

MELNIKOV HOUSE
. . . which would overthrow the old architectural order as wholly as the Soviet experiment had overthrown the old social

order. It further realised that engineering might serve ends other than functionalism — or that the functional and the fantastical could be the same.

DYNAMO METRO

Anything was possible, so long as it broke with the past and shed the burden of history.

Few projects got off the ground in the '20s. After revolution, after civil war, there was economic chaos which militated against building. And by the end of that decade Stalin's power was absolute.

Eras which dispense with aesthetic caution are rare because they are feared by authority. And authority would not be authority were its knee-jerk reaction not to inhibit artistic liberty while claiming to grant licence.

FROM THIS POINT OF THE FILM ONWARDS WE DISPENSE WITH IDEALISATION, WITH FORMAL SET-UPS THAT SHOW PLACES AS THEY MIGHT BE SEEN IN POSTCARDS. INSTEAD: INTRODUCE LADA AND MOSKVA CARS, TRAMS, BILLBOARDS INC. BIG BROTHER, PEOPLE IN UNIFORMS. EXTERIORS. PROTRACTED SEQUENCE OF RECOGNISABLY RUSSIAN ARCHITECTURAL BACKDROPS. ONION DOMES, BRIGHT COLOURS, NARYSHKIN BAROQUE, ETC. SERIES OF SHACKS, HUTS, RIVER BANKS, HEDGES, BURNT-OUT BUILDING NEAR THE EGG — NON STALINIST LOCI — PLACES WHICH ARE EVIDENTLY UNPLANNED, UNCONTROLLED

Dictators by definition abhor what they cannot control, what they have not dictated.

So artistic or architectural revolution and political revolution are seldom happy bedfellows.

THOUGHT BUBBLE OUT OF JM'S HEAD: TASTEFUL SOCIALIST REALIST DEPICTION OF STALIN IN BED WITH ARCHITECT — ARCHITECT IS BEING BUGGERED BY STALIN. MELNIKOV'S BUS GARAGE

The masses are not to be won over by avant-garde experiments which are, evidently, exclusive — and which are, usually, the endeavours of that section of the bourgeoisie which does not consider itself to be bourgeois.

By the early 1930s modernism's radical programme had been quashed by Stalin's reactionary cultural revolution. Architects were warned about the dangers of uniformity, and modernism was declared un-Soviet. The Constructivists had either emigrated or had swallowed their pride and adopted the populist idioms that the state now sanctioned.

The coercion of the proletariat can only be achieved by populist means; that is to say, by means with which it is already familiar.

FAST FOOD STALL IN COUNTRY OR OUTER BURBS

By cakes and ale . . .

JM, TRUNK ONLY, HOLDING OUT LOAF OF BLACK BREAD TO CAMERA — WHICH TILTS UP TO SHOW HIM WEARING CLOWN MASK

By bread and circuses. . .

JM WITH NEWSPAPER 'THE SUN BACKS BLAIR'; OCTOBER SQUARE. VAST HALF-NAKED WOMAN ON ELECTRONIC BILLBOARD

By tabloids and pop music.

SCRIPT 2

THOUGHT BUBBLE: SOCIALIST REALIST NON-FLASH-FRAME OF BLAIR WITH NOEL GALLAGHER, WILSON WITH BEATLES

Political revolutions will only succeed if they espouse aesthetic conservativism.

VOLGOGRAD CENTRAL AXIS, PAVILIONS BESIDE RIVER. OTHER COLUMNS AND CAPITALS AND PILASTERS

From the Renaissance till the Enlightenment, it was taken for granted that the classical orders, notionally based on human proportions, descended from God and were invested with his authority because he had created the first human being Adam in his image.

Whether or not an untutored populace appreciates this underlying conceit — almost certainly not — the apes of God have persistently resorted to classicism and to its variations like the baroque — demob happy classicism — to embody their authority.

GERMAN TROOPS DOING THEIR SILLIER CEREMONIAL WALKS

Such authority was more willingly accepted — at least more easily exercised — in Hitler's Germany than it was in Stalin's Russia. Germany was a fraction of the size, was linguistically unified, was ethnically more or less homogeneous — save for Jewry which provided the internal enemy which is *sine qua non* of a dictatorship: the bourgeoisie fulfilled the same role for the Soviets.

MOSCOW PARK. CISTO PRUDNIYJ PUBLIC GARDENS. GOTHS, WRECKED DRINKERS

And then there is the German aptitude for ordered compliance. Russia, on the other hand, gave us Bolshevism and, hence, the word bolshie — which signifies the very opposite of compliant. Many Russians were too bolshie to subscribe to Bolshevism.

Nietzsche's notion of the Will to Power stipulates that the majority who do not seek power enter into a submissive pact with those who do. But Nietzsche was a German. His observations were based in familiarity with the mores of his compatriots. Engels and Marx were also German . . .

The familiar question: could Marx have lived in a Marxist state? Only if the old shaman had himself been the despot. Which is what he'd have wished. Despite his sentimental optimism about the proletariat, he was a paternalist: 'They cannot represent themselves — they must be represented.'

RUN-DOWN, NEAR-DERELICT VILLAGE NEAR VOLGOGRAD OR MAGNITOGORSK

Perhaps it was the wrong country that adopted Marxism, a country unsuited to it. Marx himself certainly thought Russia too backward, too agrarian and too lacking in concentrations of population . . . Russia was certainly inured to despotism. But that did not make it an ideal testing ground for the comradely, hypocritical despotism of Bolshevism — which was described by the apostate Marxist Peter Struve as 'imported foreign potions mixed with native rotgut'.

Of course, Germans drink . . .

FULL FRAME OF VODKA BOTTLES, MAYBE VODKA DEPARTMENT OF MOSCOW'S BIGGEST SUPERMARKET — OR A VODKA WAREHOUSE

Joe Building

Russians, however, *over*-drink.

Shelter, food and sex apart, the most basic human instinct is intoxication — the achievement of an altered state. In this regard Russians are superhuman.

RED SQUARE/KREMLIN BILLBOARD. JM POSED SO THAT HE MIGHT BE IN THE SQUARE

The Kremlin was a pissocracy: in his later years Stalin was the people's piss artist — an Oliver Reed gone to the bad.

Incidents of vodka-fuelled humiliations were routine. Men with the power of life and death over the planet's inhabitants were Badered day and night.

CAMERA MOVES OFF JM ON TO KREMLIN AND THEN INTO REAL KREMLIN FROM A DISTANCE

The Kremlin — a political/religious/ martial/social compound — did not observe the two-year-long prohibition of alcohol that it imposed on the rest of the USSR.

STEPPE. JM SHOWING VARIETY OF INTOXICANTS IN THEIR PACKAGING. MOOING NOISES OFF

And nor did anyone else observe it. It was unenforcable in a country whose people will do anything to get out of its head. To seek oblivion. Dry-cleaning fluid, glue, rubber, de-icer, dentist's alcohol, industrial alcohol, fertiliser, shoe polish . . . Russians will imbibe anything.

STEPPE. JM TURNS TO SEE HERD OF COWS BEARING DOWN ON HIM

Reindeer . . . Heavily disguised reindeer, admittedly. Genetically modified reindeer.

The hallucinatory fungus *Amanita muscaria* is so valued that two reindeer can be exchanged for one fruit body. The fungus is dried, powdered, drunk. The drinker's urine is subsequently ingested because the renal process increases the toxin's potency — and will increase it over and over again.

To many people in the north of Russia this interesting recreation was, understandably, preferable to compulsory participation in the great Soviet adventure.

Social utopia — no.

Psychotropic utopia — yes.

PLAN OF APARTMENT WITH DOTTED LINES DIVIDING IT UP INTO TINY UNITS

New mankind needs new forms of habitation. At a micro level this meant the horrors of collective living, of strictly apportioned space in apartments and houses requisitioned, or stolen, by the state, which now owned all property.

MOSCOW'S OUTER RING. EIGHT LANE BOULEVARDS. DRIVING SHOTS. MIX DAY AND NIGHT

At the macro level it meant new forms of city. Which caused a dilemma — for the city was, according to Engels who had witnessed at first hand the conditions of England's nineteenth-century industrial cities, an exclusively capitalist construction. Engels was, as it happens, wrong.

It is unsurprising that Soviet Russia — forged in nineteenth-century thought — should have been enthusiastic about the escapist late nineteenth-century solution to urban problems: the Garden City, devised by Ebenezer Howard, whose anti-urbanism was as virulent as Engels's.

SOKOL GARDEN VILLAGE OFF LENINGRADSKI PROSPECT OR PROZOROVKA GARDEN SUBURB

The garden city was the sort of solution that bravely turns its back on the problem, buries its head in the sand and attempts to start from zero.

Ernst May was another German, the city architect of Frankfurt where he had applied lessons in planning — though not in design — that he had learnt as a young man working on Hampstead Garden Suburb.

MAGNITOGORSK ARCHIVE

He was invited to Russia to build entire garden cities: Stalin had ordered that 700 such settlements should be created. The first, the model for those that would come after, was to serve the steel works of Magnitogorsk in the mineral-rich Urals.

VOLGOGRAD. BELCHING CHIMNEYS, TENT CITY POSTCARDS, BARRACKS POSTCARDS

A hell of noxious mills and toxic furnaces provided work for tens of thousands of people who subsisted in the most rudimentary barracks and tents. It was the wild east, a pig-iron Klondike. There was also, inevitably, a forced labour camp. In that regard it resembled Nazi Germany. But in its pollution and eco-hostility it could not have been further away from that most green of regimes.

Everything was against May. Bureaucracies, budgets, builders, schedule. And his neglect to design for the harsh climate.

MAGNITOGORSK MODERNISM. ARCHIVE

But beyond these, there was the matter of his aesthetic unsuitability. It is bewildering that he was appointed. May's undecorated, uninflected blocks were denounced as unsuitable for Heaven on Earth, as examples of capitalism's patronisation of workers. This was how the West treated its workers, by giving them hutches, machines for living in. May was dismissed. He said later that he was lucky to escape with his life.

MOKHOVIA STREET APARTMENTS AND/OR PHOTOS OF THEM, MOSCOW

The dictatorship of the proletariat scorned the architecture that had been high-mindedly devised to represent egalitarian socialism. It demanded instead mock palaces.

Here was the people's will in stone.

The people's will was the sixteenth-century Venice of Sansovino and Palladio.

ST MARTHA AND ST MARY

Thankfully for their own sake, architects have throughout history been more than willing to bend to the will of the people — that's to say, to the will of dictators.

Take Alexsey Shushev. When he was thirty he was a vernacular pasticheur: this convent church is not medieval, it was designed only a few years before the revolution.

Joe Building

THE PEOPLE'S COMMISSARIAT FOR AGRICULTURE (NARKOMZEM BUILDING)
He became a Constructivist . . .

KOMSOMOLSKAYA METRO
Soon after he became a neo-baroque decorator . . .

Such lackey hacks abounded in Stalin's Russia, earning their
Order of the Heroes of Anilinguia First Class, accepting dachas
and motor-cars from the beast whose achievements they hymned.

Now, such versatility was regarded with mistrust in the West
from the beginning of modernism — because modernism was a sort of
creed, an architectonic faith. But it was a faith that flourished
in democratic societies — it was a manifest of artistic freedom.
Which its denigrators would do well to bear in mind.

A work of art or engineering, architecture or craft, made in
a totalitarian state is a facet of that state, a glorification of
that state. It is achieved by the maker obeying state thought.

Or, terrifyingly, by the maker *second-guessing* state thought —
which, under the Soviets, was capricious, ever-changing. Second-
guess wrongly and you might forfeit your career — at least.

FULL FRAME OF NEWS-STAND WITH GARISH MAGAZINES. JM'S HEAD IN CORNER
A dictatorship has to suppress imagination — for it is the enemy
of obedience. During a paper shortage — caused, of course, by
Soviet incompetence — Lenin pronounced that paper used to print
poetry was paper wasted.

STALIN POSTER HELD IN FRONT OF JM'S FACE — HANDS OUT OF FRAME
Stalin was a different sort of philistine. He approved of poetry —
so long as it was doggerel that praised him.

> *Oh Great Stalin,*
> *Oh Leader of the Peoples,*
> *You brought man to birth,*
> *You fructify the Earth . . .*

LOWERS POSTER
This is a creation myth.

**STALIN IN GREAT COAT, BOOTS, CAP, VIGOROUSLY MASTURBATING ON THE
STEPPES: SHOWN FROM BEHIND, OF COURSE**
No wonder he was furious when Roosevelt told him that he was known
as Uncle Joe. It was an insult. It relegated him. A mere Uncle?
When you are the Living God who spreads his seed to fructify, the
literal Father of the Nation . . . **[ORGASMIC GRUNT FROM STALIN]**

**ARCHIVE. STEPPE COVERED IN SNOW/SLUSH/SPERM: ENTIRE FRAME FILLED BY
SLUSH**
. . . who embodies the history of the Nation, who personifies the
Nation — who understands its variety because he is, in his own
words, a Russified Georgian Asiatic. An heir to Genghis Khan. To
Tamburlaine. A warlord of the steppe. An outsider to Moscow . . .

Though to Le Corbusier, Moscow was itself 'the carcass of an
old Asiatic village'.

VDNHK. GILDED FOUNTAIN. TABLE BORROWED FROM CAFE
It is a common characteristic of rulers that they should be

outsiders, immigrants — from provinces or dependencies or semi-detached neighbouring states.

JM PRODUCES POSTCARDS OF THE FOLLOWING
One thinks first, of course, of Andrew Bonar Law, Britain's Scottish-born Canadian-naturalised prime minister, one thinks of New Labour which is a purely Scottish construct, one thinks of Catherine the Great, a German, of the Corsican Napoleon, of the Austrian Hitler . . .
You do not need to belong to a nation to foment nationalism in it.

POSTCARD OF STALIN AS VULTURE PEERING THROUGH RIBS OF SOME CREATURE. VOLGOGRAD MUSEUM, STALIN RUG
Maybe only an outsider can take a nation's temperature.
This Russified Georgian Asiatic — Lenin called him simply Asiatic — understood a glaringly obvious point that Lenin had ignored and that Trotsky had denied: namely that this greatest and most terrible of social experiments, this experiment forged in nineteenth-century thought, this experiment which he lead, was being conducted on the *Russian* people. And that it could only succeed if it took into account the Russianness of Russia.

ST MARON OR ST NICHOLAS IN PYZHI. CHURCH DWARFED BY HIGH RISES
The Eastern Orthodox church could not have been more Russian. But it was sectarian; thus a rival to be quashed. One of the Council of People's Commissars' earliest acts, within three months of the revolution — though that is, of course, approximate — was to abolish the Julian calendar. The Eastern Orthodox had stuck to it for the reason that it was not the Roman Catholic Church's Gregorian calendar — which even Protestant Britain had subscribed to since 1752.

A DARK WOOD. JM WANDERS THROUGH IT
So thirteen days were famously lost: hence the anniversary of the October Revolution occurring in November.
Where did those days go? People searched for them. They felt they had been robbed. Their life had been shortened. It was said that the days had been buried in a wood. But when you search in a Russian wood you do not find days.

[UNCOVERS PART OF SKELETON IN UNDERGROWTH] CLOCKS ON WALL SHOWING DIFFERENT TIME ZONES. BRIEF PAN FROM MOSCOW CLOCK SHOWING THE HANDS AT 3.35 TO KETTERING CLOCK SHOWING HANDS AT 3.25
Russia had broken with its past and had come into line with the West. It shared its measurement of time with what is now, laughably, called the international community. And so it remained till 1929.

JM TALKING TO CAMERA THROUGH THICKET OF BARBED WIRE
That was when Stalin's paranoiac isolationism began. This was when the rest of the world outside the Soviet Union effectively ceased to exist. That was the pretence, the delusion. So all tokens of degenerate capitalistic cosmopolitanism had to be annulled — tokens such as a shared calendar. But the Julian calendar could not be returned to because of its religious associations.

Joe Building

COPIES OF FIVE-DAY WEEK NEWSPAPER
So the Soviet calendar was devised. It was based on a five-day week. It caused such confusion that it endured for only two years before it was scrapped and replaced — by a calendar based on a six-day week . . .

SOUTH VOLGOGRAD. AIRCRAFT TURNED INTO BAR/RESTAURANT. CUTAWAY TO SOVIET WRISTWATCH
Under both these systems there existed a denomination of time exclusive to the USSR. Here was Soviet time. It was unlike anyone else's time. It was nationalist, hermetic, exclusive, insular time. These were the properties Soviet architecture would henceforth aspire to.

NORTH VOLGOGRAD. DAM ON VOLGA. JM ON TOP/BESIDE. CAMERA MOVES AWAY DURING PTC. SHOT IS HELD ON WATER. SHOT IS REVERSED SO THAT WATER FLOWS THE OTHER WAY. THE SKY DARKENS RAPIDLY AS THOUGH NIGHT HAS FALLEN. THEN THE TREES AND SKY GRADUALLY TURN INTO NEGATIVE. TERRIBLE GRINDING NOISE
Stalin in his omnipotence could amend the unities at will. He could change the very rhythm of life, he was more powerful than the moon's cycle, he could vanquish tides. He could control the elements. In a folk tale, a state-sponsored folk tale, of course, he altered the Earth's orbit. He was a greater force than nature.

SOUTH VOLGOGRAD. ARCH ACROSS CANAL/RIVER JUNCTION FROM BOAT
He created inland seas. His slave labourers died in their thousands digging bloated canals more ostentatious than utility demanded. Their function was to prove the state's might. Like the great dams they were feats of symbolic hydraulic engineering.

VOLGOGRAD. PYLONS AND SUBSTATIONS
The electrification of the USSR was so much a cause of enjoined nationalistic celebration that parents were encouraged to call their baby girls Electrification.

JM AS SMALL GIRL WITH PIGTAILS APPARENTLY UNDERGOING ECT
It became a popular, patriotic given name along with Barricade, Metallurgist, Oncologist, Rheostatist, Galvaniser, Slideruler, Foundryhero and so on.

A MILITARY PARADE IN RED SQUARE. MIX TO CLOUDS TURNING TO ICE
Like many autocrats before him, Stalin determined to control the climate. Unlike them he partially succeeded through cloud seeding to ensure that the days of great parades were clear of rain.
  Apartments were given Italianate features and ice-cream was made widely available in an attempt to persuade Muscovites that their climate was Mediterranean.

CLOUDS FROM AIRCRAFT WITH AIRCRAFT SHADOW
Weather modification was also a Nazi preoccupation and one that the USA, inheritor of Nazi science, has covertly researched. It is, of course, a means of controlling the production of crops and, thus, population.

SOCIALIST REALIST PAINTING OF CORNFIELDS, HAPPY PEASANTS, ETC.,

SCRIPT 2

[THROUGHOUT FARM SEQUENCE THE ACTUALITY IS CRUDELY CONTRASTED WITH THE REPRESENTATIONS]

Infamously, Stalin had other methods of achieving those goals. There were the preposterous Five Year Plans — the first of which was self-congratulatorily achieved in less than five years . . .

COLLECTIVE FARMS, BARRACKS, ETC.; VOLGOGRAD SILOS; THE NIELTONES: 'AUNTIE MASHA HAD A FARM'

There was the brutal communisation of agriculture which occasioned the unprecedented enforced migration of peasant farmers into collective farms . . .

COLLECTIVE FARMS. SCULPTURES/PAINTINGS GLORIFYING AGRICULTURE

Seventy years after serfdom had been abolished under Tsar Alexander II, the Soviets, nostalgic for the perceived certainties of the nineteenth century, reintroduced it under another name.

Beyond this measure was the murder of thousands of kulaks, the land-owning peasants who were calumnised as enemies of the people even though most were little more than smallholders, but they were smallholders with just enough to eat and were thus condemned as bourgeois.

TRACTORS, COMBINE HARVESTERS AT WORK OR IN FRONT OF TRACTOR DRIVERS; LUNCH PAINTING

The mechanisation of agriculture became a Soviet fetish.

Tractors and combine harvesters were both the tools and, more importantly, the symbols of a new agrarian order. An order which resulted in a series of devastating famines — frivolously euphemised as 'problems of grain distribution'.

STEPPE. JM JOINTING A TAME RABBIT, FRYING IT. SOUNDTRACK: 'I GOT A BRAND-NEW COMBINE HARVESTER' IN RUSSIAN

One result of famine was cannibalism. Children were eaten by their own parents. Women craved pregnancy in order that they might give birth to food and that their lactate might be turned into curd cheese.

I find that the recipes best adapted to cooking human are those normally applied to rabbit and pork.

COMBINES PARKED UP IN FARM ON STEPPE

These famines were on a biblical scale. They were, of course, man-made like the majority of famines. They manifested the Living God's wrath or his negligent incompetence. He was aware of their extent yet was appallingly indifferent to their consequences.

CUTAWAY TO STALIN AS BABY

Stalinism was bizarrely infantile. There is something infantile about anthropomorphic tractors and zoomorphic combines. There is something infantile about murdering any real or supposed enemy. It is the politics of the playground, conducted in the moral void of untutored childhood.

TURN PAGE TO TERRIFYING ROUGED PANTOMIME DAME OR TENNIEL ILLUSTRATION

Stalin had a webbed foot and a withered arm. He was a creature of terrible fairy tale. He built grotesque fairy-tale castles fit for murderous witches and monstrous ogres and creatures from nightmares' sumps. A popular Soviet song of the years of the Great Terror included the line: 'We were born to make fairy tales come true.' Stalin was an invention of Lewis Carroll's come true: The Red Queen — 'orf with their heads'.

Only fairy tales — which shun the constraints of naturalism — can represent an enormity such as Stalin.

JM SITS AT A TABLE WITH SHOT GLASS AND BOTTLE OF VODKA. CAMERA VERY SLOWLY PULLS OUT TO REVEAL ON TABLE TOP A DOZEN OR SO MODEL BUILDINGS Starting from zero is also infantile. Building a new world for a new form of mankind is infantile. What would a brand-new world's buildings look like? The answer is simple. As with the calendar so with architecture — the buildings that Joe built are defined by what they are *not*.

JM POURS VODKA OVER BUILDINGS AND SETS LIGHT TO THEM They are not the old world's buildings. Yet they are not like those of anyone else's new world.

CONSTRUCTIVIST BUILDING. OR JM AGAINST WHITE WALL WITH SHADOW PLAY — INCHOATELY SINISTER Not modern. Stalin's buildings do not belong to any school of Modernism with a capital M. Though, of course, they were, inevitably, modern in the proper sense of that word. Whatever was created in, say, 1933 was at that moment modern no matter what it appeared to be.

SAME CONVENTION. TRACK ROUND A POLYCHROMATIC DOME Not Russian. Stalin's buildings are not Russian even though they allude to a Russia that might have been. And they are perforce Russian because Russia is where they are sited.

LIFT FROM *JERRY BUILDING*. VOLKISCH HOUSE WITH RUNIC INSCRIPTIONS Not Nazi. They are — god forbid: rather, living, now dead, god forbid — not Nazi because totalitarian genocidal National Socialism was, of course, a world away from totalitarian genocidal Soviet Socialism. Hitler's moustache was nothing like Stalin's.

SPLIT SCREEN OF THE TWO MOUSTACHES MAGNIFIED Although neither of the great dictators could swim . . .

SPLIT SCREEN CONTINUES: HIMMLER AND BONES, BERIA AND SUE LYON AS LOLITA The desk murderer Heinrich Himmler had nothing in common with the desk murderer Lavrentia Beria.

The one commissioned furniture made of human bones, the other merely raped prepubescent girls.

SEVERAL OVERWEIGHT DOGS OF DIFFERENT BREEDS WHICH ARE RUNNING. THE DOGS — MAGRITTE STYLE — ARE MADE OF BRICK, OF ASHLARED LIMESTONE, OF ROUGH RAGSTONE, OF BREEZE-BLOCK, ETC. SELECTION OF BARKS BY THE NIELTONES They are not reactionary, not bourgeois, not capitalistic running dogs in brick and stone because . . . Well, because they are constructed in service of a regime which opposes reaction, bourgeois opulence and capitalism's exploitation of its workers.

HOTEL ROOM. TAPED MOUTHS SHOUTING, STRUGGLING TO BE HEARD, MUFFLED ECHOES (SORT OF FRANCIS BACON)
  Not, not, not . . .

GUNSHOT NOISES. UKRAINE HOTEL, MOSCOW IN DISTANCE. JM SHOWS POSTCARDS OF FOLLOWING: CAPITOL BUILDING, LINCOLN, NEBRASKA. BOSTON AVENUE CHURCH, TULSA. CHURCH OF CHRIST THE KING, TULSA. TERMINAL TOWER, CLEVELAND, OHIO. RUTH ETTING SINGING 'TEN CENTS A DANCE'
  They are not American. The Soviet Union looked enviously at American industry, hired American engineers, may even have pursued an inverted version of the Great American Dream — but nonetheless officially America was regarded with extreme suspicion.
  The architect of this world-class hotel was arrested after having been overheard speaking English — which he had learnt in America.
  A chamfer lathe operator was imprisoned for boasting of having owned a car when he lived in America before the revolution.
  A Soviet skyscraper was deemed to exhibit different principles to an American skyscraper. An assertion that was made without qualification or explanation.

FAR EAST PAVILION AT VSKHV, MOSCOW AND PLOSTCLAD REVOLIUTSII METRO. SCULPTURE OF BORDER GUARD
  The Soviet Union's borders were sealed. Not that its citizens could find them. The absence of detailed maps was a means of preventing escape, and the soldiers who guarded the borders were regarded as heroes. They were honoured for preserving the integrity of the state. Thus the state could claim among other things that it was culturally autonomous.
  This was, of course, a lie — just as the projects which would never be built were lies.

VDNKH. JM AT TABLE OR ON BENCH. PROPS: NEST OF NAKED RUSSIAN DOLLS
  The Soviet Union was a prison which notoriously contained within it further prisons — like a Russian doll. But even the highest security prison is not entirely insulated against the world outside it . . . And Russian dolls are, anyway, originally Japanese and arrived in Russia only a few years before the Revolution.

MONTAGE OF DETAILS AND ENTIRE BUILDINGS — ALL IN DIFFERENT SCALE — IMPOSED OVER EACH OTHER, THROUGH FADES, WIPES, DIMINUTIONS OF IMAGE. SO A DOOR HANDLE SEEMS AS BIG AS AN ENTIRE BLOCK. E.G., APARTMENT BLOCKS. RED ARMY THEATRE. VOLGOGRAD PLANETARIUM. VOLGOGRAD RAIL STATION. SEVERAL STYLES OF RUSTICATION. SEVERAL STYLES OF FRIEZE. SEVERAL STYLES OF LOW RELIEFS. DOOR SURROUNDS, ETC. JM IN SOME SHOTS TO CONFUSE SCALE — DWARFED BY A CHANDELIER, HUGE AGAINST A SKYSCRAPER
  Stalin's architecture is not old, not modernist, not Georgian, not Nazi, obviously, not bourgeois, not baroque, not American, not Asiatic.
  It is, rather, all of those, simultaneously. There is a dogged conviction that more is more, an absolute faith in muchism. It is as though a building's status is determined by the sheer quantity of disparate elements that it can incorporate.

Joe Building

Elements which it hasn't borrowed but which it has stolen. That is only appropriate for a criminal regime. Like a thief, it of course denies that it steals. Architectural plunder was one of its lesser crimes.

The coarse amalgam of idioms is achieved with indiscriminate abandon. In this regard Stalin's architects were strangely prescient. They anticipated the world-wide reaction to modernism in the 1970s.

FREE-STANDING CAPITAL OUTSIDE ARCHITECTURAL MUSEUM. JM BESIDE IT

So — Joseph Stalin: glorious godfather of postmodernism . . .?

That's not so far-fetched as it might sound. Sure, postmodernism attempted to make a virtue of levity, it was allegedly knowing, jokey, allusive, witty and so on. But above all these it was populist, often tiresomely populist.

So much for the free-market responding to popular taste. It took the free-market over forty years to achieve what Stalin achieved by decree overnight. Namely, that architecture must be designed for the people, not for other architects.

HOTEL UKRAINE. SINGLE PAN INCLUDING TILTS TO SHOW THE PLETHORA OF MOTIFS

There was method in the architectural stratagems employed under Stalin.

There was a manifest tendency towards hedged bets. Here are buildings which appear to have been created in an attempt to avoid official disapproval (FLASH OF MAN BEING SHOT THROUGH HEAD) by including as many officially approved motifs as possible. This is what is meant by designing for life.

GRUTAS PARKAS/STALIN WORLD

Stalin exercised power for so long not least because Stalinism was whatever he said it was — which was seldom what it had been yesterday or would be tomorrow. He ruled by murderous caprice and malevolent improvisation. He was a fleetfoot who flatfooted his rivals — and everyone around him was a potential pretender. 'Law upon law: he forges them as though they're mere horseshoes,' wrote Mandelstam.

In the psychopathic playground of the Kremlin what was keenly sanctioned one week was a crime against the people the next. This applied to architectural aesthetics as to everything else.

VDNKH. CHANGE OF RHYTHM FROM ABOVE. DETAILS SHOWN INDIVIDUALLY. BUCRANIUM CAPITALS ON MEAT PAVILION; SHEEP CAPITALS; PIG RELIEFS; FRIEZES ON VOLGA REGION PAVILION; STONE FLOWER FOUNTAIN, ETC.

So the architect — the author — grants equal importance to every element. The sub-plots, so to speak, vie with the plot. The decorative footnotes dominate the structure to which they're appended. Ornamental parentheses expand across entire surfaces.

This strange form of egalitarianism — probably the only form of egalitarianism to be practised in the USSR — has its precedents in art nouveau and equally in nineteenth-century academicism. Not merely *architectural* academicism.

MUSEUM/GALLERY. GENUINE SOCIALIST REALIST PAINTINGS SHOWING JOYOUS COLLECTIVE WORKERS, STEEL WORKERS, ETC.

Socialist realist painting — which is neither socialist nor realist — also derived from nineteenth-century academicism. From a sort of painting decried as literary: which is a calumny on literature. And which is decried as bourgeois — which is a calumny on the bourgeoisie.

And it is, of course, accessible. Visually accessible.

VOLGOGRAD MAMAYEV HILL. EXHORTATIONAL WRITING — LITERALLY IN STONE — INCISED SLOGANS. JM READING

It *had* to be visually accessible for in the mid-1930s more than half the population was still illiterate. The Soviets brought literacy to their empire so that new readers could learn of their achievements.

But by giving people access to knowledge, a tyranny conspires in its own destruction. The greater the number of people that can read, the more thorough the controls that must be exercised over what they read. Of course, the earliest writing was an instrument of demagogic coercion. So the Soviets belonged to the most ancient tradition of state lies.

The proletariat, as Stalin realised, is responsive to vast-scale hyper-kitsch and is supposed to be grateful for it — because it is literal, it is unambiguous and it is invested with a single religious or religiose meaning.

The subject matter of such painting is paramount. There must be a story or parable whose moral or message must be clear. There must be no ambiguity. There is no participatory engagement with the work. It is a one-sided compact. The passive spectator receives the straightforward message — in a state of gratitude.

POSTER OF STALIN

'Thank you Comrade Stalin for a Joyous Life. Life is Much Merrier now. Thank you — it might be added — for *obliging* us to be happy. To be unhappy is to be an enemy of the state. Happiness is our duty. Therefore we are happy.'

THE NIELTONES: 'HAPPINESS' (KEN DODD) IN RUSSIAN. SPLIT SCREEN. DAVID'S NAPOLEON ON REARING HORSE USED AS ADVERT

Socialist realist painting's other sources include the work of painters such as Jacques Louis David who had pandered to Napoleon's triumphalist narcissism.

What did Napoleon say to himself when he crowned himself . . .

YAKOVLEV'S ZHUKOV ON REARING HORSE

Vasily Yakovlev's rendering of Marshal Zhukov is sumptuously inane: this is the victor of the most mechanised war in history, and here he is in the flaming ruins of Berlin on a white charger . . . The anachronism is laughable. The intended heroism is, unwittingly, mock-heroism. The reversal of the swastika is formulaic superstition.

VOLGOGRAD. VAST STATUARY. BOTTOM OF MAMAEV HILL. EXTERIOR PLANETARIUM

Public architecture aspired to the same state as debased painting.

Joe Building

That is, to be literary, to be read. The gestures derive from
hagiographic baroque statuary of the counter-reformation. The
minimal difference is that the subjects are Stakhanovite tractor
drivers and fecund collective-farm girls — that is, the saints of
the new society, whose actual life was no doubt as miserable as
that of Rome's saints and whose death every bit as terrible.

**VOLGOGRAD. MAMAEV HILL. SCULPTURES IN SQUARE OF HEROES (BESIDE POOL).
WITHOUT REVEALING MOTHER RUSSIA**

The purpose of socialist realism was to proclaim the triumphant
march of the revolution, and to encourage that revolution which
would lead to further revolution and thence onwards forever . . .

MAMAEV HILL. SCULPTURES. MOTHER'S GRIEF, MEMORY OF THE GENERATIONS

It was, of course, a lie. Fiction is a lie that tells a truth.
Socialist realism was a lie that tells a further lie. It was
institutionalised mendacity, removed from the subjects it
purported to represent. The Soviet Union after its first decade was
anything but revolutionary. It produced art and architecture which
resisted change, let alone the chimera of progress. Its paramount
characteristic, once it was established, was paralysis.

**JM IN C/U. SHOT FROM A GREAT DISTANCE. THE FRAME BEGINS, ALMOST
IMPERCEPTIBLY, TO WIDEN. THIS SHOT MUST SEEM TO GO ON AND ON AND
ON . . . EVENTUALLY JM IS REVEALED AS STANDING BESIDE THE PLINTH
OF MOTHER RUSSIA. SHOT CONTINUES TO WIDEN UNTIL WE SEE THE ENTIRE
STATUE. AND THEN WE SEE THE SHANTY-LIKE DWELLINGS BENEATH IT. JM
BECOMES TINY SPECK**

Socialist realism's grip on reality was as frail as that of Norman
Rockwell or reality television.

Bad art for propaganda's sake. It is spin made stone — the
organised social lie as Kenneth Rexroth said of Woody Guthrie.
The bigger it grew, the more extravagant its dimensions, the more
apparent became its vacuity.

Buildings, harvests and expeditions were dreamed up solely in
order that they could be gloriously represented.

The battle of Stalingrad was obviously not dreamt up. It was
horribly real. The Soviet Union, our ally, suffered most and won
the greatest of victories. It was the central, most crucial battle
of the Second World War, the Great Patriotic War, and had the Red
Army not prevailed the consequences would have been unthinkable.

Its terrors and its heroism are captured on film and have been
recreated in literature.

But socialist realism cannot treat with the actual. Presented
with an immense subject its only response is to match epic valour
and epic squalor with epic scale — and epic emptiness. Millions of
lost lives and the squalor of battle and its alleged glories are,
supposedly, commemorated by this vast figure.

Mother Russia is meant to signify sacrifice, and the indomitable
spirit of the nation whose soldiers and civilians thwarted Nazism,
and the military virtuosity of Zhukov, and the perspicacity of the
great war leader Stalin — who did not actually visit the battle
zone.

Kitsch may be defined as the gulf between a professed, intended sentiment and the bloated expression of that sentiment. This is kitsch. Mother Russia is huge and void, lavish and arid.

DETAILS OF MOTHER RUSSIA — AS C/U AS POSSIBLE, INCLUDING GAPING MOUTH
La Voie Sacrée at Verdun is profoundly affecting. So too is Lutyens's elegaic monument to the fallen of the Somme. But this is merely a monument to monumentalism. It is not a conduit to remembrance.

If, however, we ignore its purpose — what it is meant to be for — then it is certainly one of the world's most spectacular freaks — a petrified possibly trans-sexual inflatable warrior doll — often mistaken by western sex tourists for the Memorial to the Scorned Mail-Order Bride.

ROAD OF SHACKS BENEATH MONUMENT. MUD, RUSTING CARS, ETC.
It shuns humility, it is mute about the battle's privations and degradations, yet it says everything about the collective caste of mind that considers such a monument appropriate, that glorifies itself and glorifies war while pretending to be glorifying the dead, and that treats ghosts and bones better than living beings.

It is axiomatic that the size and number of a regime's monuments are in direct proportion to the indignities suffered by that regime's subjects. Just as the number of people in uniform is in direct proportion to a society's repressive uncouthness.

The cumulative effect of so much XXXXXL statuary is that exterior space becomes a public gallery: a roofless museum of exhortation and bloated congratulation — it is the proletariat which is being congratulated, the proletariat knows however that its triumphant achievements are only possible because of the Living God . . .

MOSAIC STALIN, VOLGOGRAD PLANETARIUM, VOLGOGRAD MUSEUM BUST
The Living God whose representation was everywhere, idealised representation. Stalin — the self-proclaimed Man of Steel — was as vain as an ingenue. This God was a starlet that never ages. His raiments, calculated for immediate recognition, match his buildings: they become ever more pompous, more imperial.

TREES NEAR NORTHERN RIVER TERMINUS
His fingers are bloated maggots. He leers through cockroach whiskers. He thinks with his bones, he feels with his brow, he tries to remember his human form.

Mandelstam had the measure of him — and would, as a consequence, die having been stripped of *his* human form.

The tyrant reduces the persecuted to his level while maintaining the fiction that he is a benevolent pastor, wise teacher, generous father — hence a strain of building which is carefully tainted with frivolity. Hence carrot buildings, infected with the mad gaiety of a fairground — a fairground, however, at which attendance is mandatory.

MOSCOW STATE UNIVERSITY
The tyrant is also claimed to be the personification of the nation. And of its law. Hence stick buildings which are the law — an

Joe Building

agglomeration of whims, a bloated statute book — turned to stone.

MOSCOW STATE UNIVERSITY AND OTHER 'SEVEN SISTERS'

In the matter of the relationship between patron and architect, Stalin turned back the clock to an imprecisely pre-Renaissance time and to an imprecisely non-European place — to the storybook world of Akbar and Kubla Khan. He was like a deity of the steppes with blood and yak butter on his hands and fermented mare's milk in his beard. A deity who decrees that his yurt must be replicated in marble.

The brand-new world that Joe built was a brand-old world . . .

As the dotard Stalin crept into murderous second childhood, so did the buildings he commissioned increasingly recall those of his distant infancy in the nineteenth century.

The structural primacy that had given constructivism its name was overturned. Structure was disguised, hidden, buried beneath encrusted decoration, just as Stalin's deformed body was hidden and disguised by uniforms and retouching.

Architecture became ornament. Ornament became architecture.

MOSCOW/VOLGOGRAD. VAST EMPTY SPACE. HUGE SKY TO MATCH THE SPACE

All around buildings there was disposed abundant space — the better to display each ornament, to set it off like a jewel. A jewel is a token of wealth and power.

Urban space is not passive. All absolutists have exploited its aggression. It is oppressive and threatening. Its paradox is that it incarcerates. It uses scale rather than manacles and chains. It diminishes humankind.

MOSCOW. WIDE EMPTY BOULEVARD, JM IN CENTRE

The fretting about the form that a non-capitalist Moscow might take was in practice resolved by its being partially transformed into a city of such vast spaces — voids of oppression. Which are also happily apt for grandiose ceremonial, for triumphalist marching, for imperial pomp, for martial theatre, for the staging of the state's narcissistic opera . . .

MOSCOW. CASINOS ON LENINGRADSKAYA. PUSHKINSKAJA. LAS VEGAS STYLE NEON. DARK BACK STREETS. EMBANKMENT. BRIDGES. PAINTED-OUT ARMY LORRY WINDOWS. BIG BROTHER BILLBOARD

Its current character remains, after more than half a century, indebted to Stalin — but hardly in the way which he foresaw.

Moscow is a city that is akin to London — it changes at every corner: it changes style, it changes scale, it changes race, it changes mood, it changes purpose.

Unlike Paris it is, architecturally, marvellously impure.

Unlike Rome it does not offer a wearying diet of masterpieces: far from it . . .

JM LOOKS OUT OF CAR TO SEE HIMSELF IN CROWD BESIDE ROAD SNIFFING FUMES. LADA AFTER LADA IN SHOT

Its vast, empty, melancholic, formerly imperious spaces are still vast — but full: of Ladas and people . . . mostly Ladas, and the enticements of leaded petrol. Still imperious, though.

SCRIPT 2

LONG LENS JUXTAPOSITIONS OF BUILDINGS OF DIFFERENT AGES

A new city has squatted within the old. Stalin's statues are no longer here — the great airbrusher was himself airbrushed within a few years of his death.

Khrushchev's denunciation of Stalin in 1956 was presaged just over a year earlier by his denunciation of ornamentalism. A sign of how much buildings were regarded as expressions of Stalin's will and how much appearances mattered. And how much Stalin's disappearance mattered to Khrushchev.

But Stalin's Metro and his buildings and his 1935 arterial road plan are inescapable. And they are wonderfully contaminated, gleefully compromised.

STREET MARKETS. HYPERMARKETS. GUM STORE. YELISEEV FOOD STORE.JM WALKING

Stalin had no more idea than anyone else what a socialist city might be. Engels's equation of urbanism and capitalism was, as I say, often wrong.

In Moscow, however, it was predictively right. Stalin's Moscow was a stage waiting for the anarchy of capitalism to happen. There is, nowhere in the world, a greater Asiatic bazaar hoping to be taken for a European one.

Trotsky, too, was wrong when he wrote of 'the proletariat's congenital incapacity to become a ruling class'. Throughout the world the proletariat has done just that — through consumerism. The proletariat rules us all — its populist taste is universal law.

VERITÉ FOOTAGE OF MAY DAY RALLY

Stalin may have turned a country into a council estate littered with death camps and monuments to his tyranny. But he is increasingly celebrated. The sentimental longing for the days of his leadership is shared over generations and social classes. The tiny minority of Germans who express a similar respect for Hitler in this way are subjects of obloquy.

MONOLOGUE: NOSTALGIC DOTARDS

Stalin is a remarkable person. Soon his monuments will be everywhere. They were removed but they will be put back — If we had such a leader today our lives would be much better. We would have a unified state — the Soviet Union!

CONTINUED VERITÉ FOOTAGE

The dispossessed and the ideologically right-on march under red flags waving pictures of Chairman Joe. They apparently yearn for an era when comradely enamel medals were awarded for super-fertility and reaping feats, for digging and foraging and grassing up your fellows; to live under the Soviets was to be at school forever.

MONOLOGUE: OLD WOMAN

Our Motherland is the Soviet Union. I love her and I am proud of her. I cannot be sold for dollars! The Soviet Militia protects me. And what's happening now is our Motherland is being torn into pieces and destroyed completely.

CONTINUED VERITÉ FOOTAGE — CRAWLING MAN WEARING PUTIN MASK

Joe Building

They exhibit a dissent which their hero would never have tolerated — but which the later liberal tsars did, to their posthumous regret no doubt. Had their secret police been as single-minded as the Cheka and the NKVD, Djugashvili would not have lived long enough to become Stalin . . . Ulyanov to become Lenin . . . Bronshtein to become Trotsky. Never trust three pseudonymous moustaches . . .

MONOLOGUE: MIDDLE-AGED MAN

Since childhood, Stalin was like my religion. Time passed and I became a deputy manager and from being my religion I came to a real understanding of Stalin. He is the greatest knight of the twentieth century, the most honest man in the world.

BACK TO STREET

Communism is now a choice . . . which probably means that it isn't communism.

EIGHTH SISTER SKYSCRAPER

Just as marching in honour of Stalin is not taboo, neither is building in his image. Again, no German would, sixty years on, contemplate mimicking the Hitlerian manner. But Joe's buildings are everywhere in Moscow. The architecture of terror and purges and genocide is just a style choice, it is stripped of its terrible associations, served up in a spirit of willed forgetfulness.

NINTH SISTER SKYSCRAPER

But these buildings leave us in no doubt about the kind of architecture that betokens both the last truly powerful caste in Russia and the last proud moment in Russian history: the victory in the Great Patriotic War six decades ago — though it often feels as if it was six weeks ago.

CATHEDRAL OF CHRIST THE SAVIOUR

Of all the buildings made in Russia since the Soviets' demise, none is so symbolically loaded and aesthetically questionable as the Cathedral of Christ the Saviour. This is a precisely detailed reconstruction of the cathedral which stood on this site and which was demolished to make way for the Palace of the Soviets. It's an immutable law of humankind that when one religion dies those with an appetite for credulousness will seek out another.

Let us turn back the clock to a never time and never land where nothing changes, to the illusory solaces of stasis, to the illusory solaces of an older superstition . . . which will, in the cyclical way of things, itself eventually be succeeded — by a second coming?

CUT BACK TO PARADE AND FLAG-WAVING CROWD. CHANTS OF 'NATION! MOTHERLAND! SOCIALISM! NATION! MOTHERLAND! SOCIALISM!'

END

3
BLAND AHOY!

Power, Politics and Insipidity's Triumph

THE DISMAL PROFESSION

THE RIBA'S ANNUAL conference in 1979 was striking for three reasons, all of them regrettable. The first was the remarkably flimsy, intellectually fraudulent, historically ignorant 'keynote' address by Tom Wolfe, which he would subsequently expand into his captious polemic *From Bauhaus to Our House*. The second was the alarming passivity of the audience, which either didn't realise that it was being collectively insulted or was, by then, so inured to deprecations of architecture that it could not stir itself to respond by, say, walking out.

The third was the sartorial dreariness of that almost entirely male audience, which was almost entirely dressed in the drab suits of high street tailoring. Was I in the presence of some sect that worshipped at John Collier – 'the window to watch' – or at Hepworths, or Burtons? No, this was architecture's rank and file. And although it may be foolish to judge by appearances it is even more foolish not to judge by them. A familiar and reassuring figure of my childhood and adolescence had vanished: the flamboyantly bow-tied, floppy-haired chap with deafening tweeds and yellow socks. (Actually, I'm not sure if they all wore yellow socks, but a friend of my father did.)

That generation of architects – now dead, and by the time of that grim conference mostly on the point of retirement – dressed as though to express their self-proclaimed status as artists. Or aspirant artists. Or, at the very least, as arty folk. By 1979, however, architects had come to think of themselves as technicians, as members of a profession, as businessmen. An architect with the temerity to regard himself as an artist was by then a laughable and immodest dinosaur. That is the way it has, for the most part, remained these past twenty-five years. Twenty-five years that have thus been atypical in the history of architectural endeavour; twenty-five years that will, with luck, come to be seen as a strange hiatus.

The Dismal Profession

At a typically lavish bash to launch his tectonic bildungsroman *Place*, Sir Terry Farrell made a speech that was quite the opposite of that which one might expect on so convivial an occasion. It was bereft of jocular niceties. Instead it was an impressively passionate call for a reassessment of the way architecture is taught and, just as important, of whom it is taught to. Now, this was not a matter of pedagogic hair-splitting. Farrell's incontestable contention is that architecture's self-image – abetted by the skeletal purity of much current work – is increasingly that of a branch of civil engineering. His argument, based on his experience as an astonishingly precocious painter of neo-romantic landscapes, illustrator, pasticheur and photographer, is that a background in the visual arts is of incomparable value to the future architect. The corollary is that the current emphasis on mathematics and physics is misplaced, though hardly surprising given the mistrust of art unless it is 'practical' or 'vocational'. The qualifications demanded for entry to architectural courses are, simply, wrong. And schools are complicit: they discourage pupils who are not strong in maths in the belief that this is the primary gift required. Which would no doubt interest, *inter alia*, Michelangelo (sculptor), Inigo Jones (stage designer), Sir John Vanbrugh (playwright), Sir Edwin Lutyens (doodler), Le Corbusier (painter/sculptor), Pancho Guedes (sculptor and the only member of this list to be a 'qualified' architect).

These people have changed the way we think of buildings. They have done so through that combination of imagination, memory, intuition, risk, bloody-mindedness, whimsy, obsession, practicality and craft that we deem art. And which obeys its own rules – that are unquantifiable, and so impervious to bureaucratic scrutiny and professional control. In the mid-nineteenth century, opponents of architecture's being granted the status of 'profession', such as Norman Shaw, foresaw the problems posed by attempts to regulate an art as if it were a discipline like the law. There is today an obvious irony in the insistence on irrelevant educational qualifications. For there has never been a better moment to be innumerate yet to conceal that failing and render it utterly irrelevant.

(2005)

FUCK E**LISH *ERIT*GE

THE TWO MAST HOUSES just within the Victory Gate of Portsmouth Dockyard are raised above the water on piloti. They are structures of remarkable grace, clinker-built, painted the palest green. They are vast, as they needed to be. Their survival in an industrial site devoted for a century to the servicing of mastless vessels is a matter for celebration. The use to which the more southerly is put is a matter for obloquy: the Mary Rose Shop is a repository of tawdry, insipid tat. It's the sort of stuff to make me wince – a dismal, timid inventory of mediocrity. Bad taste is forgivable. It's no taste which is so disheartening.

Mr Tudor Rat is 'a soft toy, very cuddly, and fun to have around'. (Watch the kiddies' buboes grow. Contract Weil's disease.) 'A rich burgundy leather fan-shaped jewellery box inspired by the detail on the central boss from a Mary Rose gun shield.' A Henry VIII 'collectors' thimble. A 'Collectable Mary Rose Henry Bear'. 'Henry VIII and Elizabeth I pots. Take off their heads and store your treasures.' Mary Rose tea towel. Mary Rose coasters. Mary Rose cups. Mary Rose paperweights. Tudor Rose cross-stitch kit – comprises scissor keep, needle case, spectacle case. 'Glossy gusseted carrier bag with rope handles.' Jewellery 'inspired by the collection of Anne Boleyn'. The mug punter surveying this dross is treated to pyped musicke from distante tymes.

Now this crock of olde shyte might be deservedly ignored were it not for its enjoyment of the imprimatur of English Heritage. This preposterous quango – composed, apparently, of the blind and the bland – has the temerity to oppose, say, Renzo Piano's London Bridge Tower yet sanctions the debasement of marine archaeology, naval history and the magnificent dockyard itself by encouraging the sale of crass souvenirs which are no doubt 'accessible'. It is all too evident what sort of England English Heritage wishes to preserve (or create):

an England fit for beige cardigans and Country Casuals, an England that is ancient, deferential and dreary.

Directly across the road from the Mast Houses stands a fine, hangar-like storehouse completed on the eve of the Second World War. It is threatened with demolition and English Heritage will not recommend that it be listed.

That's bad enough. Its arrogantly negligent attitude towards the nearby Tricorn Centre is far worse. The Tricorn is a building of national importance, but English Heritage has again indulged its preening antiquarian prejudice. You know the mindset: if it's Georgian it must be good, if it is of the '60s it's not merely expendable but unworthy of consideration. The Commission on Architecture and the Built Environment (CABE) is, mercifully, equally parti pris but on the other side of the fence and may yet still save it.*

There is no surer means of pandering to philistine populism and currying favour in middle England than by uttering the words 'concrete monstrosity of the '60s'. The proposition that all copies of *The Naked Lunch* should be burnt and every print of *Rocky Mountain and Tired Indian* should be shredded is unthinkable, the stuff of yesterday's dystopian nightmares. But buildings contemporary with those works are regarded as acceptable targets.

We properly despise the mid-twentieth century's destruction of Victorian buildings (also 'monstrosities') and the late twentieth century's destruction of art deco buildings ('vulgar'). We know better now.

No, we don't. Apart from being of the '60s and of supremely sculptural concrete, Rodney Gordon and Owen Luder's Tricorn Centre is also a (former) shopping centre. Which further relegates it: buildings are subject to a taxonomy of use that disregards their aesthetic worth and determines whether they will be listed. A temple is deemed to be a higher form of building than a temple of commerce even if that temple of commerce is the most thrilling brutalist work to have been made in Britain.

---

* (It didn't.)

The Tricorn's detractors – and they are many – presumably have no objection to Portsmouth's latest shopping mall, Gunwharf Quays. This is a place that is offensive precisely because it dreads giving offence. It is architecture as frozen apology for itself. Its meekness is mitigated only by the Spinnaker Tower, a 'millennium project' which is showing signs of dereliction even though it is nowhere near complete. I concur with the waggish suggestion that the city would have been better represented by a 150-metre-tall OMO packet: if you don't get the allusion read *Pompey*.

In the depths of the mall itself you might be anywhere. There are outlets and factory shops selling, presumably, last season's styles or seconds though that doesn't quite explain what the contents of a Cadbury's factory shop might comprise and I didn't have the stomach to investigate. Alongside these are a carousel, a pork and bap stall, the usual chains and a branch of Gieves & Hawkes, a tailor which had a long presence on The Hard in Portsmouth in its capacity as a naval outfitter. This branch is predominantly civilian. Aspirant tars can get their togs at a fascinating shop called Baum in Queen Street: the tropical kit includes Northampton bench-made white canvas shoes at a bargain price.

I didn't experience the tropical microclimate of Gunwharf Quays because I was freezing most of the time I was in Portsmouth. Still, it must exist.

Everyone between the ages of fifteen and twenty-five wears clothes that suggest a heatwave. This is not occasioned solely by a desire for display – or so a Novocastrian tells me. The mindered clubs and bars which youth frequents are apparently so hot that the very minimum of garments is required. On Saturday nights queues stand in line and shiver as the Spithead Bite whips up off the water. Towns on keels slip past in the night. Gosport twinkles enticingly. (No one has ever written that before.) The harbour lights shine like jewelled fruit gums. They're gaudy, much too gaudy: I think I know a body that can sort that out, dim them so that they cause no displeasure to the fastidious English eye. Vessels may collide.

(2004)

AMATEUR

WHEN JOHN VANBRUGH, soldier, spy, herald and playwright, re-invented himself as John Vanbrugh, opportunist architect to the nobility, Jonathan Swift guyed him as 'Vitruvius the second' and wrote cattily: 'Van's genius without Thought or Lecture/ Is hugely turn'd to architecture.'

These mocking observations were, presumably, prompted by Swift's suspicions about dabblers – though both Christopher Wren, astronomer, and Inigo Jones, stage designer, might equally, at the outset of their architectural careers, have been charged with untutored tyroism. But their œuvre was already made when Swift pronounced. He had no means of anticipating that Vanbrugh's magnificently sullen works would ascend to even greater heights.

The architectural amateur has existed in many guises, most notably as folly builder, but he – and it invariably is a he – has existed with decreasing frequency since the trade or craft was dignified as a profession in the mid-nineteenth century. That elevation to the status of conspiracy against the laity has militated against the professional employment of even the most enthusiastic hobbyists. They have thus been restricted to designing for themselves.

The 1st Earl of Lovelace had the fortune to inherit both a country house and the entire village of East Horsley outside Dorking where, during the 1850s and '60s, he was able to practise his coarsely picturesque art with prodigal abandon. Amateurs are, typically, more restricted: building is, after all, a costly enterprise. Le Facteur Cheval's Palais idéal in the Drôme, François Michaud's sculpturally adorned houses in the Creuse and Lucien Favreau's borderline demented bungalow in the Charente are the products of lifelong obsession, hyperbolic essays in naive or outsider art, bricolage pursued to its crazed end.

In the early 1930s the painter Meredith Frampton designed the one building of his life, a house for himself on the West Wiltshire Downs, a French-influenced curiosity with a mansard roof and *oeil-de-boeuf* windows. One might have said that it was, in England, sui generis, had not it echoed the one building that the painter and critic Roger Fry designed, a decade and a half earlier: again, a house for himself, on the edge of Guildford. Neither, evidently, pays any heed to professional architectural fashion. Nor does the writer Wilfred Scawen Blunt's house in Sussex – a William and Maryish conceit constructed at the height of the Gothic revival.

The amateur is in these instances his own client. Patron and architect are one. If there is a single quality that defines amateurism it is bloody-minded whimsy.

I have excited the wrath of a former fashion designer called Wayne Hemingway by suggesting that his employment as an architectural consultant to Wimpey on a housing scheme in Gateshead is a ploy to lend that volume builder a dash of daring. If Wimpey and its peers really want to improve their product, I went on, there are people whom it might be an idea to commission: they are called architects.

Shaw omitted to add that professions are conspiracies against other professions, as well as against the laity. This truism is tiresomely, repetitively, demonstrated by the construction industry's factionalism and territorialism. The mutual loathing of builders, architects, quantity surveyors and civil engineers is an endless entertainment: the letters column of *Building* magazine is a morass of recriminatory spite and pitiful whines.

It is unsurprising, then, that Wimpey should enlist an enthusiastic amateur with a name and without, yet, the ballast of interdisciplinary prejudice. Wayne's bells-bulls-balls namesake once averred that 'it is a man's prerogative to change his mind'. What has caused me to change mine, to a degree, is not having this ex-clothie describe me as 'out of touch' (ouch!) but, rather, enjoying the opportunity to peruse his Gateshead project – albeit in miniature, at a RIBA exhibition, Coming Homes – without his mediation of it.

His whining, self-pitying, nagging Lancastrian exegeses on telly, in print and in numerous hard-hitting interviews – 'Wayne, can you transfer your incomparable genius for patch pockets to defensible space?' – have occluded the qualities which he has brought to the project. Which may not be those Wimpey bargained for. These qualities are not those of the old-fashioned amateur – whimsy is notable by its absence.

They are, rather, the qualities of the old trad professional, of the almost extinct genus of architect who thought first about the purposes of a home and its immediate environs and designed it accordingly. Yet Wayne affects to berate the shibboleth of form following function. Of all the projects displayed his is the most conservatively modernist, the most rooted in the humane modesty of the post-war new towns: this is where volume builders have got to.

But . . . If only . . . There are so many modern (as opposed to modernist) ideas and prototypes kicking around, seeking more than parochial commissions and not finding them. Modular prefabrication; properly maintained high-rises; 'mobile' homes which take advantage of the lax planning regulations which govern those strivingly twee coffins for the living with leaded lights, dimple glass and pitched roofs but which have been sleekly reworked with an orthogonal eye.

Britain is blessed with professional architects who can achieve something beyond another Stevenage. Nothing stands in their way save crass politicians, educationally subnormal planners and moderately talented amateurs seductively ushering them back into the recent past.

(2002)

MAMMARIAL

AT FIRST GLANCE – and you are here glancing from high across the Arno, where the roads twist through naves of stone pines above the Piazzale Michelangelo – the Duomo appears to be straightforward synecdoche, an architecturally applied instance of the literary conceit where a part stands for the whole. Much of the rest of the structure is obfuscated – by other buildings, by the thick breath of half a million tourists, by the exhalations of half a million cars – but the dome is unextinguishable. It is the cathedral. It stands, too, incidentally, for the city of Florence.

This first glance, this first paragraph, is entirely misleading, absolutely wrong. The word, duomo, does not signify an inflated cupola, or any other vaguely hemispherical structure on top of a building. It derives from the Latin domus – a house (whence domestic, whence domicile, whence, presumably, Domestos), and it means, specifically, god's house. Or one of god's houses – god is famously over-housed. The Italian god may be housed down the road at Sienna, say, in a striped cathedral without a dome, with only a campanile tower reaching towards heaven – this house is still a duomo. So is Milan Cathedral, of which the word was first used: Milan Cathedral is the only major gothic cathedral in Italy and even the Victorians at their most promiscuously eclectic failed to top off Gothic buildings with domes.

The point is that the word carried a sacred connotation before it carried a geometrical one. It described a holy space before that space was defined by the characteristic form signified in French by dôme and in English by dome – a usage which is less than 350 years old, which is contemporary, then, with the earliest appearances of the dome in English building, in the form of lanterns on houses derived from Netherlandish models. But these were slight affairs.

The first and greatest of the few major English domes is that of St

Paul's, which is evidently a sacred building – though it wasn't evident to me at the age of six when I first witnessed it in its sooty sublimity; I couldn't understand how it could be a cathedral, for cathedrals had spires like Salisbury's, and pointed arches. Even to an infant's eye, it obviously wasn't a musty repository of ancient mysteries and incomprehensible superstitions: the fearsome, jealous, tyrannical wrath-monger called god couldn't possibly live here. St Paul's was worldly. Even though I didn't yet know the word worldly, I could discern the building's swagger and urbanity and temporal grace. But for all those qualities, St Paul's is a church. It is a monument to unreason – even if the particular brand of unreason is tempered and Protestant rather than high-tar and Catholic. Its precursors are, of course, exclusively Catholic: the baroque was a Catholic idiom. It was the tectonic propaganda of the counter-reformation.

And the dome was the paramount part of the baroque's kit. Wren's baroque may not be agitated classicism in a distorting mirror and it may be ornamentally chaste, may indeed be infected by sobriety, by a certain coolness and compromise and level-headedness, but the fact remains that London's supreme building wears Roman (and French) clothes which have been only marginally toned down.

This should please the Christian Bomber Blair, whose fondness for Rome is well known. The Millennium Dome will inevitably be presented to the world as a technological marvel, a fantastic opportunity showcasing all that's wonderful about something or other . . . ad nauseam, ad vomitum. But even Blair probably has enough Italian to know what duomo means – and Richard Rogers, who is half Italian, certainly has. The Dome, the Dome, the Dome at Greenwich, the Dome that all sentient people are bored to the teeth with, is a covertly religious building. It is the First Church of New Britain and its dedication is to St Toni Scrounger.

Anything with a spire on it would have been too self-proclaiming. But what a spire means in England and in northern Europe is what a dome means in the southern half of the continent: they'll get on-message down there. They'll understand in Zaragoza and Turin that it's the Saviour-of-Europe's first Grand Projet.

Is there just the one message to be got? Is the dome as a form so bound to its sacred past that it cannot escape the meaning that institutionalised superstition has dumped on it? If it has difficulty escaping that burden, it is because its most singular quality is its aptitude for abeyance. There is no other comparable architectural device which is so prone to flit in and out of the common gamut: there was the Pantheon, Rome; then there was Hagia Sophia, Istanbul; then . . . well, save Innuits' igloos and Puglians' trulli, there was nothing – there was a few centuries' hiatus when, because mankind had forgotten its technology, the dome disappeared from the mainstream until Brunelleschi set to work in Florence.

The Renaissance prompted a habit in western culture of perpetual subsequent renaissances, of serial revisits. A style, a manner, a subject may have been dead for 3,000, 300, 30, 3 years before it is exhumed, re-displayed, re-invented – the only certainty is that it will happen, one day. Nothing ever goes away for ever. Reincarnation is the norm. Yet the dome appears only fitfully.

Given that copyism comes all too easily to architects, it might have been expected that after St Paul's domes would enjoy a vogue. But Wren's own domes at Greenwich Hospital, Vanbrugh's at Castle Howard, Hawksmoor's mausoleum in the grounds of that house, Archer's pavilion at Wrest Park and Gibbs's Radcliffe Camera don't make a summer. And, anyway, the (more or less) baroque in general was short-lived in these islands.

Despite its baroque associations, the early Palladians didn't entirely abjure the dome – witness Chiswick House, witness Mereworth (actually, don't bother – the grounds are guarded as zealously as those of an Ian Fleming villain). Later in the century the makers of picturesque landscapes appreciated the eye-catching properties of domed temples, domed churches, domed mausoleums (Stourhead, Nuneham Courtney, Hawkstone, Brocklesby, etc.). It can be persuasively argued that follies have a secondary function beyond that of occasioning delight and adding accents to a landscape: they are also working models, stylistic exercises, try-outs for idioms that may subsequently be employed for utile structures.

Thus began the Greek revival. And sham castles proved so popular that milord and milady wanted to live in one. But once again the dome failed to make the leap. Sure, both S. P. Cockerell's Sezincote and Nash's Brighton Pavilion, the two great late-Georgian essays in the Hindoo style, do possess domes of a sort. But these are buildings which never really get beyond the state of follydom. They exist in an exotic cul-de-sac. And while their authors, and such contemporaries as Soane and James Wyatt, did design other domes, they are invariably peripheral to the main composition. They are the punctuation marks, not the text.

And so it was throughout most of the nineteenth century: it wasn't that domes were structurally inhibiting – successive technologies were making them ever easier to achieve in ever greater spans – but that they were aesthetically disregarded. University College, London; the old Bedlam, now the Imperial War Museum; Haileybury College; the wretchedly feeble National Gallery – they look as though their architects hoped that their domes wouldn't be noticed. The Albert Hall, the largest domed building of the century in Britain, is more than 200 metres in circumference and possessed of a wonderfully impressive interior – but from without, its dome is so shallow as to be unnoticeable.

It is a matter of some regret to devotees of high Victorian perversity that none of the really wild men who flourished between the mid-'50s and the mid-'70s – Teulon, Pilkington, Lamb, etc. – brought his exhilarating insensitivity to bear on domed structures. For what might have been, one has to look to Piedmont and to the apparently certifiable work of Alessandro Antonelli at Turin and Novara: this frighteningly bizarre architect shared his English contemporaries' impatience with notions of beauty, harmony, proportion and so on. He went for energy, clashing masses, terrible shrieks, hyperbolic geometry. His domes are piled on top of teetering drums; they are stretched so that the soothing roundness of the mammarial form is distorted like the exhausted teat of a veteran wet nurse; their ugliness beggars belief.

In England from the mid-1890s, more than 200 years after St Paul's was begun, Wren's example was at last followed. There had been isolated outbreaks of baroque in the '50s: Wellington College, Leeds

Town Hall, the immediate surrounds of Victoria Station – all of them enthusiastically francophile, which is odd, given the dedication of the first. But then all things architecturally French were at the height of fashion at the very moment when the paranoiac Palmerston was building fortresses to keep that nation's armies out.

The baroque revival which extended well into Edward's reign was, with rare exceptions, based on English precept – which meant that there wasn't that much to base it on. There are many more baroque buildings of the turn of the nineteenth and twentieth centuries than there are of the turn of the seventeenth and eighteenth. And countless among them sport domes: the combination of red brick, grey stone dressings and green oxidised domes is ubiquitous. There is a dome beneath Justice on the Old Bailey which has the temerity to stand round the corner from St Paul's; there is a deep dome on Methodist Central Hall and there are several shallow ones on Westminster Cathedral; there are domes in Cardiff and Lancaster and Delhi and Belfast and Pretoria and Edinburgh and Calcutta, and there would have been a dome in Liverpool had Lutyens's Catholic cathedral not been interrupted by war, then ignominiously abandoned in favour of Paddy's Wigwam.

Actually there are domes in Liverpool, on the Mersey Docks Building at Pier Head. The dome was evidently no longer confined to sacred buildings, nor even to buildings such as courts, governmental offices, town halls and so on which are required to fulfil a representational or symbolic function. Domes had become commonplace, if not quite common: they were attached to stores, public libraries, theatres, billiard halls, mansion flats.

Indeed, by the time they went out of fashion in the years immediately preceding the First World War, they had acquired an almost frivolous etiquette. And after that war, just about their only manifestation was on the caramel faience of Lex garages which sprung straight out of Dulac's or Pogany's mildly oriental illustrations in the first age of mass motoring, when touring at forty miles an hour was an adventure rather than a chore and the buildings associated with it were suitably escapist.

Mammarial

When an architectural form becomes as ubiquitous as the dome did in England in the first decade of this century, its versatility and tendency to crop up in the most unexpected places means that, for the duration of its popularity, it carries no particular associational significance – it can mean anything: temperance, hedonism, bellicosity, mercantile vanity and so on. Retrospectively, of course, we forget the billiard halls and the kursaals. We think of imperial splendour, Britannic might.

That is what Edwardian baroque domes speak of a century on: when we race past the Ashton Memorial beside the M6 on the way to the Lakes, we do not, despite the structure's size and bombastic exhortation, remember the noble dynasty of linoleum barons it commemorates. We're more likely to reflect in a generalised way on the risible confidence of the epoch it belongs to, on the almost touching hubris it displays and on our terrible knowledge of what came next. A knowledge which, quite unjustly, vitiates mighty domes, turns them into cheap rhetorical gestures. That's hindsight for you.

After the war, the English dome duly went into another long hibernation. Architectural fashion had swung towards timid gentility, towards the straitened respectability of Quality Street Georgian and the cosmetic reassurances of the mock-Tudor. Grand domed gestures were for the Belgians: Liège has two examples of the domed expressionist church, and the northern Brussels skyline is dominated by the unloved and hilarious basilica at Koekelberg.

The grandest domed gesture was that of Hitler's poodle Albert Speer, whose Reichshalle would have been the largest dome in the world, more than twice the height of St Paul's: Speer possessed an infantile faith in superlatives and an indefatigable appetite for bigness. He believed that a building's merit was in direct proportion to its vastness. Had he been homosexual, he'd have been a size queen.

With such a history of patronage by the church, by the British empire, by the Third Reich and by Harrods, it is hardly astonishing that the post-Second World War modernist establishment in England considered the dome off limits, reckoned it a form unworthy of consideration. And it wasn't condemned simply by cultural and political association, but by its intrinsic property – a dome is curved, it doesn't belong to rectilinear geometry, and that geometry was paramount throughout the third quarter of this century. The only acceptable dome had been the Dome of Discovery at the Festival of Britain – which wasn't really a festival of Britain but of (mostly) imported Scandinavian tics, and the dome wasn't really a dome at all, but a Dutch cap. The most spectacular proposed English dome of that era, Pier Luigi Nervi's for the Pitt Rivers museum in Oxford, was never built.

Now this dome, The Dome, the First Church of New Britain, is also a bit of an IUD, but it improbably derives from the Festival of (old) Britain. It belongs to a different tradition. During the years of the straight line's global hegemony, there were always oddballs knocking about non-orthogonal notions: some, like Bruce Goff, even built, prolifically. Buckminster Fuller didn't. But that has hardly lessened his fame and his influence. Still, the trick as an architect is to get the thing commissioned, to turn it from perfection on paper into flawed actuality. The first great Rogers building, the Beaubourg, could not have existed without the wheezes and manifestos and drawings of the Archigram group. But the members of that group are distinguished by seldom having got anything off the ground, certainly nothing on the scale of the Beaubourg. They have been more energetic, however, about reminding the world of the Beaubourg's debt to their projected Fun Palace.

The Dome bears a kindred debt to Fuller's techno-utopian models. The difference is that 'Buckie' (it's the fate of the 'guru' to incite the familiarity of his followers) is not around to bitch about where the idea came from. Does the fact that the inspiration for this cable-net tent is, let us say, more than borrowed necessarily make it the work of a great artist? The syllogistic corollary to T. S. Eliot's questionable dictum is that the more derivative a work, the greater its begetter.

At first glance – and you are here glancing from among the sugar refineries at Silvertown, from the heights of Honor Oak, from across allotments near Maze Hill – the Dome is magnificent and otherworldly. It might be fresh in from Sizewell or Sellafield, if not from the planet

Tharg. And it will remain magnificent. On a telly programme with the only unindoctrinated Labour MP, Bob Marshall-Andrews, at the beginning of last year, I suggested that the only thing to do with it was to leave it empty because that volume of enclosed space was potentially uplifting. David Hockney came up with the same idea a few months later – so they *do* steal, these mature artists: that's the proof, Eliot was right. It doesn't really matter, of course, that the Dome will initially be used to propagate the cult of the Islington messiah. It really doesn't matter what it's filled with – good jokes, like Ron Mueck's fifteen-foot-tall representation of a constipated, squatting William Hague, or meretricious junk selected by committee.

Politicians come and go – some sooner than others. And as for the meretricious junk – who now remembers what was actually exhibited at the Great Exhibition? No one. But we all know what the Crystal Palace looked like and we all rue the night of 30 November 1936.

(1999)

THE ABSENTEE LANDLORD

GOD IS, OF COURSE, a human construct, an invention, a fiction, an irrational absurdity. But it's an absurdity which is perpetual, immortal, protean; an absurdity whose existence it is absurd to deny because God is omnipresent, God is in believers' minds or souls or hearts or wherever it is they know he is. This is a fantastically successful invention. If ever an idea had legs; this one has run and run and is still full of puff.

God has been doted on since man – in his bewildered ignorance and reverential wonder – conceived the notion that all of this, all of us, must have been made by something else, something bigger, by an omniscient grease monkey who fitted the parts together. And who now services them in return for songs, sacrifices, oblations, genuflection – but who visits plagues, earthquakes, typhoons, famines, buboes and hideous skin conditions on those who fail to sing to him, who neglect to thank him.

This is not a sophisticated conceit – which is what accounts for its appeal to hundreds of millions. God is a universal solution. He is the one in whom faith is taken as a mark of earnest, of moral probity, etc, rather than as evidence of delusion or a symptom of insanity.

Churches – the buildings – cannot be denied. And whether one regards the conceit that they are built to glory as tosh or transcendentally sublime, it is indisputable that they are efficacious propagators of the faith. And the fetishes they contain – the cross, the stations of the cross, the host, the representations of the son in glass and the mother in stone and of light's vanquishing of darkness – all concentrate the beguiled mind and the fiction called the soul.

The idea of the church in England was self-evidently till very recently inextricably bound to the Church of England, the Church of Convenience founded to sanction a king's polygamous lubricity and his

Bluebeard appetites. Its moral basis may be infirm but its architectural basis was as strong as the rock which is Peter's name. It annexed an incomparable store of parish churches, abbey churches, minsters, cathedrals, the bulk of which are in the several and overlapping pointed arch idioms that are lumped together as Gothic.

Ask a child to draw a church and it is a rudimentarily Gothic church that will be ideogrammed: here's the church and here's the steeple, look inside and see the people. That is what a church looks like. It looks Gothic. The nave houses the congregation (if there is one) and the steeple or spire points to the sky, which is where god lives in climatic insouciance wearing funny clothes and a beard which obfuscates his irate face. He lives there in defiance of gravity, which is a post-god discovery.

Even in the century-and-a-half's hiatus between the Restoration and the Regency, when classicism in its various mutations and perversions was absolutely predominant, the church did not abandon the Gothic. The Gothic, after all, was its trademark, its logo. It built classical and baroque structures and then proceeded to give them Gothic silhouettes. It added vertical emphasis in the form of spires. It created architectural mongrelism in the service of self-advertisement. It was more concerned to maintain the continuity of easy recognition than to adhere to classical tenets. A church had to point to God.

After that hiatus, after the years of the Gothic survival, came the years of the Gothic revival. The orgy of church building which characterised Victoria's reign re-created the Middle Ages before her subjects' eyes. God had excelled in the Middle Ages, the Gothic centuries. That had been god's moment, and here he was again, he was back and dominant after the enlightenment. God is a goth, which is not to make some frivolous link between the idea and Black Sabbath, but to suggest that the appetite for the sort of belief which is dignified by the word God is most wholly fuelled by the Gothic.

The Gothic is conducive to pious irreason. The Gothic revival was founded as much in a pseudo-morality as it was in an aesthetic preference. It fixed for all time – that is, till the present day – the association between a particular architecture and the church. And it fixes the church in the dark ages and in the decor of fear. The Gothic is often crazed, often suggestive of architectural psychosis. It complements and heightens the mysteries, the terror, the morbidity of the passion. It never lets us forget that Christianity is a death cult.

And the Gothic is an agent of de-secularisation, it is an architectural means of achieving sacredness. It made schools, law courts and railway stations into quasi-religious buildings which were thus invested with added authority: we thanked God for selective education, for the wisdom of judges, for making the trains run on time.

Atheism, unlike Christianity, possesses no architectural type or decorative style which is peculiar to it. It has no need – there is nothing to represent or celebrate. Only a void can stand for a void. Atheists do not worship. Atheistic genuflection is connected to shoelaces or sex.

Still, had atheists built they could have built at no more propitious a moment than during the middle years of this century, when nearly all architecture had broken with western Christian precedent, when architecture was so ideologically determined that it was itself a cult or religion. Atheists, quite properly, didn't build. But the church, amazingly, did.

The Gothic hegemony had lasted almost a century. It extended into the 1930s. Between then and the outbreak of the Second World War a number of churches were built that dispensed with the revivalism of, say, Giles Gilbert Scott's Liverpool (Anglican) Cathedral, but not with Scott altogether. The example of Scott's first design for Battersea power station is evoked in N. F. Cachemaille-Day's work: his buildings are sacred cinemas. They derive, too, from northern German expressionism and from Albi Cathedral, which strikes the unwary as an Odeon *avant la lettre*.

But Cachemaille-Day's work is atypical. His churches were merely swallows. Summer had to wait till after the Second World War, after the Holocaust. If the Holocaust didn't extinguish faith in god, nothing ever will. How could faith persist and the church build in the wake of that enormity? Easily, is the answer. Concern about the culmination of 1,945 years of Christian anti-Semitism was not the church's priority. The true purpose and instinct of the church, as of every other institution and organism, is to perpetuate itself. The church exists to create its future, to demonstrate through its buildings and utterances that it is a pedagogic force, a moral force, a cultural force, that it is relevant, that it belongs to today. It is, consequently, acutely fashion-conscious.

The first post-war architectural fashion was that which we now associate with the Festival of Britain, the festival of borrowing from Scandinavia. The Festival style was whimsical, airy, short-lived. Post-war Britain was poor. Building was licensed. And by the time licences were revoked the Festival style was out of fashion. Parish churches usually still are poor. They are forced into a conservationist stance. Thus they are valuable repositories of forgotten crazes and half-baked ephemera which have elsewhere been excised or amended. There can be few examples of the Festival style, and of the early '50s spectrum, so complete as Christchurch in the Coventry suburb of Cheylesmore. (The Royal Festival Hall itself was made over in the mid-'60s.) Conventional wisdom has it that compromises produce banality and insipidity. First-generation post-war churches fascinate because they are the products of compromise, not despite that condition. They are gay and trashy and cheap looking. They display a skittish lack of earnest. But then the old fearsome god had proved impotent in the face of systematic atrocity and industrialised annihilation, so perhaps it was time for a revised version, for god the unmighty, for god the ever such a nice bloke, for god the softy, god the timid.

In churches such as St Paul, Lorrimore Square, just south of Elephant and Castle, or St Andrew, Stevenage, we can discern the first steps on the road to the cheery new hell, the hell which isn't such a bad place after all, the hell which provides its clients – for that's what they are – with therapy and drop-in facilities before they are adjudged fit to pass on to the next station-stop in that great afterlife of theirs.

The dedications should be to St Cosy and St Cardigan. Anglicanism was getting very chummy indeed. Pulpits were stripped of pomp and, you might think, of authority. They were no longer boxes to thunder from. They were made of parquet and laminates and were apparently designed by G Plan. They might have been expressly wrought to discourage fulmination. St Cosy is a sunny place which prompts no thoughts of death, judgment, heaven, hell. This was the church which was distinguished from other branches of the social services only by the Dave Clark Five collars worn by its operatives.

The church underestimated the appetite for its own product. It misunderstood its unique selling proposition. That appetite has come to be satisfied by fringe cults, nuthouse sects, wobbly temples, worship shacks. The church got it hopelessly wrong. It wasn't less mumbo-jumbo that its punters craved, but more. For all their stylistic variations, the churches of the '50s displayed certain invariable mannerisms. The detached campanile, a feature it shared with fire stations, doubled as a spire, a vertical sign – the trademark was no more shunned than it had been in the eighteenth century. And interiors still comprised nave, choir, baptistry and so on.

The compromise was the result of those spaces being rendered in a manner that was admissible to architects who were ill at ease with copyism, but yet lacked the killer instinct to do away with the spatial forms themselves. So the Gothic template was adhered to but Gothic devices were shunned. This was modernism in a corset – which for those who subscribed to the holy church of modernism was no modernism at all. Modernism was extreme, fundamentalist.

A similarly resolute indecision characterises the sacred art of the '50s and early '60s. It was all hedged bets. The sitting-on-the-fence school predominated: semi-abstract, sort of representational, trapped in the gap between personal expression and liturgical propriety. When Coventry Cathedral was consecrated in 1962 it enjoyed a popular fame unmatched by any post-war British building with the possible exception of Lloyd's. It became a place of pilgrimage: queues formed around the rebuilt city centre. Despite the ostensible purpose of the building, it was not holy queues it attracted. It was a symbol of renewal. It was a tardy monument to the qualified victory in the war which had cost Coventry its former cathedral.

There is certainly qualification in Epstein's sculpture beside the entrance. St Michael, the dedicatee, is above Lucifer – but Lucifer is merely down, not out. He isn't defeated, merely planning the economic miracle. Cantilevered out from the steps beside this work is a viewing platform. Even before the interior is broached, the pilgrim is invited to behold this phoenix in the form of a radiogram as an art gallery, as Basil Spence's museum of expressly commissioned work on holy themes. It is a time capsule, rather akin to the collection at the Royal Holloway College that was put together exclusively during the years that that Loire château was being built in Surrey.

Spence created the space and opportunity for artists in several media – incised stone, an apparently basketwork ceiling, Graham Sutherland's dominant tapestry, a descending flock of coat-hangers in the choir, outsized bobbins posing as candelabra, a *faux naïf* mural by Steven Sykes, wonderful stained glass by John Piper and Patrick Reyntiens. What, at the time, looked like contrasting styles seems all these years on to be of a piece. It is a collective tour de force, but a pre-modern tour de force. It is the last major church in an 800-year-old tradition. It may be a gallery but it at least pretends to be a cathedral.

The baroque was the propagandist style of the counter-reformation. It dazzled in the service of Rome. It fought protestantism with showiness, cleverness, richness, virtuosity. Four hundred years on, Rome addressed the schism by trying to out-prod the prods. The second Vatican council of the early '60s responded to the Liturgical Movement – which was by then a century old, and which advocated a return to the forms of worship supposedly used by the early, primitive church. It decreed that all that was not essential (whatever that meant) was to be eliminated from new places of worship and the hierarchical division of clergy from congregation was to be quashed.

Not to be outdone, the Church of Convenience arrived at the same conclusion – they work in mysterious ways, these competing liturgical engineers. Holiness, henceforth, was to be expressed through form rather than through icons. The churches were making liturgical concessions to the architectural ideology of the time – an ideology which anathematised ornament and the western figurative tradition.

This was a prodigiously weird carry-on. It wasn't merely a case of the churches sleeping with the enemy, but of marriage, impregnation. It was as though the meat trade had commissioned new abattoirs from vegans. It was as though the churches had abandoned Bach and Fauré in favour of John Cage and Cornelius Cardew. What next? Dedications to St Herod the Cleanser and St Judas the Supergrass?

During the years of its cultural and moral centrality, the church was an unparalleled architectural force. It led, it owned the initiative. Now it was cravenly following extreme secular fashion. It was not the patrons but the artists who called the shots. This gave rise to buildings which were exercises in pure form. They were like student projects (i.e. fantasies) which somehow got built. Modernism regarded itself as something more than a mere style. It was a faith. So here was one faith supposedly serving another but in fact grabbing the chance to prove itself a moral and pedagogic force.

This was cuckooism on the grand scale. It was the glorious parasitism of confident art upon wavering religion. Churches started to come in all shapes. There were bunkers and ships. There were churches that looked like silos, churches such as St Andrew, Livingston, in Lothian, whose entrance was hidden, churches with swervy roofs and hyperbolic paraboloid roofs, churches that aped giant ammonites. The faithful must have had to work hard to convince themselves they were attending church at all.

The new language of sanctity was a desperate esperanto. Every church had its own idiom. There was no norm, no consensus. The only rules were that churches were not allowed to resemble churches and that figuration was off limits. Where once everything had had to look like something else, now everything had to look like itself.

Otherwise, ecclesiastical modernism was unconstrained, especially by practical demands. God, like the Queen, doesn't go to the bathroom. So god's houses need no bathroom. Johnny Architect was presented with what he had always longed for – carte blanche.

This bewildering licence was just one of a series of own goals which included the New English Bible, the revised prayer order, the vernacular Mass, Honest-to-God, agnostic clergy, vicars with guitars trying and failing to reclaim the best tunes. The attempt to appeal to the modern world by coming into line with the modern world annulled the church's appeal – which was nothing if not pre-modern. The ascendancy of theatre in the round hadn't escaped the church's notice; here was a device which would involve the congregation, a crudely material and literal device which assumes that physical proximity to the celebrant and to the host occasions a greater chance of divine access. Richard Scott's St Thomas More in the Birmingham suburb of Sheldon is typical of this trait.

Much of the ecclesiastical work of the high modern years prompts awe. But that awe is more likely to be aesthetic rather than religious; it is not different to that which is incited by a grand hangar or concert hall – the potency of a building is quite distinct from its nobility (or baseness) of purpose. Efficacy as a church rather than as an architectural conceit derives as much from a lack of ambiguity as from any specific spatial or iconic devices.

Scott's bizarre Our Lady Help of Christians in another Birmingham suburb, Tile Cross, may look east to Orissa (or somewhere) and west to Las Vegas, but it is unmistakably a church. It does not share its space with a youth club or a knitting circle. It makes it emphatically clear that it is a single-use building and that that use is holy. Further, it implicitly makes the point that such spaces should not be shared, should not be let out to all-comers with a good cause. The spiritual is separate from the temporal, no matter how worthy the temporal cause may be.

The last of the modern megachurches, Clifton Cathedral, seems embarrassed by its function. Sure, its stations of the cross are coarsely graphic and don't flinch from the horror of Calvary. But the building would rather be the National Theatre. Its brute, board-marked concrete and jagged discords come straight off the South Bank.

An architecture which proclaimed itself as democratic – which, if it means anything, means that it was mostly made in the service of democratic institutions – is incapable of addressing the problem of a publicly visible hierarchy. The bishop's throne is pitiful. No pomp. No majesty. No sign that the celibate who occupies it pretends to do so at god's behest. There is no modern vocabulary to embody such an ancient and anachronistic idea. Abstractions cannot be represented by abstract architecture. Abstractions require figurative metaphors, physical analogues, explanatory devices.

Baptism does not require that sort of intercession because it is an idea which is made visibly manifest. The initiation by water is literal. This superstitious immersion may be a social rite, but it is a social rite which is performed in churches and which swells attendances. Baptisteries are thus the focus of otherwise hardly used churches. They are commensurately richer and more detailed than the buildings which contain them. Baptism will doubtless eventually be performed in specialist lidos, water palaces, geyser centres. This is not an outrageous speculation. It is the only vaguely religious ritual apart from weddings and funerals which is still observed.

Fifty years ago cremation was used to dispose of less than 10 per cent of British bodies. Now that figure is over 75 per cent. Crematoria are the true churches of the late twentieth century. They're places where the fire is real and not one of hell's props. They are no more for those suffering egress than baptisteries are for the mewling dunkee. It is the witnesses that these places must address – and in the case of death that means the bereaved, the stricken, the weeping, the disorientated.

This may be dealt with, architecturally, head on – by suggesting the momentous physical change that has afflicted the unknowing protagonist, by suggesting that the body will be swallowed, consumed in the building's fearful maw, transported to another place. Such in-your-face grief-management may compound the offence caused by the sense of loss. It may heighten the indignity prompted by the funerary production line and by the bathos of the New English patois.

On the southern edge of Edinburgh, at Mortonhall, Basil Spence designed a crematorium chapel which is dignified, grave, earnest, solemn. It is capable of catalysing the illusion of transcendence. It encourages eschatological curiosity. It is almost unique among modern British sacred buildings in displaying a fitness for its purpose.

(1997)

NEO-GEORGIAN

WE DO NOT dress like Milord Foppington or Mister Petulant. We do not eat salmagundi or posset. We do not hurl our faeces into the street. We do not travel in horse-drawn vehicles. We do not write with quills. We do not suffer tooth extractions without anaesthesia. We do not hang rustlers. We do not utter mild oaths like stab me vitals! We can tell the time. We can read. Our water is clean. Our life is long.

Yet the pull of a particular past is apparently so potent that many of us – most notably the very rich – want to replicate it, if only cosmetically. Such people as the banker Robert Gillespie and the interesting entrepreneur Nicholas van Hoogstraten want to live their long life literally immured within a token of the past. Their beau ideal is a neo-Georgian palace in the country. New, yet old. Now, yet then. Centrally heated, yet corniced. Double-glazed, yet sashed. Steel-framed, yet sort of Vanbrughian or sort of Palladian or sort of Adamish.

The aspiration to all this is so commonplace – witness the tens of thousands of lesser neo-Georgian villas – that its sheer dotty oddness passes us by. It's an oddness which is exacerbated by its straitenedness and its randomness. We're not talking here about going the whole hog. The thing outside with a wheel at each corner is called a car: its bodywork does not mimic that of a stagecoach. The tiny thing outside with a wheel at each corner is called a Smart car: its bodywork does not mimic that of a Sedan chair. The roaring gleam ascending into cloud is not shaped like a Montgolfier. And inside: a fridge evidently fulfils the same function as an ice house but its form makes no allusion to the fact.

The gamut of objects which we infect with the pretence that they come from another time is limited to furniture, frames, cutlery, chinaware. Domestic objects, all of them. Home is where the past is. Neo-Georgian is an approximate epithet for an approximate school of architecture. The Georgian to which neo- is affixed signifies the long century and a half from the Restoration till the Regency, a period during which domestic building, whether urban or bucolic, enjoyed a harmonious evolution. There existed a broad consensus about the form a house should take. Thus a house of, say, 1826 will manifest its direct descent from one of, say, 1682: kindred proportions, kindred symmetry, kindred reliance on local materials. By the former date, however, the consensus was becoming frail due to the aesthetic lure of the picturesque. And the consensus would soon be entirely ruptured by the 'moral' thraldom to medievalism and by the increasing ease with which materials might be moved about the country.

Still, that protracted period of domestic classicism gives its apes a great deal to draw on. But why do they want to draw on it? The positions of client-consumer and architect-supplier are disparate. Which comes first: the client's whimsical wishfulness and desire to make a 'statement', or the eagerness of a small yet hugely self-confident tribe of architects to beat against the current? I have down the years admiringly 'consumed' Pope, Swift, Gray, Cowper, Hogg and countless others who wrote when buildings were 'Georgian' – but I wouldn't want to imitate them even if I had the capacity. The very idea is preposterous. So would be that of a current composer who sought to imitate Handel or Haydn. Who, today, sees any point in painting in the style of Gainsborough?

All that any exercise in willed anachronism can capture is mannerism, surface. This is a lesson that has never been wholly learnt by architects (and whoever it is that confects Queen Anne telly cabinets), nor by their clients. It is worth noting Marc Girouard's observation that the earliest essays in neo-Georgianism were made, in the 1870s, by architectural amateurs: architect and client were one, there was a touch of folly about these first creations. But follies have always been stylistic try-outs, the concept cars of the built environment.

And so the idiom gradually took off professionally and, by the time of Edward VII's death and the accession of George V it was a recognisable and increasingly loud component of the architectural babel. Not that loud is an appropriate adjective. The appeal of neo-

Georgian – I'm guessing – derives from its supposed quietness and modesty, from the presumption that it summons up good taste and Englishness. The facts that the earliest 'Georgian' buildings, those which are actually Caroline, are almost wholly Dutch in inspiration, and that subsequent refinements were made in imitation of French, Italian, German and Greek precedents, are happily forgotten in pursuit of the chimera of architectural nationalism.

Its champions will forever make the Jesuitical point that a neo-Georgian RAF mess of 1937 designed by the War Office's W. Ross is of that year and no other and is no more or less literally modern than a contemporary modern movement building designed by Gropius and Fry. That is an easy paradox with which it is impossible to sympathise.

This is make-believe of the most insipid order; it is wholeheartedly half-hearted, no matter how convinced its begetters are of its probity. The very idiom appears to defeat those who embrace it. Beginning with Lutyens there is an ignominious history of architects who have enfeebled themselves by putting their imagination in abeyance while they played an ancient tune and played it weakly. It is that quality – imaginative bereavement – which characterises neo-Georgianism: good taste means, in this instance, no taste.

(2002)

POSTMODERNISM TO
GHOST-MODERNISM

MODERN ARCHITECTURE DIED on 16.05.1968 in Canning Town. We all know this. An explosion in the kitchen of an eighteenth-floor flat caused part of the recently completed Ronan Point to collapse. According to an age-old tradition, celebrated in music hall song, in east London gelignite (notoriously unstable) is kept in the fridge.

Modern architecture died on 16.03.1972 in St Louis. We all know this. The Federal Department of Housing demolished the sixteen-year-old Pruitt Igoe project, designed by Minoru Yamasaki, author, too, of the Twin Towers of the World Trade Center.

Modern architecture died on 16.10.1973 in Kuwait. We all know this. OPEC, greedily vindictive in its animus against Israel's allies, cut the production of crude oil and raised its price 70 per cent: by early 1974 that price had quadrupled.

Each of these versions, with its fatidic 16th day of the month, has its supporters, DIY pathologists who gleefully proclaimed the absence of signs of life. Calendrical coincidence is not however maintained by 03.05.1979, the day that Margaret Thatcher attained power. She would proceed to drive the final nail into modernism's coffin by dismantling the economic and cultural apparatus that had supported it. No architectural idiom can survive without the armature of patronage. British modernism was the material emblem of the cross-party, Butskillist, welfarist consensus that had endured since 1945. Indeed, during the period which Labour, at the 1964 election, decried as 'thirteen years of Tory misrule', the governments of Churchill, Eden and, especially, Macmillan had sanctioned the construction of an unprecedented volume of social housing. The welfare state was noblesse oblige nationalised and welfarism was la pensée unique, from which French locutions no decent person dared dissent in public. Its monuments served (and proclaimed) the democratisation of tertiary education, the provision of cradle to grave health care and of a decent dwelling, the primacy of high culture, the beneficence of the nation.

With, early on, the exception of a few soon-to-be-culled 'wets' (Ian Gilmour, Francis Pym, Mark Carlisle, etc.) Thatcher's administrations were composed of self-made garagists, by-their-bootstraps mini-tycoons and nabobs of estate agency who exuded an unmistakable whiff of Poujadisme. Had they even heard of it these people were no more sympathetic to the idea of noblesse oblige than they were to the activities of trade unions. Institutionalised philanthropy was not susceptible to the new god of The Market.

Nor, according to the polls, was it what the public sought. The outgoing prime minister, James Callaghan, spoke memorably of a 'sea-change', thus revitalising that expression of Shakespeare's and ensuring that if he is recalled for anything it is for his impotence in the face of electorate's will; though had he summoned the nation to the polls a few months previously, in the autumn of 1978, he would almost certainly have been returned to office. Would British architecture, consequently, have taken a different course?

Probably not. The exercise of suffrage is curt. The process of architecture is tortuously protracted. The same architects tend to flourish no matter what the regime; it is not a pursuit of the morally fastidious. While political power changes hands overnight, an architectural scheme conceived under one government's tenure will very likely not come to fruition till its successor is firmly ensconced.

Thus Mrs Thatcher's accession followed close upon the completion of two architectural landmarks: Neave Brown's Alexandra Road and the Hillingdon Centre. The former would turn out to be the triumphant culmination of a tradition of inspired social housing, the end of the line. Whereas the Hillingdon Centre, although it too was a *public* building, and had been planned as long ago as the first oil crisis, might have been commissioned as the bespoke herald of a cultural 'sea-change' commensurate with that which had occurred at the ballot.

These civic offices were uncannily predictive. But this was prophecy as apostasy. In the eyes of the modernist establishment, represented by the *Architectural Review*, the unblemished practice of RMJM and its chief designer, Andrew Derbyshire (b. 1923, author in 1960 of New Zealand House) had committed an act that was ideologically heretical, aesthetically treasonable. Hillingdon was not even akin to the decorative, mannered, late-modernism of, say, Richard Seifert (1910–2001) or William Whitfield (b. 1920), which at least paid lip service to the holy orthodoxies.

Here, in the wilds of suburban Middlesex, was a suburban town hall composed, apparently, of several dozen suburban villas and suburban bungalows which had thrown their keys into the centre of the room and which were now conjoined in cosily elephantine abandon. Like any suburban orgy (think South Ruislip) it was more comical than sexy: it broke so many rules and was so wholly divergent from the precepts of canonical modernism that it was revolutionary – in the snuggest, homeliest, most carpet-slippers way. It was the architectural equivalent of Benny Hill or Sid James: coarse, matey, blokeish, undemanding, unthreatening, *accessible*. Inadvertently and fortuitously, then, just the ticket for a decade when we would be enjoined to get on our bike in search of greed (which, we were governmentally instructed, was a good thing) and to make contact with our inner barrow boy.

Populism is the provision by a prosecco-drinking elite – which far from considering itself elite considers itself *subversive* – of Lidl own-brand tannic plonk to the masses who, devoid of the elite's exquisite taste and compendious knowledge, are incapable of appreciating anything better. Anything better would be wasted. Populism is supremely patronising. It is sneeringly presumptuous: the masses are reckoned to be uncultivable, unimprovable, stripped of free will. They are slow-learners or no-learners who are to be held in pens of perpetual infantilism. They can be expected only to respond to what they already know. They are forever condemned to belong to the ignorance community. The proudly vaunting philistinism which has afflicted Britain for three decades found its first architectural expression at Hillingdon.

Here was a building whose components, if not their wacky aggregate, were familiarly banal, instantly recognisable, utterly comprehensible. These properties, in a startling variety of guises, would define the architectural mainstream for years to come. Architecture abandoned the analogues of atonality and twelve-tone serialism in favour of compositions worthy of a tectonic James Last. High seriousness was buried beneath an avalanche of toytown rustication, inverted diocletian windows and distended columns. The modernist hegemony was replaced by a pluralistic dressing-up box. The word 'irony', shorn of its literary and rhetorical meanings, was indiscriminately employed to signify smug knowingness and coyly daring playfulness. Jokiness became de rigueur. Just as battalions of unfunny comedians ensure a sycophantic response by laughing at their lame gags *before they tell them,* so do self-proclaimingly 'witty' buildings announce themselves with hyperbolic clashes of scale, a dubious chromatic sensibility, a willed illiteracy, a children's entertainer's garrulous importunacy. Grown men – and they were nearly all men – went back to the playground. And to their history books. This was ultimately the architecture of contrition. Architects knew how much they had been despised. They were now eager to please. They were penitents: some equipped with nursery colours and elemental shapes, others with servility and forelock.

The clamorously bruited diminution in state spending during Margaret Thatcher's administrations was more notional than actual. It was cut by less than 5 per cent. Revenue was excised and books balanced by the privatisation of utilities and by the sale to their tenant of council houses under Michael Heseltine's 1980 Housing Act. ('Selling the family silver' was, actually, Mrs Thatcher's own wary summation of her predecessor Harold Macmillan's misgivings. He himself did not use that precise expression, though, having been instrumental in building 300,000 council homes per annum between 1951 and 1963, he might have entertained doing so). These sales – what kind of *right* is the right to buy? – were effected for the short-term gain of electoral advantage. They were socially calamitous, the more so given the drastic reduction in the creation of new public housing. The private and housing association developments of the 1980s were in signal contrast to both the towering megastructural schemes of the '60s and the impenetrably dense, labyrinthine low-rise projects of the early '70s.

Just as the exhilaratingly harsh Gothic monolith of the 1850s and '60s had been succeeded by the genial warmth of the misnamed Queen Anne revival, so did this new domestic revival conjure up a cosy sense of homeliness. The standard was set by Jeremy Dixon (b. 1939) who, a few years previously, had been co-designer of a regrettably unbuilt, confidently modernist, prismatic glass pyramid for Northamptonshire County Council. Now, in a redbrick purlieu of Notting Dale, he had designed a stylistically disparate but equally startling terrace for a housing association. It was a work of unmistakably reactionary inspiration which drew on a generalised Dutch vernacular and, specifically, on De Stijl. And he had built it on a street, a form of thoroughfare which modernists had scornfully cast aside for it connoted both insanitary Victorian slums and despised interwar suburbia.

One of the besetting problems of British and, indeed, European modernism had been the relationship of the building to its immediate environs. Architects are not necessarily urbanists any more than composers are conductors or playwrights directors. The slavish crazes for tall buildings surrounded by parks and for cluster blocks accessed by novelty walkways, experimental alleys or untested tunnels had created ill-defined, semi-public, vulnerable spaces that were awkward, illegible and threatening: deftly crafted projects such as Whicheloe Macfarlane's High Kingsdown in Bristol were the exception. The North American ideal of adherence to the building line had been largely rejected. With a revival of such building types as the terrace and the semi, it was inevitable that streets would return, along with curvy 'closes' and less accommodating cul-de-sacs.

Dixon's terrace would prove massively influential. That is to say that its design was sedulously ripped off. Architects' plagiarism is shamelessly bereft of cunning. The London Docklands Development Corporation, a Heseltine-appointed quango responsible for the redevelopment of the banks of the Thames downriver from the former Pool of London, was statutorily sanctioned to apply but the laxest planning stipulations. Height and area only; 'aesthetic control' was forbidden as an illiberal contravention of the market's power. Thus the LDDC smiled beneficently on meretricious hackery. The area became a sprawling, infrastuctureless showcase for nicked ideas, secondhand wheezes and all the typologies that ever fell off the back of a lorry. It was a neat summation of London itself – a brick and stucco Klondike, ever mutating, ever chaotic, joined up only by chance and proximity. The most accomplished copier of Dixon, other than an entire generation of Dutch architects who relearnt their idiolect from him, was Dixon himself. On the south-eastern perimeter of the Isle of Dogs he created an unusually coherent quarter of even more wholeheartedly Netherlandish semis and neo-Adelaide villas, laid out in crescents and squares, terraces and avenues. In less sure hands this civil, understated urbanism was liable to teeter into insipidity: witness Arup's Lloyds TSB beside the Floating Harbour in Bristol, a crescentic building partially derived no doubt from that city's ponderous Council House, and one which is desperate to please by giving absolutely no offence to anyone, anyone at all.

A more usual stratagem for pleasing everyone was to follow the Hillingdon route, to borrow from the past's discards, to make art out of what had been calumnised as artless – this after all is what Richard Hamilton, Peter Blake, Andy Warhol, Mel Ramos, Roy Lichtenstein and countless others had been doing for years. But architecture is a tanker that turns torpidly even though its crazes and fads shift with the speed of skirt-lengths: thus many buildings are out of date by the time they are finished; the bigger the building, the longer the construction and the greater the gap. Architecture is also predominantly practised by people who are still reckoned young at the age of forty or fifty: this is the age at which they tend to at last get their chance. The extraordinary flowering of pop music in the '60s was achieved by war babies and baby-boomers. The architecture contemporary with that music was the work of their parents' coevals. The modernist asceticism and straitened purity of these architects were there to be snubbed by filial insolence: one generation reacts against its predecessor. Baby-boomer architects embraced self-conscious impurity, high colour, loud ornament, unashamed facadism, burlesque frippery, (sometimes desperate) jollity, knowing vulgarity and, equally, unknowing vulgarity.

The cleverest and nimblest was Piers Gough (b. 1946), the most inescapable and ostentatious Terry Farrell (b. 1939). Both of them were incorrigible show-offs. Gough once described his work as 'B-movie architecture'. That merely gets part of it. He certainly displayed a fondness for obscure West Coast architectural subcultures such as Borax and Googie, garish idioms which provided the décor, even set the mood of countless *films noirs* and saturated Technicolor melodramas. But Gough was nothing if not catholic in what he stole and from where: Viennese jugendstil, Regency gewgaws (he was brought up in Brighton), Home Counties arts and crafts, nautical motifs, the Belgian seaside, arterial road Metroland, officers' mess neo-Georgian, Wrenaissance, Edwardian free style, the Amsterdam school. And so on: any source that was marginally tainted or not quite respectable. Gough's invention was boundless. His method often appears to have been that of collage: stylistically disparate and temporally diverse components are juxtaposed. They elide. They collide. At his best – The Circle and China Wharf near Tower Bridge, the de Baroness shopping centre in Breda, Cascades in Docklands, the Black Lion development in Brighton – he has created, and continues to create, buildings that remain attached to the retina, images that carry an ineradicable emotional charge.

Terry Farrell's unmistakable work of the '80s and early '90s is vast, hefty, glaring, squat, brooding. London's skyline may not be altered but its ground level most surely has been. Like Gough and Dixon, Farrell had looked back, but to less definable topoi and eras, to unknowable generalised sites stored in his architectural back-brain, big sites, of course. Unlike them (and most American post-modernists) he was unmoved by the picturesque, by sweetness and light. His lumbering set pieces respond in kind to the modernism he had briefly embraced at the start of his career, behemoth to behemoth. They have the scope and ambition of megastructures if not their style. Alban

Gate bestrides London Wall with brutalist confidence. Embankment Place is not *at* Charing Cross, it *is* Charing Cross; it derives from some dream of North American heavy engineering. The laughably unsecret MI6 building at Vauxhall Cross is a brutal temple to necessary surveillance. It is formidable, overwhelming, undeniably powerful. Farrell embraced the cult of sullen ugliness with the gusto of such high Victorians as Pilkington and Teulon. He too possessed the drive and muscular vitality known then as 'Go!'.

No former modernist was more penitently opportunistic than James Stirling (1926–92). In the late '50s and '60s he had enjoyed, among his fellows, an unrivalled reputation. Among those unfortunate enough not to be architects, among mere *lay* persons exposed to his buildings' manifold peculiarities, he was rather less fêted. His constructivist Leicester University engineering building was faintly praised by Anthony Burgess as '. . . functional. We can . . . admire [it] without being moved'. This was mild damnation in comparison with the rebukes that became routinely vituperative. Calls were made for the demolition of his history library at Cambridge within a couple of years of its completion. The Florey Building at Queen's College, Oxford, fared little better. His buildings, like their bombastic maker, looked tough but were perpetual invalids, basket cases. By the end of the '60s Stirling had become a by-word for extravagant failure. Throughout most of the '70s he taught. When he returned to the actual practice of architecture he had, though he of course denied it, transformed himself into a fully fledged postmodernist – it was akin to witnessing a man in late middle age dancing to the latest hits. His Neue Staatsgalerie at Stuttgart mixed Schinkel's classicism, fairground forms, wavy glazing, massive masonry (which, daftly, was considered rashly akin to Albert Speer's), brash colours, bunker-like voids. Its exceptional incoherence did not prevent his career taking off again. A science centre in Berlin followed: it was as if de Chirico, another artist who famously lost his way, had devised a cylindrical cassata, baby blue and baby pink. He received commissions from Harvard and Cornell. At last, after than more than a decade, he was welcomed back to Britain; at least he was welcomed by the burgeoning and sycophantic architectural media which didn't even entertain the idea that its esteem of him – and his esteem of himself – was inflated out of all proportion to his achievement. Charles Jencks, the taxonomically inclined cheerleader for postmodernism, described the Clore Gallery as 'the greatest game of contextualism played to date . . . An architecture for all seasons'. The ostensible purpose of this extension to the Tate was to house the gallery's collection of works by J. M. W. Turner. The building manages the improbable double of being both trite and overweening. Its reception by those who were not groupies, not parti pris was, quite properly, unanimously hostile.

> MARINA VAIZEY These fun colours are not fun; they are shocking in the wrong sense. They are vulgar. The most subtle colourist in British art is defiantly shouted down.
> GAVIN STAMP An architect is showing off at the expense of England's greatest painter.
> JOHN McEWAN A well-upholstered bomb shelter.
> RICHARD CORK The entire scheme had been conceived as Stirling's work of art, rather than as the most fitting vessel for Turner's long-abused bequest to the nation.

And so on.

Stirling died prematurely during a routine operation in 1992. His last building, No. 1 Poultry, in the City of London, was not completed till six years later. By which time it looked not only oafish but hopelessly retardataire. Fashion had changed. Architecture, no matter how much architects deny it, is in absolute thrall to fashion.

Postmodernism was, however, far from exhausted. While the epithet itself may have been buried some time between Black Wednesday and New Labour's *machtergreifung*, the retrospective process that it signified has continued to flourish as the architectural norm. Instead of looking back to the Belle Epoque, or to primitive classicism, or to Tudor palaces (the bizarre Richmond House in Whitehall) or to Wurlitzers (Barclay's Bank in Lombard Street) or to Anton Furst's Gotham City in *Batman* (the 'gothic' Minster Court in Mincing Lane) or to the Beaux Arts (Canary Wharf), architecture has, for the past decade and a half, contented itself with energetically recycling the modern movement in its many guises. Most – but certainly not all – of modernism's stylistic history has been plundered while the (possibly chimerical) ethical apparatus attached to it has been cast aside as an irrelevance. The quotes and allusions and citations and 'homages' which characterised '70s and '80s postmodernism are just as commonplace today, save that they tend to be occluded, they are deployed discreetly. They have been *synthesised*. Synthetic modernism or neo-modernism or IKEA modernism (Owen Hatherley) or CABE-ism (Rory Olcayto) is as revivalist, as 'historicist' as neo-Georgian or the old English style or the baroque of 1900–1910. But unlike these idioms it is loath to admit it: it adopts the pose of the 'modern Gothic' of the 1860s which disclaimed medieval precedent. It pretends to originality. And its omnipotent, ubiquitous global practitioners possess both a PR machine and – hardly different – a deferentially anilingual architectural press to support their vain pretensions.

The volume and influence of that press increased with the advent in 1983 of *Blueprint*, a cleverly positioned glossy newspaper which sought to make architecture attractive to a public which the worthy, rather ingrown *Architectural Review* and the cultist, jargon-splattered *AD* could not reach. If architecture was for the people then magazines about architecture ought to be for the people. Further publications appeared: *Wallpaper, Icon, Frame*, etc. They were not perhaps the most strenuously critical of journals. The national press, too, had begun to pay attention. The broadsheets, as they then were, having largely neglected the subject save to report social catastrophes and crumbling concrete, appointed architectural correspondents, starstruck panderers who sucked up to and repeatedly eulogised the same big names – whom they themselves had helped to create as big names by sucking up to them in a vortex of mutual dependence. These people were seldom inclined to challenge the claims of the globetrotting heroes whose vehicles and aircraft they had been privileged to travel in.

In the late 1970s Philip Johnson (1906–2005), the wizened Talleyrand of New York architecture, then in his Chippendale phase, described his former pupil, the future Baron Foster of Thames Bank

(b. 1935), as 'the last modern architect'. Good try. But wrong. Norman Foster was, rather, the first neo-modern architect. Or, anyway, the first neo-modern architect to get the timing right. It never does to be too far ahead of the game. It was the (literally card-carrying) communist Douglas Stephen (1923–1992) who had in 1965 been the earliest British architect to return to the forms of early modernism, when he adopted the rationalist idiom of the Lombardian fascist Giuseppe Terragni for a development of private flats on Campden Hill in Kensington which retains a delightfully baffling timelessness. Stephen's marvellous David Murray John Tower (1975) in Swindon also looked back, to a 1920s platonic ideal of a tall building. It's like the realisation of an unmade project glanced in a magazine – a far from uncommon instance of fiction made architectural fact, of 'paper architecture' given three-dimensional life. Foster's Willis Faber Dumas building in Ipswich (1975) struck those who insouciantly came upon it as some unrecorded prodigy of the 1930s, specifically as an unknown work by Owen Williams (significantly an architect-engineer) whose black vitrolite *Daily Express* plant in Manchester Foster must have known as a child. Douglas Stephen went on to design a few largely unsung buildings in London.

Norman Foster (and his 1,400-strong practice under its succession of names plus a brigade of civil engineers in the small print) went on to design many of the most dotingly lauded and slaveringly mediated structures in the world. Foster belongs to a small cadre of masters of the architectural universe who dump lumps of prestigious tectonite with promiscuous abandon. He is globally prolific, globally honoured. Astana, Barcelona, Berlin, Bordeaux, Frankfurt, Hong Kong, London, Madrid, Millau, New York, Nîmes, Rotterdam, Singapore . . . And Cosham, Finnieston, Gateshead, Norwich, Swindon, Woking. Whether they are orthogonal or geometrically delinquent or blobbily biomorphic his work is characterised by sleekness, precision and jewel-like elegance. It is often glacially beautiful. These are machines for working in, for studying in, for creating in, for shopping in. And for living in – for living a life that's tidy, costive, ordered, rational, *fosterian*. His buildings may be emotionally affectless, heartless, but they are exemplary. They process and condition their users as though they are revenants from the age of architectural determinism; although they are made for vast corporations and questionable dictators they retain a dilute utopianism, the hint of a will to improve the race.

Foster and his fellow milord, his partner of long ago Richard Rogers, are, due to the media's and the curatocracy's adhesive pairing, as inseparably conjoined in the collective conscious as Chang and Eng. Foster's neo-modernism had its provenance in buildings made before he was born, and in the moral modernism of Pelican books, and in Frank Hampson's Dan Dare strip in *The Eagle* (whose oldest reader he must have been) and in the early, Sarasota period, oeuvre of his (and Rogers's) Yale teacher Paul Rudolph. In contrast, Rogers's and Renzo Piano's competition-winning scheme for the Beaubourg looked back only a decade to the sometimes fantastical, 'unbuildable' paper architecture of the Archigram group, to Cedric Price's proposed Fun Palace for Joan Littlewood and to the forms if not the materials of Paul Rudolph's design for Yale Architecture School.

A telly documentary which followed Price and some members of Archigram to the newly completed Beaubourg marvellously captures those originators' mix of pride, resentment and amused astonishment at the barefaced chutzpah of Rogers and Piano, both of whom would also make the stratospheric leap to become global attractions, big pulls, the people to call in when, in defiance of economic discretion, gestural engineering and landmark beacons and iconic icons and 3-D logos and lighthouse rebrands and vision hubcaps and regenerative torches are required to bolster megalopolitan vanities. Chipping Camden gets a bus shelter by Calatrava, Chipping Norton responds with a stocks by Gehry.

Global architecture is trophy architecture, underwritten by base tribal primitivism, corporate one-upmanship, nationalistic boastfulness. It was always thus – the height of medieval spires, the vastness of fortresses, the lavishness of palaces. The clock may be digital but it has been turned back. Architecture is once again for show and swank, mere building is once again for utility and shelter. This is the historic norm. The years of philanthropic welfarism were abnormal.

Unlike Foster, Rogers and Piano have been little imitated, no doubt because they share an unpredictability which is rare at this level of architecture, whose practitioners repetitively stick to a jealously forged 'signature' that renders them bankable, reliable, recognisable – Gehry, Liebeskind, Calatrava, et al. Rogers's Lloyds building(s) in the City of London, the law courts in Bordeaux and Antwerp, the Hesperia hotel in Barcelona and the Beaubourg are entirely different from each other. What they have in common is that they are among the most viscerally thrilling structures of the past 50 years; his boorish Knightsbridge flats for oligarchs and petro-midases and his Millennium Dome are not. Piano has given London both the unmissable Shard and possible proof of daltonism at Central St Giles.

A multi-teated amputee reptile with a prothesis: forty years after the heyday of Archigram its founder member Peter Cook (b. 1936) demonstrated, in collaboration with Colin Fournier (b. 1944), that he retained that group's power to astonish with yesterday's novelty. The work in question was a civic art gallery, in the Austrian city of Graz. Future Systems had existed for a decade and a half before it received, in 1995, the commission to build the media centre at Lords, a chunk of 1960s product design immeasurably magnified. David Chipperfield's most significant projects are in Spain, Germany and China. Will Alsop built the startling offices of Bouches-du-Rhone, department 13, at Marseille and a large ferry terminal beside the Elbe at Hamburg-Altona before he secured commissions in Britain. The Olympics Aquatic Centre is Zaha Hadid's first major building to be completed in the country where she has lived and practised for almost two thirds of her life.

After he had completed the Beaubourg Richard Rogers was so short of work, so broke, that he considered moving to America to teach. So it goes on. It is impossible to deny the received idea of Britain's tawdry, risk-averse unwillingness to take a chance on 'new' talent. Especially given the generalised refusal to acknowledge that there is actually nothing remotely new or audacious about the notion of the avant-garde, particularly when it tends to feed predominantly on the avant-garde of many decades ago, merely saucing the dish with sustainastic, sustainabulous splashes of green piety and chromatic discord. And given that London is the most cosmopolitan, most polyglot city in Europe – the home of the lingua franca where the lingua franca is seldom heard – it is odd that there should exist, alongside its resistance to architecturally indigenous daring, an equal disregard of foreign architects. Piano, Legorreta, Nouvel and Herzog de Meuron are rare exceptions.

The mainstream of British architecture over the past decade and a half is notable for its blandness. Because the media, in any field, thrives on the atypical, and represents the atypical as typical, that which is actually quotidian, ubiquitous and commonplace is overlooked, save when its physical dimensions plead for attention. Synthetic modernism is merely a stage of postmodernism, it offends only by its cringing fear of offending. There is something of Uriah Heep about the mood and tenor of the monuments that New Labour, its PFIs and its quangos bequeathed to Britain. The most bellicose prime minister of the last 150 years and his slimey mates were mere inheritors of the long-planned Falkirk Wheel and Jubilee Line stations which opened under their questionable tenure: these, anyway, like the Millau viaduct were essentially works of civil engineering decorated by architects. What was more commonly built under that corrupt regime was block upon block of 'luxury' flats which pleaded to be liked – in the early, needy manner of the Christian Bomber Blair. Block upon block – in London's Docklands and on Salford Quays, beside the Clyde, at Cardiff Bay, Ocean Village and Brindleyplace, everywhere – was decorated with strips of oiled wood, terracotta-coloured tiles, translucent bricks, riveted Corten, titanium scales, random fenestration. But mostly it was just glass and funny angles. The gamut of devices was straitened.

The one idiom from the modernist past that was not exhumed was brutalism. It was presumed, no doubt, to be liable to scare the people. It was all too likely, as one its practitioners, Owen Luder (b. 1928), put it, to say: 'Sod you.' Luder and his designer, Rodney Gordon (1933–2008), could only watch from the sidelines as their masterful exercises in the sublimity of sculptural concrete were demolished to make way for an architecture of cosmetic vacuity and feeble consensus. An architecture which was all but official, governmentally ordained. This was the architectural monoculture sanctioned by CABE, the jobs-for-the-boys quango whose middle-of-the-road attachment to synthetic modernism and avoidance of controversy left little space for the two extremes.

The one extreme represented by the 'new urbanism', by the Prince of Wales (b. 1948) and Quinlan Terry (b. 1937), a classicist whose devotion to pastiche is almost as fervid as Norman Foster's. The other by the genuinely modern rather than the 'modern': i.e. that which is novel, which has not been done before, which does not obey rules evolved by an extant culture, which exhibits properties that are, perforce, yet unknown, which returns primacy to the imaginative artist and removes it from the people, the consulted community, who know only what they have already seen, what already exists. It is the role of the artist to produce what does not already exist and to exorcise the persistent ghost of the day before yesterday's architecture.

It is not difficult to surmise which of these two extremes is more likely to have flourished when the architecture of Coalition Government by Trustocrats is assessed and exhibited, here, some years hence.

(2012)

VICTORIA DIED IN 1901 AND IS STILL ALIVE TODAY (2001)

PART ONE

TITLES. HEAVY RAIN. ARTIFICIAL-LOOKING, LAID ON POST-PRODUCTION.
CAMERA TRACKS IN TOWARDS A WALL COVERED IN TATTERED VICTORIAN
POSTERS. IT TILTS ACROSS THEM SLOWLY. PANS UP AND DOWN. THEY SHINE
IN THE RAIN, THEY ARE TORN TO REVEAL OLDER POSTERS BENEATH THEM.
THEY ALL RELATE TO DRUGS, MADNESS, RELIGION: ADVERTS FOR PROPRIETARY
BRANDS OF LAUDANUM AND PAREGORIC, EXHORTATIONS TO REPENT, THE WAGES
OF SIN, MADAME GLEET, THE DOLLYMOP'S GIFT TO SIR JOHN THOMAS, BEWARE
FRENCH PRACTICES, BEDLAMITES, ETC. EVENTUALLY THE CAMERA COMES TO
REST ON: 'GOD, POX, LAUDANUM' THEN IT TRACKS HORIZONTALLY ON TO:
'VICTORIA DIED IN 1901 AND IS STILL ALIVE TODAY'. . . CREDITS.
RAIN RUNS DOWN FURTHER POSTERS FOR THE CREDITS. CAMERA TRACKS AND
PANS TO FIND 'WRITTEN AND PERFORMED BY JONATHAN MEADES'. A RED BUS
MOVES ACROSS FRAME OBFUSCATING POSTERS. JAMES SMITH UMBRELLA SHOP,
NEW OXFORD STREET. HEAVY RAIN. A RED BUS MOVES R. TO L. TO REVEAL
FRONTAGE. JM ON EDGE OF FRAME

Our Queen and Empress, Victoria Regina, is supposed to have
departed this earth a hundred years ago, to have slipped away to
be reunited with her beloved Albert and his cock ring.

Can we be sure of this?

It feels as though she is all around us. The dead, of course,
are always with us: our friends, our parents, live in our personal
memory, the famous (and the infamous) live in the collective
memory.

FAMILY PHOTOS IN FRAMES — OF APES, ANCESTOR PORTRAITS

But Victoria inhabits a further space. She is so much with us that
we don't notice her. We take her for granted.

**VICTORIA STATION, ROAD, PUB, PARK, SPONGE, PLUM TREE, VICTORIA CHIP
SHOP**

We take trains from Victoria. We walk down Victoria. We drink in
Victoria. We eat Victoria. We climb Victoria.

PALL MALL CLUBS, LAW COURTS STRAND, BURSLEM TOWN HALL

Our institutions are Victorian.

Our traditions are Victorian — indeed the very idea of
tradition is Victorian.

Our sense of nationhood, of our relation to the state is
Victorian.

Our cities are the cities which her protracted age bequeathed
us.

TOWER BRIDGE

Our shorthand for London is Victorian.

JM LOOKS UP, HOLDS OUT HAND TO FEEL RAIN

Still reigning. Still reigning after all these years.

**JM WALKS OUT OF FRAME. VICTORIA (CHRISTOPHER BIGGINS) EXITS SHOP AND
PUTS UP UMBRELLA SO THAT FACE IS OBSCURED. WAVES TO HAIL MODERN TAXI**

SCRIPT 3

CAB. CAB STOPS. SOUND OF HOOVES AND CABMAN'S 'WHOA!' SIMULTANEOUS WITH HER APPEARANCE: ELO 'RAIN RAIN RAIN'. CAMERA FOLLOWS CAB, PANNING WITH IT. VOICE OFF:
My god — him too! Quick!

CAMERA PANS BACK IN A BLUR TO SHOW ELVIS PRESLEY COMING OUT OF SHOP WITH UMBRELLA, AS MANY BOXES OF BURGERS AS HE CAN CARRY
They *are* always with us.

STATUES OF VICTORIA: WIPE FROM ONE TO THE NEXT. N END OF BLACKFRIARS BRIDGE (NEARBY: CITY OF LONDON SCHOOL AND TEMPLE), 15 CARLTON HOUSE TERRACE, FLEET STREET NEAR LAW COURTS, GLASGOW GREEN FOUNTAIN, ETC.
When the present queen succeeded to the throne in 1952, children of the post-war baby boom were optimistically styled New Elizabethans. It was an epithet that didn't stick.

In contrast — by the time of the Great Exhibition in 1851 Victoria's subjects had begun to refer to themselves as Victorians and to the multitude of things that surrounded them as Victorian. She became a noun, an adjective . . .

VICTORIA STATUE IN WINCHESTER CASTLE — THIS IS THE MOST FANTASTICAL AND GROTESQUE OF ALL STATUES OF HER (BY ALFRED GILBERT)
. . . an icon and, apparently, a malevolent toad.
These appropriations of her name by a people and an age may or may not have been deferential. The point is that such coinages were *made*.

TRACKING SHOTS OF EIGHTEENTH-CENTURY TERRACES AND EARLY NINETEENTH-CENTURY TERRACES. THEY WIPE INTO EACH OTHER. QUEEN ANN'S GATE. COWLEY ROAD, CAMBERWELL, EDINBURGH NEW TOWN, REGENTS PARK, HIGHBURY FIELDS, PARAGON, BLACKHEATH
Victoria's succession coincided with the dissipation of an architectural consensus which had existed for 150 years. From soon after the Restoration to soon after the Regency there was a unanimity about what buildings should look like. A house of the 1830s spoke, so to speak, the same language that a house of the 1680s had. The idea was that a terrace should appear as a unity.

TRACKING SHOTS OF VICTORIAN TERRACES: THE GARDENS SE22, NIGHTINGALE LANE SW4, THE CHASE SW4, ONSLOW SQUARE SW7, ETHEL TERRACE, MORNINGSIDE, ETC. SHOW IN CHRONOLOGICAL ORDER. KELMORE GROVE, OAKHURST GROVE, SE22
During the six decades of Victoria's reign the notion of the terrace as one grand construction was gradually thrown out. And so was any attempt at consensus. Individuality was all. There was perpetual revolution. Or, rather, revolutions.

IN FRONT OF FACTORY GATES
There was the Industrial revolution — continued.

JM IS OBLITERATED BY PUFF OF SMOKE AND SOOT. COBBLED STREET — JM PTC SOOTY FACE
There was social revolution.

JM IS OBLITERATED BY MOB OF CARDBOARD CUTOUTS WHO RUSH IN FROM ONE SIDE AND TRAMPLE HIM. COBBLED STREET. JM PTC AS HE PICKS HIMSELF UP

Victoria Died in 1901 and Is Still Alive Today

There was political revolution.

JM OBLITERATED AGAIN BY MOB RUSHING IN FROM OTHER SIDE WITH FLAGS. LEGS AND FEET IN FRONT OF HEDGE: JM'S AND MAN, WOMAN, CHILD AND PIG IN MUDDY COUNTRY CLOTHES AND BOOTS
There was demographic revolution.

BUCOLIC VOICE:
This the way up Lunnun then?
JM:
It is indeed. That's a fine specimen — should fetch a fair price . . . and you'll doubtless get a few bob for the pig too.

HURSTPIERPOINT, WEST SUSSEX. HEAVILY ARMED CAMOUFLAGED CLERGY SINGING 'ONWARD CHRISTIAN SOLDIERS'. THEY PASS JM
There was religious revolution.

PARODY OF WARHOL'S MAO WITH VICTORIA AS SUBJECT — THE ENTIRE SCREEN IS FILLED WITH MULTIPLE VICTORIAS AS IN THE MAO SILKSCREENS
There was cultural revolution.
Victorians wanted to start from zero. They selfconsciously defined themselves as belonging to a new age. They thought of themselves as new. They belonged to new social classes.

MONTAGE OF VICTORIAN INVENTIONS. LIFT ASCENDING. NEW THINGS SHOVED IN AT EACH FLOOR BY UNSEEN HANDS. JM AMIDST THEM. CLOCKS TICKING. NOISE OF STEAM
They were surrounded by new things.

NATURAL HISTORY MUSEUM, OXFORD. JM WEARING APE MASK
They were subjected to new thoughts.

ST PANCRAS INTERIOR. JM IN RABBIT MASK WITH WOMAN AND TEN CHILDREN IN RABBIT MASKS
They bred new people at an extraordinary rate.

CAMERA TILTS DOWN TO SHOW DEAD RABBITS (REAL ONES) ON FLOOR.
There were stillbirths, however.

SUBTITLE: 'FEWER THAN SIXTY ANIMALS WERE INJURED IN THE MAKING OF THIS FILM'. MONTAGE OF:
a) PECKFORTON CASTLE, CHESHIRE
They built new buildings. Victorian architects fretted publicly about the need to create a distinct and specific Victorian architecture, an architecture fit for their age.
But the architecture fit for their age had also been fit for other ages.
Their new buildings were also old buildings.
They belong to the centrally heated Middle Ages.

b) ST PANCRAS
They belong to the age of steam and chivalry.

c) LANCING COLLEGE CHAPEL, W SUSSEX
They belong to the age of gross superstition and applied science.

d) KINGSWOOD HOUSE, DULWICH, WITH BULL TETHERED OUTSIDE AND GIANT JAR OF BOVRIL

. . . to the age of battlements and Bovril

e) GLASGOW — GREEK REVIVAL CHURCH: WELLINGTON CHURCH, UNIVERSITY AVENUE; NEAR NAKED BOYS DRESSED IN GREEK LOINCLOTHS HOLDING LYRES, AND HEAVY PETTING ON STEPS

They belong to the age of gaslight and ancient Greece.
They all belong to the age of Victoria.

f) MEMORIAL HALL, 14 ALBERT SQUARE, MANCHESTER

Which was also the age of Venice . . .

g) HARRINGTON GARDENS, SW7. DUTCH WOMAN IN VERMEER/VAN EYCK DRESS APPEARS FROM DOOR:

Schpliff? You want good schpliff?
. . . and of Vermeer

h) ENTRANCE OF WANDSWORTH PRISON

. . . and Vanbrugh.

TEMPLETONS, GLASGOW GREEN

It's as though not content with colonising half the globe Britain wanted also to colonise the past — its own past and everyone else's past too. It coveted time as well as space. Anything seemed possible in that first glorious moment when all the old certainties dissolved.

Britain in the nineteenth century was land of the future as the USA was in the twentieth century, but its buildings are three-dimensional oxymorons.

ST FRIDESWIDE, NOTLEY ROAD OSNEY, OXFORD

The might and potency of Victorian architecture derive from this preposterous desire to be in two places at once, in two centuries at once, and from the tension between these two irreconcilable goals — which were pursued with absolute conviction, absolute irreason.

This was dressing up for real — if you pretend to be something for long enough you become it. This was serious fantasy contaminating primary reality, *usurping* primary reality. This was the manifestation of a pathology.

JM GOES OUT OF DOOR

Nothing was done whimsically. When Alice Liddell carved this door she did it in absolute earnest — just as Lewis Carroll wrote in absolute earnest.

ST PANCRAS INTERIOR. JM RETURNS AS WHITE RABBIT: OFF-WHITE RABBIT WITH MYXOMATOSIS AND NICOTINE-STAINED TEETH, A NASTY COUGH AND WHEEZE, RUBS CROTCH PERSISTENTLY. HOLDS TIN OF LAUDANUM MARKED 'DRUGS'

What you lurking about here for then, my babas?

DRINKS FROM TIN

Fancy a tipple any of you?

OFFERS TIN

Ooh I couldn't half do with a spot of executive relief — I don't suppose . . .

CUT TO JEFFERSON AIRPLANE: CHORUS OF WHITE RABBIT LOOPED — 'FEED YOUR HEAD, FEED YOUR HEAD'
ALTON CASTLE, STAFFS (SHOT FROM LODGE INITIALLY, THEN CLOSE BY), CHEADLE CHURCH, STAFFS

The Victorians didn't borrow from the past. They stole. They grave-robbed.

Not, they claimed, for aesthetic reasons. Not because it looked good. But because it would *do* them good. For spiritual reasons: supposedly. They deluded themselves just as feng shui's pitiful adherents delude themselves today.

Augustus Pugin was the catholic architect and propagandist who made what should be acknowledged as a great English invention. He understood that his essentially philistine compatriots were loath to be appealed to on aesthetic grounds.

So he devised what might be called The Pietistic Excuse or The Mask of Righteousness:

It is not, he realised, sufficient to tell the English: I design this way with pointed arches and tracery because I happen to like the geometry . . . No, I design this way because pointed arches are holier than round ones and spires will take us closer to god than round arches and domes will. Aesthetic preference has to be disguised as moral and sectarian necessity.

PETERSHAM HOTEL, NIGHTINGALE LANE, RICHMOND. CAR PARK. RED VW WITH PRIEST IN DOG COLLAR. WHITE CITROËN WITH SEEDY WHITE RABBIT. FRONT ON TO CAMERA. CAMERA TRACKS FROM VW TO CITROËN. IN VW PRIEST PLAYS WITH ROSARY. IN CITROËN A HEAD BOBS UP AND DOWN ON RABBIT'S LAP

Pugin's invention is specious — it's like justifying a preference for red cars over white cars by stating that red cars are decent while white cars are degenerate . . . But you never can tell -

SWIFT WIPE TO KING'S CROSS RED VW KERB-CRAWLING BESIDE WAIF ON PAVEMENT. JM IN DOG COLLAR LEANING FROM WINDOW. PIXELATE FACE AND REGISTRATION NUMBER

> *Hey little boy — d'you wanna make a bargain?*
> *Come for a ride in my Volkswagen?*
> *I am forty-four and you are only ten,*
> *But I'm the kind of man that likes you little men.*

LANGFORD PLACE, NW8 — SPIKY, CREEPY GOTHIC HOUSE. NIGHT

An inanimate object such as a car or a building cannot possess moral or immoral properties, for those properties are human and a building is merely a construction of brick, stone, glass, iron and so on.

But humankind can *invest* a building with such properties. Humankind can infect a building by association, by propaganda, by persuasion. We can say: This looks creepy. This *suggests* evil. And all because of an over-attenuated gable and a window like a diseased eye.

CAMERA GOES THROUGH WINDOW INTO DARKNESS TO SHOW A DISEASED EYE AND THENCE ON INTO BRAIN. OUTSIDE OPTICIAN IN MACCLESFIELD. JM WEARING OPTICIAN'S TESTING SPECTACLES

Or is there more to it than that? Is there a part of the human brain which decrees that these very shapes are the shapes of perfidy. In which case — why were they so lavishly employed?

KINGSTON CLINIC, EDINBURGH. NIGHT

So far as architecture was concerned reason went to sleep throughout Victoria's reign. Or maybe reason was *put* to sleep.

VICTORIA V/O

> *Sleep tight my little princes . . .*

VELVET PILLOW GRIPPED BY VICTORIA'S HANDS — SHE IS GREEN-VEILED: THEY ARE LIKE CLAWS. FILLS FRAME AND COMES OVER CAMERA AS THOUGH SMOTHERING IT. OUT OF THE DARKNESS — A FAINT IMAGE OF A FROCK-COATED MAN SEEN THROUGH THE GAP BETWEEN HEAVY CURTAINS, THIS GAP COVERED BY A MUSLIN CURTAIN. HE IS INITIALLY BENDING FORWARDS AS THOUGH TO SMOTHER CHILDREN, THEN HE SLOWLY STRAIGHTENS HIMSELF WITHOUT TAKING HIS EYES OFF THE UNSEEN CHILDREN. UNEARTHLY SOUND.

JM PTC WEARING VULTURE MASK

Victorian architecture explicitly acknowledges that the human appetite for all that is inexplicable is greater than the appetite for reason.

KEBLE COLLEGE CHAPEL, OXFORD

Like the baroque of the counter-reformation, the Gothic was a weapon. It was an instrument of anti-Enlightenment zeal and of a powerful, reactionary church which preyed on the fears of an uneducated, superstitious populace susceptible to nonsensical ritual and absurd mysteries.

And he certainly does move in mysterious ways: the chapel at Keble was paid for by guano — bird shit.

LOUD SCREECHING OF GULLS. FLASH FRAME OF GULLS ON TIP — THE FRAME IS FILLED BY THEM. CUT BACK TO JM COVERED IN BIRD SHIT

KEBLE COLLEGE, OXFORD

The Gothic was also a means of turning the secular sacred, of contaminating daily life with religion, of making god omnipresent in the most literal way, of rendering holy a place of learning — forever: materials that won't weather or mellow suggest a yearning for permanence.

MONTAGE OF RAGSTONE CHURCHES — COURTFIELD GARDENS SW5, ONSLOW SQUARE SW7, REDCLIFFE SQUARE SW10, ST BARNABAS, PIMLICO ROAD SW1

At any given period 1 per cent of architects are goats, and 99 per cent are sheep. The majority merely follows fashion irrespective of what the fashion is supposed to represent.

The same goes for fashions in ideology, political thought, religious conviction and, of course, clothes.

What was the bustle if not a device to emphasise the sexual primacy of the buttocks? It was thus an aid to contraception. It promoted anal penetration. And no one, as we all know, ever got up the duff from a meat suppository. But is that how the hundreds of thousands of women who wore them thought of them . . .

ANIMATRONIC BUSTLE WHICH MURMURS 'BUY ME AND STOP ONE'.
ELVETHAM HALL, HARTLEY WINTNEY, HANTS

Fashions are also occasioned by the means of their production. Things come into being not because they are needed but because the technology exists to create them.

Again — the mania for revivalism and retrospection can be regarded as an extreme, pathological, form of escapism. The quotidian reaction to quotidian experience was to attempt to dress it up, to disguise it — anything to avoid the actuality of the real. It is as though an entire nation was involved in trying to hide something from itself. Architecture's role was to sweep actuality under the carpet, to euphemise, to lend a literal front of propriety and respectability. Respectability was more than a social nicety, it was a moral imperative.

LOW BACKWARD TRACKING SHOT OF JM'S FEET PEEPING OUT FROM A TABLE CLOTH.
BARCLAY BRUNTSFIELD CHURCH & BRUNTSFIELD LINKS — PEOPLE GOING TO CHURCH ON SUNDAY. JM DRINKING FROM LAUDANUM BOTTLE — SIX-PACK OF RING-PULL CANS WITH LEGEND DRUGS OR LAUDANUM OR L. JM AS MR HYDE AND DR JEKYLL. PTC IN UNISON

'Dr Jekyll and Mr Hyde' may be read as a representation of this ur-Victorian imperative. As a child Stevenson could walk out of the front of his parents' house into the aggressive rectitude of bourgeois Edinburgh.

Or he could walk out the back into the close where the family's servants would make extra money by letting the rooms above the stables. They let to mumpers . . .

JM BEGGING ON EDINBURGH STREET — ALBERT MEMORIAL — SITTING IN GRAND ARMCHAIR (HIDDEN CAMERA). JM REFRAIN TO PASSERS BY:
Can ye spare a florin for a cup of tea, Jimmy.
. . . They let to scaldrum dodgers . . .

JM STILL BEGGING, NOW WITH SOILED BANDAGE OVER HEAD AND ONE EYE PATCHED. JM TO PASSERS-BY:
Can ye not see it's a running sore I got and pox of the head. Spare us a florin for a cup of tea.
. . . They let to bludgers.

JM IN ALLEY. HITS PASSING BUSINESSMAN — GILES GORDON — OVER HEAD WITH SPADE AND TAKES HIS WALLET AND BRIEFCASE. BLOOD EVERYWHERE. JM ADDRESSES UNCONSCIOUS FIGURE:
I gave you a chance to gimme a florin. No prisoners, Jimmy.

SUBTITLE: 'DO NOT ATTEMPT THIS AT HOME — BUT IF YOU DO, WEAR A CONDOM'.
MEWS. COUPLES FUCKING. OUTDOORS ORGY. PEOPLE SHOOTING UP, DRINKING, ETC. JM WALKS BETWEEN THEM

Thus the back of the Stevensons' house was a thieves' pantry. It was the very obverse of respectable. The opposing worlds existed within yards of each other. The nether world preyed on the visible, official world — which tried to deny its sordid sibling's existence even though it knew it was everywhere. Sounds familiar?

Here's Long John Silver or, rather, W. E. Henley, Stevenson's inspiration for that celebrity pirate:

SCRIPT 3

Madame Life's a piece in bloom
Death goes dogging everywhere
She's the tenant of the room
He's the ruffian on the stair.

ST THOMAS'S HOSPITAL, LONDON SE1. WAIF WITH SYMPTOMS OF THESE DISEASES, PILING UP ON TOP OF EACH OTHER

The ruffian was always on the stair. Death *did* go dogging everywhere. Victoria's reign began and ended with smallpox epidemics — and in between cholera and typhus manifested with extreme regularity. There was a protracted outbreak of bubonic plague in the 1850s. Diphtheria, dysentery, encephalitis were all virulent. There was yellow fever, scarlet fever, blue fever, mauve fever — any colour fever. A lot of people were in a lot of pain a lot of the time.

TINY COFFIN — WREATH THROWN ON IT.

ST THOMAS'S HOSPITAL. JM SWIGGING FROM LAUDANUM TIN

Diseases were, of course, blown by malevolent zephyrs, carried on miasmas, vectored by fallen angels, spread by satanic bodies. Or maybe diseases were retribution, god's wrath.

The utility and modest ornament of hospitals betrays their low status in the architectural hierarchy. And, evidently, the sick's low status too. The sick are generally powerless and the architectural pecking order was based on power — whether that power be ecclesiastical, mercantile, aristocratic or political . . .

So how did a building display high status and power?

JM MOVES OUT OF FRAME TO REVEAL: PALACE OF WESTMINSTER (PARLIAMENT) ON OTHER SIDE OF RIVER — ZOOM ACROSS WATER

Ostentation mixed with piety always helped. An explicit show of *plenty*. Plenty of everything. Plenty of size. Plenty of ornament. The primitive conviction that more is more.

INTERIOR OF THE PALACE OF WESTMINSTER

And within — the more like a fence's lair the better.

Which is apt here since governance is a euphemism for legalised theft, for self-sanctioned peculation, for the redistribution of wealth — among cronies.

The material expression of temporal power was doggedly literal. The adoption, so early in the reign, by the architecture of public polity of this robber-baron mentality set an eagerly followed example.

INTERIOR OF CARDIFF CASTLE

At both personal and public levels status was in direct proportion to the energy expended on ensuring that no space was left unfilled, no surface left blank. Status was in direct proportion to the quantity and richness of acquisitions, to the aggregate of *things*: and there were more things in the world than ever before.

ST PANCRAS INTERIOR. VICTORIA AT PHOTOCOPIER PRINTING OFF SCORES OF PICTURES OF HERSELF: THEY ARE SPEWED OUT (SPEEDED UP SLIGHTLY) AS THEY FALL ON TO FLOOR COVERED BY HERALDIC FLAG

Victoria Died in 1901 and Is Still Alive Today

There were more ways of making things. There were more ways
of copying an original and making multiples: oleographs,
Stevengraphs. There were more people to make them. There were
more factories and works to house machines to make anything and
everything. Britain mass-produced for the world.

SILHOUETTE OF FIGURE WITH SMALL AXE

Where did the Sioux and the Cheyenne get their tomahawks, scalping
for the use of?

**FIGURE IS LIT TO REVEAL ROY WOOD* IN INDIAN HEAD-DRESS AND WARPAINT
WITH TOMAHAWK AND SCALP IN FORM OF A PERMED WIG**

They got them from Birmingham

RW — STRONGEST BRUM ACCENT POSSIBLE:

Why don't thou drop by for a cup of tie and a pow-wow?

**GUNNERA AT ELVETHAM, BICTON MONKEY PUZZLES, RHINEFIELD RHODODENDRONS,
BIDDULPH, STAFFS**

The more you owned and displayed the higher your position in
society, *and* the better and holier your soul — apparently. There
was so much that might be possessed. And not merely manufactured
things but exotic flora, exotic fauna.

The equation of temporal plenty with godliness was indeed a
symptom of a medieval mindset.

EAST HORSLEY TOWERS, SURREY

Perhaps that should be a *neo*-medieval mindset. No matter how much
Victorians might have wanted to live in the fourteenth century,
they didn't. They lived in the nineteenth. A medieval house built
in 1861 belongs to that year — not to 1361. The 1361 that 1861 was
interested in was a fiction. 1861 wanted 1361 on its own terms. It
wanted 1361 to be Victorian, to be the 1361 they'd have created
then had they had the good sense to be Victorian.

LICHFIELD CATHEDRAL, WEST FRONT. JM SUCKING PIXELATED LOLLY IN FRONT

And just as the present was infected by the past, so was the
past infected by the present. The past was there to be remade
in the Victorians' image. A medieval cathedral is likely as not
an on-site recreation of that cathedral by Victorian restorers.
It is a copy of itself mediated by the taste of several hundred
years after it was initially built. The west front of Lichfield
cathedral, for instance, is a creation of the 1860s. It was
medieval in a manner suitable to Trollope's generation.

BIG BEN: POSTCARDS, MODELS, HP SAUCE BOTTLES — ROW UPON ROW OF THEM

All Britain was being re-branded. Big Ben became one of its logos:
it came to stand for London, hub of the Empire, capital of the
most powerful nation on Earth. It stood also for the paramountcy
of clock time: obey parliament's clock as you obey parliament's
laws. Until the industrial revolution, sun-up and sun-down had

---

* Mr Wood was booked for 2 p.m. at a studio near London Bridge. At 2.30 he
still hadn't shown up. When phoned he said that he had been on the way to
Stafford Station when he met a chum and they were still in the pub. Sorry. I put
on the head-dress.

sufficed for the great mass of the populace. Precise horological knowledge had been vouchsafed only to the few. Time had been a luxury item. Factories and railways made it a necessity. A fob watch became a mark of respectability, responsibility.

HIGHBURY — CHURCH AND CLOCK TOWER, THE LATTER IN FOREGROUND
A borough, too, lent itself respectability by the construction of a clock tower. Time-keeping was a civic virtue.

PORT TALBOT STEELWORKS — POSED SHOTS OF BARDS, HARPS, DECLAMATORY SPEECHES IN WELSH, WELSH COSTUMES. JM SHOUTING TO MAKE HIMSELF HEARD
Big Ben was named after Benjamin Hall, the plutocratic Welsh MP who oversaw the rebuilding of the Palace of Westminster. For which he was rewarded with a baronetcy. His wife Augusta, Lady Llanover, was equally busy — creating Welshness, a form of Welshness appropriate to the new age. The profits of heavy industry were spent on the construction of a bogus past: myth systems, festivities, rituals, romantic national costume, national songs, national dishes. None of this simply happens. It doesn't come out of the ether. It has to be created because it is needed: the newer a society, the more obsessed it will be by its supposed roots, by its presumed ancestors, and the more anxious it will be to track them down like a foundling looking for its mother.

LULULAND, BUSHEY, THE FRONT DOOR
The man who created the bardic costumes and thus the image of Welshness was not himself Welsh. He was a German, living in Hertfordshire, in a house designed by an American: he was the painter Hubert von Herkomer and this is all that remains of that house, Lululand, the only English work of the Bostonian H. H. Richardson.

CAMERA MOVES PAST JM THROUGH DOOR INTO THE BRITISH LEGION CLUB BEHIND — PEOPLE PLAYING BINGO, ETC. EXTERIOR CARDIFF CASTLE
The Anglican creators of Welshness did not devise a specifically Welsh architecture. They imported English idioms. There was, of course, a chasmic gap between romantic notions of Welshness and the Welsh themselves . . .

CLASSICAL CHAPELS: MERTHYR TYDFIL, PORT TALBOT, PONTYPRIDD, THE HAYES, CARDIFF
. . . who already had a specifically Welsh architecture, and a specifically Welsh building type — the chapel.
The chapel was the emblem of the real, non-conformist Wales — which was not preoccupied by fancy dress and harps. This Wales embraced the classical survival rather than the Gothic revival.

CASTELL COCH
The Gothic revival in Wales was a form of cultural colonialism. It was the architecture of an immensely wealthy entrepreneurial caste. None of them wealthier than the third Marquess of Bute, a singular patron with a sense of humour. He once employed the notorious paedophile Frederick Rolfe aka Baron Corvo to run a boys' school.
The greatest beneficiary of his patronage was William Burges.

Victoria Died in 1901 and Is Still Alive Today

ROSTRUM: LES TRES RICHES HEURES AND NORTHERN GOTHIC PAINTINGS

Just as classical architects based actual buildings and landscapes on the imaginings of painters such as Claude, so did Burges build actual castles that derived from the idealised illustrations in books of hours and from northern Gothic paintings. Thus were medieval fictions made Victorian fact.

CARDIFF CASTLE

Between them Bute and Burgess suffered religious mania, opium addiction and tertiary syphilis — the trinity of afflictions that made high Victorian architecture glorious. It was religious mania that determined the styles that were acceptable. But it was laudanum and venereally founded madness that determined the manner of their interpretation.

END OF PART ONE

HOTEL DU VIN SMOKING ROOM. COMMERCIAL FOR LAUDANUM
WHITE RABBIT

Laudanum — for after shave, after shower, after anything

PART TWO

Laudanum is a suspension of opium in alcohol. It was the most widely used drug of the nineteenth century. Victoria herself used it — though late in life cocaine was her tipple of choice.

FLASH FRAME OF VICTORIA ROLLING UP A BANKNOTE WITH HER HEAD ON IT. LINES ON A TABLE

You could buy countless proprietary brands of laudanum in any chemist's or corner shop without a scrip.

FLASH FRAME OF JM PUSHING SUPERMARKET TROLLEY FULL OF BOTTLES, TAKING MORE OFF SHELVES. PA ANNOUNCEMENT: 'THERE ARE 25 REWARD POINTS ON A LITRE OF PAYTON'S PAREGORIC THIS WEEK. ENJOY'

It goes without saying that the toleration of such drugs ruined Britain, prevented it from creating the world's greatest ever empire, prevented it from becoming the most powerful nation on Earth. Today, of course, we've licked all our problems by banning everything.

Cardiff Castle shows us inside Burges's stoned brain. Inside his diseased brain too — tertiary syphilis presents with hallucinations, delusions, a gamut of retinal disorders including, eventually, blindness — it has that in common with masturbation, which is an effective way of avoiding syphilis. It also prompts a condition in which the arithmetical faculty is impaired.

THE DECOR SEETHES AND TURNS INTO BUBOES, SORES ETC.

There was no cure for the French disease — all diseases get blamed on someone else. And there were at least 80,000 female and 30,000 male prostitutes working in London in the 1860s — many of them children. Who knows how their great-great-great-grandfather made his start in life?

WALLACE MONUMENT NEAR STIRLING. VICTORIA WITH MOP AND BUCKET

SCRIPT 3

OUTSIDE KEEPER'S HOUSE

My great-great-great-grandfather may very well have hawked his brawn in the close behind Stevenson's house. He may have been a catamite to the manse. He may have gammed sailors in Leith at threepence a time. The age of consent, incidentally, was twelve, in implicit acknowledgment of life's brevity. In 1870, at a salary of £20 per annum, he became a key player in the North British janitocracy when he was appointed first keeper of this — the Wallace Monument. The monument to bogus Scottishness. And a typically Scottish structure: the crown at the top is a device which originated in England . . . But that should be no surprise.

HARRINGTON GARDENS, LONDON SW7

For tartan is originally Flemish —

DUTCH OUTFIT IN TARTAN INC. CLOGS.
ST PANCRAS INTERIOR

. . . the kilt was invented by a Lancastrian ironmaster.

FRAME WITH VICTORIA KNEELING SO THAT HER HEAD IS BENEATH THE KILT

And haggis is French.

VICTORIA APPEARS FROM BENEATH KILT HOLDING TWO HAGGISES

Victoria who did more than anyone to propagate it all was an English queen with a German husband . . .

So it's entirely apt that Mel Gibson should be an Australian.

BRAWNY SCOT WALKS INTO FRAME CLUTCHING TELEGRAPH POLE:

Will you no' toss my caber laddie?

SCOTTISH BARONIAL HOTEL IN POLLOKSHAWS, GLASGOW. KILBRIDE CASTLE

Still, unlike bogus Welshness bogus Scottishness — Walter Scottishness — created a plausible architecture which attempted to validate the romantic fabrications associated with it. The baronial style was lavish, butch, brutal, mad.

FLASH FRAME OF ROOM ENTIRELY HUNG WITH ANTLERS

It took the Victorian obsession with restless skylines to an extreme. It obviously alluded to the auld alliance — it is often wholly French in inspiration.

BLAIR DRUMMOND NEAR STIRLING

These buildings try to evoke a warrior caste. They may be luxurious, they may be second homes or aggrandised shooting boxes.

FLASH FRAME OF MONARCH OF THE GLEN — PULL BACK TO SHOW JM SHOOTING AT IT. BLOOD EXPLODES FROM IT, NOISES OF EXPIRING STAG

But they dissemble their sybaritism. They put up a convincing show of anticipating or even inciting armed attack.

BLOW-UP OF BRAEMORE HOUSE (DEMOLISHED)/FORTH BRIDGE, LONG LENS. JM IN FRONT OF BLOW-UP. HE IS IN SHARP FOCUS, THE HOUSE BEHIND IS FUZZY SO IT IS NOT IMMEDIATELY APPARENT THAT IT IS A BACKDROP

This is the house of the engineer John Fowler.

JM SLASHES AT THE PHOTO TO REVEAL THE FORTH BRIDGE BEYOND

And this is his greatest work.

Victoria Died in 1901 and Is Still Alive Today

Why would the author of this prodigy of cantilevered construction, one of the marvels of Victoria's reign, elect to live in a house of make-believe?

Well, a bridge was one thing, a gentleman's house quite another. A bridge was blue-collar. There was as much a structural hierarchy as there was a social hierarchy. The idea that this bridge might ascend the structural hierarchy as John Fowler had ascended the social one was a non-starter: the social hierarchy was a climbing frame. But structures were bound to their position in the hierarchy by their use.

GLASGOW. CENTRAL GRIDIRON BUILDING, CA' D'ORO

Urban Scotland was defiant of metropolitan English practice. Glasgow, second city of Empire, belongs both architecturally and sentimentally to northern Europe rather than to an English-dominated union. It recalls the Cities of the Baltic and, even, of north America.

NECROPOLIS

During Victoria's reign it expanded at an exponential rate. Its population increased by 100,000 every ten years.

CO-OP WAREHOUSE, MORRISON STREET, GLASGOW

Glasgow is built on the superhuman scale common to small nations with a point to prove — the silicone enhancement of the nationalistic impulse.

GLASGOW TENEMENTS/FLAT BLOCKS: BALMORAL CRESCENT OFF QUEENS DRIVE; FOTHERINGAY ROAD AND TERREGLES AVENUE; VARIOUS STREETS IN PARTICKHILL (E.G. LAUDERDALE GARDENS) AND IN MARYHILL

It's tall because its inhabitants learnt, as England's were unwilling to, that you pay for the privilege of living in a city by living vertically. Glasgow's Victorian tenements were once as reviled as 1960s tower blocks. But if *any* sort of building is left unmanaged it will deteriorate, especially if the inhabitants bubble coal gas through milk before drinking it.

JM DOING THIS IN SOME BAR:

Backlands Bree. It really fucks your head.

SUBTITLE: 'DO NOT TRY THIS AT HOME: MILK WHICH ISN'T TRANSFORMED INTO CHEESE CAN BE TOXIC'
CALEDONIA ROAD CHURCH; ST VINCENT; MORAY PLACE; 200 NITHSDALE ROAD; THE ENTIRE THOMSON OEUVRE . . .

What Glasgow had most of all was Alexander Thomson — Greek Thomson. It is, of course, unlikely that a revenant from ancient Athens would take him for a compatriot.

His work is unique in Britain. And because he was so widely copied, Glasgow itself is unique in Britain. Glasgow is what all British cities would have looked like had ecclesiastical medievalism not spread like a virus through their fabric and had a continuum with the pre-Victorian classical tradition been maintained. Glasgow was the last British city to engage with reason.

Classicism is essentially urban and formal. It embodies civic properties.

Glasgow owes its coherence as a city to the centripetal impetus of its classical suburban buildings: they reach *in* to the city's core rather than *out* to the country around it.

ST MARY'S CATHEDRAL, EDINBURGH/FOREIGN OFFICE, LONDON SW1
The major division in Victorian architecture is habitually represented as being between classical and Gothic. The fact that George Gilbert Scott designed both St Mary's Cathedral in Edinburgh and, at Palmerston's behest and against his wishes, made the Foreign Office classical, demonstrates merely that Scott, a principled propagandist for the secular use of Gothic, possessed a healthy respect for the commercial advantage to be gained by flexibility in those principles.

MORNINGSIDE PARK, EDINBURGH OR STEWART'S MELVILLE SCHOOL, QUEENSFERRY ROAD, EDINBURGH
Architects always have and always will move from one style to another according to a client's whims and to the demands of professional survival. Frederick Pilkington did the same — though in Pilkington's case the style was always one of his own invention, his knack was to make buildings which, whatever gamut of devices they employed, always seem on the point of uttering a silent scream.

JM STANDS BESIDE HORSE'S HEAD (FRESH FROM EQUINE ABATTOIR).
ELVETHAM MANOR, HARTLEY WINTNEY, HANTS
The same goes for Samuel Sanders Teulon, whose indifference towards the canons of harmony and proportion perplexed even his most hardened contemporaries. The combination of brutality and apparently wilful perversity that his buildings displayed was known as go!

Go! — a term first applied to horses — came to mean hard-edged, harsh, muscular, polychromatic, endlessly energetic — there is always a lot happening, much of it counter-intuitive. Such architecture begs the question *why?* Why would anyone design a house like this — and how did they arrive at such a resolution? Teulon, too, suffered tertiary syphilis which may be a clue. Was go! a symptom? Was there a syphilitic school of architecture?

OXFORD NATURAL HISTORY MUSEUM, THE PARKS
Go! transcends style. It has much more to do with the bloody-mindedness of an individual architect.

Gothic revival buildings can be infected with go!

WELLINGTON COLLEGE
So can baroque ones . . .

ASHMOLEAN MUSEUM, OXFORD
And early Victorian ones can . . .

PALACE HOTEL, OXFORD ROAD, MANCHESTER
. . . and those of the 1890s — by which time the word go! was hopelessly out of fashion. But the quality it signifies has not disappeared. It comes around now and again. It was everywhere in

the 1960s, which like the 1860s were also inclined towards sexual abandon.

SPLIT SCREEN: DETAILS OF THE FOLLOWING RATHER THAN WHOLE BUILDINGS
No, the real division in Victorian architecture — in all regions other than Scotland — has nothing to do with style. It is, rather, the division . . .

PRUDENTIAL, HOLBORN/HOLBORN UNION, CLERKENWELL ROAD
between hard and soft . . .

FETTES, EDINBURGH/PORT SUNLIGHT
between artifice and naturalism . . .

ROCKMOUNT CHURCH ROAD, UPPER NORWOOD/PORT SUNLIGHT
between vigour and douceness . . .

WADDESDON MANOR, BUCKS/WADDESDON VILLAGE
between the worldly and the would-be humble . . .

LAMMERBURN, NAPIER ROAD EDINBURGH/COTTAGES OPPOSITE ENTRANCE TO NAPIER UNIVERSITY CAMPUS
between the candidly industrial and the hand-made . . . the *apparently* hand-made.

MULTIPLE SCREEN: SIX FRAMES SHOWING THE FIRST OF EACH OF THE ABOVE PAIRS (I.E. PRUDENTIAL, FETTES, ETC.)
The two strains existed side by side. But it is the first category, the grandiloquent, the rhetorical, composed of materials which age gracelessly, that we think of as quintessentially Victorian, ur-Victorian.

CRIMEA MEMORIAL AND STATUE OF FLORENCE NIGHTINGALE, WATERLOO PLACE, SW1
The Victorians mythologised themselves. They were hero-worshippers.

PEOPLE KNEELING AS THOUGH AT PRAYER
Florence Nightingale's lamp which lit her rounds at Scutari was nothing like the one she holds here — but this one links her to Rome, to classical antiquity. And long before Tennyson sought to invest The Charge of the Light Brigade with dignity and nobility, William Russell had begun the sentimental glorification of cannon fodder in his eye-witness report for *The Times* . . .

READS FROM THAT DAY'S *TIMES* FRONT:
'CHRISTIAN BOMBS KOSOVAN TODDLERS'
 . . . *with a halo of flashing steel above their heads, and with a cheer which was many a noble fellow's death cry, they flew into the smoke of the batteries* . . .
This was an eye-witness account. Yet it makes no acknowledgment of the squalor or the humiliation — it's just euphemism and unwitting mock-heroism. This was another way of sweeping things under the carpet.

WEDGWOOD INSTITUTE, BURSLEM
There was such an appetite for gods that the months of the year were granted figurative form for the improvement of the Potteries.

Before the Wedgwood Institute was finished, one of its designers, John Lockwood Kipling, left to take up a post in Bombay. With him went his wife and, in utero, his son, who would be named after the place of his conception, a lakeside resort a few miles from the Potteries called Rudyard.

Rudyard Kipling embodies all that our era, in its moral superiority and manifest righteousness, finds distasteful about Victorian England. He's the archetypical DWEM. He transcends such fashions as anti-imperialism, anti-racism, anti-talent. Has our conscience-stricken government issued an apology for him yet?

CHARTERHOUSE SCHOOL NEAR GODALMING

Almost seventy years after it was written, Kipling's *If. . .* was appropriated by Lindsay Anderson as the title of his movie about rebellion in a public school — which was still, in 1968, essentially Victorian: in its mores, its cruelty, its wretched team spirit, its morbidly muscular Christianity and, unmistakably, in its architecture.

These places were part seminary, part tribal initiation, part OCTU. They were intended to instil notions of duty, obedience, valour. To harden the body and the soul. To produce character not cleverness.

It's heartening how often they failed.

Their ideal product was Newbolt Man, a creature hymned by the risible poet Sir Henry Newbolt:

> *To honour, while you strike him down,*
> *The foe that comes with fearless eyes . . .*
>
> *Henceforth the school and you are one*
> *And what you are, the race shall be . . .*
>
> *But his Captain's hand on his shoulder smote:*
> *'Play up, Play up and Play the game'. . .*

Oh the sheer, murderous infantilism of it all . . .

LANCING COLLEGE CHAPEL

The apogee of self-abnegating piety was reached in the schools founded in the 1850s and '60s by the Rev. Nathaniel Woodard. They were intended for shopkeepers and tradesmen who wished their sons to be turned into Christian gentlemen. These austere schools were emphatically not the grimpons and picks of social alpinism. Keep 'em humble might have been their motto had they not had a better one — Onward Christian Soldiers.

HURSTPIERPOINT COLLEGE. CLERGY WITH GUNS, GROUND TO AIR LAUNCHERS, HELMETS WITH CAMOUFLAGE BRANCHES AND GRASS TUFTS, A BARBED-WIRE CROSS, SINGING 'ONWARD CHRISTIAN SOLDIERS'

This hymn was written by the Rev. Sabine Baring-Gould who taught at Woodard's Hurstpierpoint. To take holy orders in the mid-nineteenth century was to enter rather than, as today, to preclude oneself from the mainstream.

Baring-Gould was an energetic and fecund man. He had twelve

children: the wife didn't wear a bustle then. He wrote a hundred
books. And he collected and bowdlerised thousands of folk songs.

**JM PICKS UP MUSICAL BOX/TANKARD THAT PLAYS 'WIDDICOMBE FAIR' SONG AND
DANCE NUMBER IN 'THE YOUNG GENERATION' STYLE: THE RUSTICS IN ONE-
PIECE STRETCH OUTFITS WITH HAYSEED APPLIQUÉ. JM IN FOCUS, BEHIND HIM,
SO OUT OF FOCUS THAT IT IS JUST A BLUR, A GREY PANTO HORSE WITH SEVEN
YOUNG RUSTICS AND TWO OLDSTERS, ALL OF THEM TRYING TO GET ON IT**

> *Tom Pearce, Tom Pearce, lend I thy grey mare*
> *For us to go to Widdicombe Fair*
> *With Bill Brewer, Jan Stewer, Peter Gurney, Peter Davy,*
> *Dan'l Widden, Harry Hawk and Old Uncle Tom Cobleigh and*
> *all . . .*

If *Onward Christian Soldiers* and public schools are all High
Victorian go!, Widdicombe Fair belongs to the later Victorian
reaction against it.

**MULTISCREEN OF EXTREME HIGH VICTORIAN EXAMPLES: ROYAL DOULTON'S,
LAMBETH HIGH STREET, SE1; MILFORD HALL, SALISBURY; GERRARDS CROSS
CHURCH**

> For twenty-five years from about 1850 the mainstream of
> architecture was uncompromisingly harsh — it exploited new
> technologies and it exploited the railways' capacity to free
> buildings from a reliance on local materials.

**ECCLESTON PADDOCKS, ECCLESTON NEAR CHESTER. DUKE OF KENT SCHOOL,
EWHURST, SURREY**

> Now this sort of architecture did not disappear. Indeed some of
> the most aberrantly dissonant buildings of the reign belong to its
> late years . . .

BROMPTON ORATORY; WESTMINSTER CATHEDRAL

> Pugin turned in his grave when the two most prominent Catholic
> churches in London were built — Brompton Oratory in the Jesuit-
> approved style of Vignola . . . and Westminster Cathedral like an
> import from Byzantium.

STANLEY HALLS, SOUTH NORWOOD

> Mr Stanley endowed — and designed — a technical school and a
> concert hall with the money he had made from the Stanley knife
> . . .

**FLASH FRAME OF TATTOOED SKINHEAD WITH BLEEDING FACE AND EYE DANGLING
FROM SOCKET. FRAME WIDENS TO SHOW VICTORIA SMILING, FEELING THE BLADE
OF A BLOODY STANLEY KNIFE. GRINS AT JM.**
**ROYAL HOLLOWAY COLLEGE, EGHAM. VICTORIA WITH PILLS THE SIZE OF EXTRA
STRONG MINTS AND VET SYRINGE, JABS HERSELF IN ARSE**

> Mr Holloway founded a college with the money he had made from
> patent remedies . . .
>
> But by the time he built it this free-booting vigour was a
> thing of the past. What happened to the spirit of sod-you-ism, to
> the wild experimentation, to the refusal to be ingratiating. What
> happened to that self-confidence . . .

LEICESTER SQUARE/PENSHURST, KENT/HAMMER FIELD, PENSHURST — 'ALL

THINGS BRIGHT AND BEAUTIFUL'

It gave way to self-consciousness and to prettiness of the sort that George Devey specialised in. Devey took the picturesque to new heights of naturalism and fakery. His houses — cottages suffering elephantiasis — give the impression of having been added on to down the centuries, of having been based on the foundations of older houses. Each of his works has a bogus history incorporated into it.

GRIM'S DYKE, HARROW

But coyness and sweetness were not enough.

At one level Victorian architecture was akin to late twentieth-century conceptual art: it required a text to lend it purpose. To make coyness earnest and sweetness serious.

Enter William Morris.

You know the type. He was a cerebral Conran **[FRAME OF CONRAN AT TABLE LAID WITH DESIGN OBJECTS]** who propagated the art of living in the Middle Ages rather than the art of living in France. He was the River Café with a beard **[FRAME OF R. ROGERS AND R. GREY WITH BEARD]**.

THE RED HOUSE, BEXLEYHEATH

He was a rich luddite who commissioned his own house at the age of twenty-four — the well is a telling touch. He was an entrepreneurial socialist who hoped to change the world with wallpaper, a manufacturer who abhorred machine production.

OXENFORD GRANGE, SURREY. DEEPEST ENGLISH LANDSCAPE, SOUNDTRACK OF CHILDREN PLAYING, VASELINE ON LENS . . .

His life was a protracted exercise in nostalgia — whose literal meaning is *the longing for a lost home*. Morris was forever trying to recreate his childhood which had been a prelapsarian commune with nature, baking hedgehogs in clay and riding a pony dressed in a suit of armour — as one does.

FLASH FRAME OF CHILD ON A FALABELLA PONY DRESSED IN ARMOUR, POSED LIKE THE MILLAIS PAINTING SIR ISUMBRAS AT THE FORD WHICH IT SEGUES INTO

We all had a childhood. What was peculiar about Morris was his desire to inflict his childhood on a nation. What was even more peculiar is that he succeeded.

Morris's ideal was the supposedly unselfconscious tradition of local building.

FLASH FRAME OF THE STONEBREAKER BY JOHN BRETT

He considered manual labour ennobling rather than demeaning — sure proof that he had never had to rely on it for a living.

The irony of his position was that his example was so appealing that he was himself an effective mass producer — of technophobic nostalgia. He turned nostalgia from being a private pining into a shared yearning that could be acquired by subscription to the Morris aesthetic.

EXAMPLES OF THE WORK/PORTRAIT OF ALL THE FOLLOWING, EACH WIPING ITS PREDECESSOR OFF THE SCREEN

Victoria Died in 1901 and Is Still Alive Today

Morris connected well. His close circle included: The architects:
Philip Webb

OAST HOUSE, CROYDON ROAD, HAYES COMMON
and George Edmund Street

HOLMBURY ST MARY CHURCH, SURREY OR ST PHILIP AND ST JAMES, WOODSTOCK ROAD, OXFORD
whose pupil he had been.
 The painters: Edward Burne-Jones (who was Kipling's uncle by marriage)

ANY PAINTING — THEY ALL LOOK THE SAME
Ford Madox Brown

PAINTINGS: *WORK* OR *THE LAST OF ENGLAND*
Dante Gabriel Rossetti — who died of chloral addiction

PAINTING OF DANTE AND BEATRICE
Rossetti's wife Lizzie Siddall — who died of a laudanum overdose.

PAINTING: *OPHELIA* BY JOHN EVERETT MILLAIS
Along with Morris's wife Jane she was one of the pre-Raphaelite muses who were known as stunners — proving that there is nothing new under the sun — or in *The Sun*.

ST JOHN, SYLVAN ROAD/AUCKLAND ROAD, UPPER NORWOOD. JM BUYS PAPER FROM WAIF. PLACARD: 'DRUG DEATH RIDDLE OF TRAGIC STUNNA'. PAGE-THREE PRE-RAPHAELITE. JM 'READING' *THE SUN* AND MOVING LIPS — THE NEWSPAPER IS PRINTED LIKE MORRIS WALLPAPER
Tragic Lizzie's ambition had been to work with animals and to open a chain of boutiques for world peace in the twentieth century.

SCREEN DIVIDES INTO FOUR. LAW COURTS, STRAND TAKES UP FIRST QUARTER OF SCREEN, SHOT FROM OXO TOWER
During the boom years of the '50s, '60s and early '70s there was a consensus that while cities might be prey to crime, disease and poverty these were not insoluble problems, that they could be tackled by the law . . .

SECOND QUARTER: LANGSIDE INFIRMARY, GLASGOW
by the provision of hospitals . . .

THIRD QUARTER: ABBEY MILLS PUMPING STATION, ABBEY LANE E15
by the provision of effective sewerage . . .

FOURTH QUARTER: PEABODY ESTATE, PEAR TREE COURT AND CLERKENWELL CLOSE, EC1
by the provision of adequate housing.

RYLANDS LIBRARY, MANCHESTER
It was in their mercantile interest to constantly improve the lot of their people. They made life better for many people who became materially and intellectually far better off than their rural peasant forebears. Cities civilise.

GUSTAVE DORÉ ILLUSTRATION
Morris — like William Cowper and William Cobbett before him — was

convinced that cities were dehumanising to the point where the
only reaction to them could be to reject them.

**KINGS CROSS. WAIF SELLING SEXUAL FAVOURS, PICKING AT BLACK BAGS,
PULLING OUT LAUDANUM BOTTLES**

The greatest human creation was deemed expendable, there were
apparently alternatives . . . Alternatives whose ramifications
would be felt all over England throughout the twentieth century
and on into this one.

England was the first country to industrialise and urbanise.

LEAFY ROADSCAPES: AUCKLAND ROAD AND BELVEDERE ROAD, UPPER NORWOOD

But, more importantly, it was the first country to react against
industrialisation and urbanisation — and that reaction has
afflicted England for a century now.

It was as though it was England's lot to make the mistakes the
rest of the continent would learn from.

**1910, 1920, 1930 SUBURBS: SOUTH EDEN PARK ROAD, VILLAGE WAY, WICKHAM
WAY, THE AVENUE, LINKS WAY — ALL IN BECKENHAM**

William Morris, utopian socialist, propagated the idea of flight
from cities, which meant in practice suburbs, which would sprawl
in their attempt to connect with the country rather than with the
inner city.

And his namesake William Morris of Cowley, the precocious
automotive engineer, provided the means which allowed the
aesthetically squalid and environmentally disastrous burbs to
spread.

PORT SUNLIGHT. MORRIS MEN

When it was initially turned into tiles and beams, Morris's vision
— traduced and misinterpreted, of course, but visions always
are — was a pretty enough sham, a sort of merry England for the
middle English and Morris dancers. Rusticity without the middens.
Bucolicism without the rural poor. As sanitised as a Baring-Gould
folk song: thus the maypole was topped with a crown rather than
with a glans.

WESTBROOK, WESTBROOK ROAD, GODALMING, SURREY

The generation of architects born in the 1850s and '60s — Voysey,
Prior, Ashbee, Mackmurdo and above all Lutyens — was, alas, one
of the most gifted Britain has produced . . . I say *alas* because
they designed with such imaginative invention and such persuasive
charm that the generations which followed them aped them — badly
and persistently.

VICTORIA IN CARDIGAN AND SLIPPERS SMOKING PIPE IN GARDEN

Innocence and lack of worldliness are no more virtues than
chastity is. Cosiness and a yearning for pre-industrial society
are banal forms of escapism. How pitiful it is that this
generation of architects should be so afflicted.

**DETAILS OF ARTS AND CRAFT BUILDING: BLACKHEATH CHURCH NEAR GUILDFORD
MUNSTEAD ROUGH NEAR GODALMING; LONG COPSE, PITCH HILL, EWHURST;
(CHURCH COTTAGES, SHERE, SURREY), DETAILS OF BRICKWORK, ETC. — ALSO A**

Victoria Died in 1901 and Is Still Alive Today

GERTRUDE JEKYLL GARDEN

They buried their heads in the sand, and cement, in the
handcrafted bricks, in the folk-woven tiles, in the artisan-
wrought metal, in the bespokely hewn stone, in the roughcast, in
the old English cottage gardens, in hollyhocks and lavender.

**HOUSES BY VOYSEY: HOLLYBANK, GRANGE ROAD AND MERRYHILL ROAD, BUSHEY,
HERTS**

They built a twee fiction — and they built it beautifully.

Voysey's clergyman father was tried for heresy. He denied god's
wrathfulness: that is, he denied the high Victorian god's essence.
He believed in a nice, soft god. The son believed in nice, soft
houses — and his vision spoke to some deep-rooted little English
longing.

**VICTORIA SITTING OUTSIDE CHEYNE WALK HOUSES READING PORNO MAG CALLED
'ONE'S SUBJECTS' HUSBANDS'**

Ashbee's father was the most prolific and energetic of high
Victorian pornographers — his pseudonym: Pisanus Fraxi. The
son reacted by espousing an ascetic and puritanical socialism,
by doing good works, by founding a craft guild, by espousing
sandalled vegetarianism.

**HOUSE ATTACHED TO BACK OF PETER JONES, SLOANE SQUARE, HANS ROAD SW1,
25 CADOGAN SQUARE SW1, OLD SWAN, SLOANE STREET SW1. ROSTRUM COVER OF
WREN'S CITY CHURCHES**

Mackmurdo didn't have a problem father. He introduced Shaw to
Wilde. But in 1900 he effectively abandoned architecture to compose
tracts advising that human society should take as its model the
beehive.

STILL OF HUMAN HEAD COVERED IN BEES

He is also credited with having made the first art nouveau design —
though this may be a desperate English attempt to stake a claim in
a movement which largely passed this country by.

KNUTSFORD, CHESHIRE

Why didn't the English adopt art nouveau when it was such a craze
in the rest of Europe? Maybe for that very reason — the English
have always wished to be out of step with Europe, even if it means
seeming backward . . . *especially* if it means seeming backward.

And much art nouveau was the rococo revived, and the rococo had
never taken off here.

Then again, art nouveau is an urban idiom reliant on machine
production — indeed it celebrates machine production.

It was the opposite of the arts and crafts. It didn't pretend
to grow out of the earth. It didn't dissemble its artifice.

The most singular concentration in England, the *only*
concentration, Richard Harding Watt's work at Knutsford, has
nothing to do with any indigenous idiom — it derives from Genoa
and Livorno and Rome.

**LUTYENS HOUSES: MUNSTEAD WOOD, MUNSTEAD ORCHARD, ORCHARDS MUNSTEAD
(ALL NEAR GODALMING), TIGBOURNE COURT NEAR WITLEY (ALL IN SURREY),
GREYWALLS, GULLANE NEAR EDINBURGH**

The young Lutyens was the greatest English architect since
Vanbrugh.

He designed 200 houses and 500 of them are in Surrey.

**ESTATE AGENTS FILMED OUTSIDE THEIR PREMISES, EACH SAYING THE LINE:
'IT *IS* BY LUTYENS — HONESTLY.' THE ESTATE AGENTS IN ONE HALF OF
FRAME, A HOUSE IN THE OTHER**

More than any one else he invented what would become the idiom of
the first half of the twentieth century.

That early work of the 1890s is delicate and gay and light. For
all its eagerness to please it *is* touched by genius and by the
very un-English quality of sheer cleverness. His multitudinous
followers patently lacked genius . . .

**'20s AND '30s ARTERIAL ROAD SUBURBS: KINGSBURY (OR EALING, WEMBLEY,
GREENFORD, ANYWHERE). JM V/O BUT IN FRAME LISTENING TO SELF ON HALF-
TIMBERED RADIO**

. . . and they weren't too clever either. They merely aped his
mannerisms, with increasing inaccuracy . . . And so caused the
aesthetic tragedy of twentieth-century England.

JM SWITCHES OFF RADIO

But it is not actually this architectural coarsening which is of
such moment. Lutyens and his contemporaries established not only a
stylistic model but an aspirational one too.

In place of the economical terrace they created the kind of
land-hungry ersatz-rustic dwelling that became the cynosure of
English daydreams.

**COWS, GOATS, SHEEP IN FRONT GARDENS OF '30S HOUSES, PEASANTS IN
SMOCKS, KIDS THROWING CABBAGES AT PERSON IN STOCKS, PEOPLE CUTTING
THEIR LAWNS WITH SCYTHES, NEWSPAPER BOY ON HORSEBACK**

No matter that the rusticity grew ever more bogus, no matter that
machine production was made to imitate handcraft.

Successive governments anxious not to offend the electorate
permitted the creation of suburbs which spawned suburbs of *their*
own which spawned suburbs of their own — and so on, without end.
This is what comes of every Englishman and woman getting a castle,
the only civil liberty the most proscribed country in western
Europe seems interested in.

FLASH FRAME OF '20s CASTLES IN BUCKS LANE, KINGSBURY NW9

The ultimate paradox of the back-to-the-land impetus is that there
is now no land left to go back to.

It is not, however, the prodigal squandering of the country, of
green fields, which is our most poisoned inheritance.

JM IN CAR PTC WAIF WITH SQUEEGEE, TOTALLY SOAPED WINDSCREEN:
Get back to when you came from.

**SCREEN MIRACULOUSLY FREE OF SOAP. URBAN DERELICTION FROM CAR.
SUCCESSIVE SHOTS MOSTLY MANCHESTER**

The greatest damage is to cities themselves: England is the only
country in Europe where 'inner city' is synonymous with crime and
deprivation. But no wonder — the cities were neglected in the race

to get out of them.

It was this insatiable appetite for escape which buggered up England.

Those countries which embraced the machine *as* the machine achieved a quality of urban life which this island with its perpetually post-protestant, perpetually post-puritanical mistrust of the city has not begun to get the knack of.

Even today, a century after Victoria died, the pitiful English desire is still for a house which is a counterfeit token of old England.

Though the people-mover we clean at weekends is not half-timbered, is not tile-hung, and its windows are not leaded lights.

MACCLESFIELD 1990S/2000 ESTATES OF 'CLOSES', LEADED LIGHTS, TUDORBETHAN DETAIL. ELVETHAM HEATH/HENLEY COURT

The ultimate architectural legacy of Victoria's reign is not the buildings that surround us. Not the clapped-out hospitals, the mercantile palaces, the ubiquitous terraces, the purposeless churches.

The true architectural legacy is not physical.

It is, rather, the collective mindset founded in the late Victorian conception — a misconception as it happens — of cities as morally and physically corrupting, and in the false promise of a better life outside them.

We greedily inherited the wrong bit of that Queen's England.

END

4

VERTICAL SAUSAGING MATRIXES

Axialise an Iconic Art Hub

ON THE BRANDWAGON

WE'RE FREQUENTLY INSTRUCTED – so often that we may half believe it – that professional football is for *the community*, that a successful team is invaluable to the mood of the people, that it raises a city's collective morale. When you drive into Southampton from the west you are confronted by a vast hoarding that claims 'Saints are the spirit of the city'. In which case – why aren't the citizens contorted in misery? Liverpool FC won the European Cup four times between 1977 and '84. So one can only conclude that Merseyside's loss, during that seven-year period, of 20 per cent of all available jobs and of 10 per cent of its population were both suffered entirely by Evertonians. Who were also obviously behind Militant, the Toxteth riots and the wall-to-wall strikes of those years.

This was when Michael Heseltine, famously, had a highly original idea, which was thirty years old and German. *Bundesgartenschauen* had been held biennially in such cities as Hanover and Cologne since 1951. The years of the German economic miracle. Heseltine put two and two together: a five-month pageant of horticultural excellence and spectacular entertainment would occasion the Scouse economic miracle. The change in West Germany's fortunes had thitherto been reckoned due to fiscal reform, curtailing the money supply, repealing price and wage controls. Hydrangeas in blossom would achieve so much more.

Three and a half million people visited the Liverpool garden festival in 1984. Yet, astonishingly, unemployment remained high and per capita income low compared with, say, Manchester and Leeds.

The Garden Festival was however massively influential in other ways. Its building presaged the squalid Millennium Dome – also, originally, one of Heseltine's whims. And it proved to be the foundation of a new industry. Or, rather, an industry which presented itself as new. The Regeneration Industry.

No one had previously heard of regeneration in the sense that it is used today. It is inescapable. What is it?

Spin made stone. Rather, spin made glass, steel, Corten and titanium with maybe just a little wood for warm eco-chumminess.

Gentrification through newbuild rather than restoration.

A payday without end for the demolition community.

Publicly funded parochial vanity.

Gestural engineering: follies for the masses.

State-sanctioned style over substance: officially ordained populism.

It's 1960s comprehensive redevelopment reincarnated and given a name with cosily spiritual connotations.

Volume building in disguise.

The twenty-first century's emulation of the clearances and enclosures.

The most telling expression of the Blair Age.

It's these and more.

It's the greatest of contemporary gravy trains. Regeneration is an ad hoc collation of enterprises, a shifting, mutating, jargon-heavy mass of PFIs, Blusky Thoughts, Endeavour Drivers, Round-ovalings, Crisis Tablet Alerts, Groningen Protocols, Nielsmartings, Creative Hubcaps, Hot Drains, Blisterquenches, Meme-trepanners, Fruit Haemorrhages, Brainswarfings, Crystal-axiommings, Jeopardy Beacon Plateaux, Traction Triggers, Infra-Sustainables, Grench Scalabilities, Quadratoastables, Holistic Leverages, Vark Incubators, Vertical Sausaging Matrixes, Core Sausaging Matrixes, Helicopter Views. Think outside the elephant. Push the box. The envelope in the room.

Regeneration is not a self-regulating discipline. Yet nor is it a commercial enterprise which must survive in the marketplace. As early as 1988 the National Audit Office recommended that the Merseyside Development Corporation be wound up because it had spend hundreds of millions to very little apparent effect.

How its effects are measured is, of course, the holiest mystery of regeneration. Cost-benefit analysis is at best an imprecise pseudo-science – and when applied to regeneration it turns into another bandwagon scratching along in the wake of the distended juggernaut.

The National Audit Office was impotent against what is essentially a *faith*, a cult.

So Regional Development Agencies multiplied and their budgets swelled. This expansion emphasises that in Regeneration, results – whatever they might be – count for little beside optimistic wheezes, flashy initiatives, loud gestures, soothing assurances of improvement . . . and the guile to procure public money in the first place.

When the names Ecclestone, Caplin, Foster, Prescott, Levy, Mandelson, Campbell, Hinduja and Mills are no longer recognised and his criminal folly in Iraq is forgotten, Blair's decade – the age of Social Thatcherism – will be recalled as much by its architecture as by its low dishonesty.

There can be little doubt about what building type will be seen as almost parodically representative of this age. A kind of structure characterised by its hollow vacuity, by its sculptural sensationalism, by its happy quasi-modernism, by its lack of actual utility. Yet it possesses two definite purposes: to be instantly and arrestingly memorable, and to be extraordinarily camera-friendly. This type of structure is the soundbite's ocular analogue: the sightbite.

The sightbite may have been an instrument of that fetish of New Labour, modernisation. The sightbite may have appeared modern. But it is pastiche modern, neo-modern, synthetic modern. Form and function coalesce. Appearance is everything. This is transgressive modernism. Ornamental modernism.

1997. The notion of a three-dimensional logo is born the same year as New Labour's *Machtergreifung*. Wrong. The notion of a three-dimensional logo as a symbol of a creed, as a sign of a superstition or ideology, as an expression of a ruler's might or a city's wealth . . . this notion began many thousands of years before 1997. It is indeed very likely older than the notion of architecture as shelter.

Like Liverpool, Bilbao was the near-corpse of an industrial city with special needs. Like Pittsburgh, the steel city on which it had modelled itself, it had decayed into a precipitous rollercoaster of rusting plant, abandoned mills, defunct furnaces, empty docks. Unemployment was high, the population was declining.

On the Brandwagon

The Guggenheim Museum, Bilbao's device for putting itself on the map, was designed by a man who would not be allowed to represent Athletic Bilbao. Who doesn't speak the language or play jai-alai or bomb Madrid government offices. Further, Frank Gehry's building makes no concession to where it is. It looks as though it has been dropped on Bilbao.

It is the world's most importunate building of the recent past. The architect has pulled every gimmick out of his sleeve. It screams deafeningly: take my picture, take my picture. Adding, *sotto*: but mind where you take it from, take it from the sanctioned side – like an ancient film star it has a good side. It bellows: look at me. But what are we looking at?

What we are looking at is an advertisement, a perpetual marketing campaign. Every time it is photographed or posted on the web or televised it is Bilbao that the world hears of. If architecture is frozen music then Gehry has devised a very catchy jingle.

What Bilbao has shown is that the word Regeneration has to be made flesh, so to speak, has to be represented materially. A faith requires figurative expression. Bilbao has also shown that regeneration breeds regeneration. That gestures breed gestures. Stasis is replaced by perpetual mutation. Edgelands must be excised. Wastelands must be tidied. Cities must become permanent building sites. New structures must constantly be made. Send for Calatrava. Send for Calatrava. Again. Regeneration craves bridges. Send for Foster. Milord came in a helicopter, for forty minutes, to see *his* Millau bridge for the first time a few weeks before it opened. It's akin to Wayne Rooney condescending to read a few pages of his autobiography.

Gehry's scheme proved to be inspirational as soon as it was published, let alone built. We should not be surprised. Architects are zealous hunter-gatherers of inspiration. In any era there is a mere handful of exemplars, each of whom attracts scores of acolytes – or plagiarists.

Regenerative Cities replicate each other's buildings. Which is hardly surprising since they all crave iconic blue-chip trackers by members of a small cadre of perpetually itinerant architects, image-makers by appointment to the world. It would hardly be in their or their patrons' interest to acknowledge that the very process of building is grossly wasteful of energy and is about as unsustainable as it gets . . .

These are the architects who are relentlessly copied, the creators of the templates for our new age's monuments . . .

To spatial delinquency . . .

To non-orthogonal geometry . . .

To the legitimisation of the crazy cottage . . .

To the inspiration of a stiff rumpled napkin at the end of a long lunch . . .

To novelty formations . . .

To magnifications of product design . . .

To the infinite capacities of software . . .

To zoomorphic and geomorphic whimsy . . .

To a wilful unorthodoxy – which has, of course, rapidly become the new orthodoxy . . .

Ultimately they are monuments to conformity.

Almost indistinguishable buildings occur in every continent. Gestural engineering is homogenising. Every building that is ever made has the potential to be a one-off. But that opportunity is missed time and again. The bespoke is being globalised.

Not only architecture but the entire apparatus of regeneration is slavish in its adherence to fashion. For an endeavour whose alleged purpose is the improvement of post-industrial cities and the alleviation of deprivation it is signally ingrown, a smooth-talking verruca.

In the very words of someone called, without irony, an urban regeneration guru: 'What we need are fashion incubators attached to art installations.'

Why should a strategy dreamed up for a Basque port be applicable to, say, a landlocked German foundry town? The assumption that one post-industrial city is akin to another is insulting – it omits all that is specific to a place. The strategy was replicated long before anyone had had the chance to assess the benefits of the Guggenheim to Bilbao. The cause was still on the drawing-board, so the effect was entirely unknowable to the calibration community.

This is evidence – as though any were needed – that what prompted Development Agencies to ape Bilbao was not the promise of an improvement in the quality of life but a covetous yearning for a flashy building.

Certainly Bilbao has made itself noticed. And nine years after the Guggenheim was opened ETA laid down its arms – which shows just how effective it has been. And then ETA picked them up again – so maybe not all that effective.

The Bilbao strategy is a two-stage initiative throttle. First comes the sightbite, the three-dimensional logo, the challenging focus, the symbolic component designed by a world-famous architect who doesn't know what country he's in today. Then it has to be decided what non-symbolic function is to be sutured on to the structure.

Regeneration remedy-creatives, guru-guys who coalface at the coalface of armageddon decisions and who insterticise tough paradigms, envision inclusions in big-thought canopies. They brainstorm like they're firebombing Dresden. They leave their synapses on the table. They go into deep thought submersion.

Then they announce that the sightbite will contain a museum and gallery and cultural logarithm with conference facilities for the regeneration community.

Why . . .?

Silly question.

Because culture is good.

Not old-style culture – Beethoven, Shakespeare, Joyce, van der Weyden. Old-style culture was not inclusive. Therefore it was elitist and very bad. But the new, accessibly accessible, fun-style fun-arts which beacon integrated modernising lighthouse revitalisation for everyone. Interactivity within the community of community. Vaulting awareness gateways. Everyone can inclusivise re-energizing. You might be so old and drooling you're old Alzheimer himself, you might be so young you're in utero. Culture will still springboard you into the happiness community.

But: we all know that every member of the SS was a pianist who could play Schubert sonatas to concert standard and had all of Goethe by heart. We all know that, on the other hand, Top Albanian Mother Teresa liked nothing better than singing along to Tony Orlando and Dawn and watching Chuck Norris movies. The proposition that concert halls, fine art, museums, galleries, theatres, public sculptures, etc., are beneficial in any way other than aesthetic is unprovable. The notion that they might be socially, morally, economically useful is, in other words, nothing more than a belief, a hope.

A hope which can, of course. be dressed up to give it substance by the application of a delightfully bogus science called Output Measurement Metrics. This hall of statistical mirrors and reflective mendacity allows Development Agencies to spend public money as they wish. It enables Development Agencies to claim that what is politically opportune possesses economically rational bases. So we get multi-faith street furniture . . . creative hub clusters . . . vibrant cycle paths . . . inclusive flagshipping lanes . . . participation footfalls . . . media villages and media village idiots . . . ethnically diverse handshake sheds . . . gender indifferent water-polo ponies . . . shia bestiality workshops . . . therapeutic tram shelters . . . and holistic pudding churches . . . These are all – somehow – reckoned to be triggers of enterprise.

The Museum of Swarf and a state-of-the-art mangotime swing bar are obviously just the inducements a CEO in Osaka is looking for when resolving where to site a new factory. Equally, are oblique forms of intervention liable to be any more effective than the direct provision of shelter to those who cannot afford to provide their own, of health care to the needy, of subsidised jobs, of an effective public transport system like Ultraspeed which would cut the time between Manchester and London to forty minutes? Add to that a seven-minute link between Manchester and Liverpool airport which would create one airport that would obviate the need for Heathrow's further expansion.

What has happened to Greater Manchester since it got two cultural sightbites within a couple of minutes' walk of each other? What are the wider social benefits of the Imperial War Museum in the North and the Lowry Gallery? Because of inclusive culture, curvy buildings and reinvigorated docks all of Moss Side's crack dens have turned into holistic dance workshops. There are no longer drive-by shootings in Hulme. The teen hooker community has disappeared from the corridors of the Britannia Hotel. The underage rent community no longer hawk their brawn at Victoria Station: most are now reading for degrees in development partnership studies. The many graduates of Tirana's legendary Polytechnic of Pimping and People Smuggling who plied their trade are all retraining as urban energisers. And I am Napoleon . . . though tomorrow I'll be Keith Harris and Orville.

Manchester may now call itself a Knowledge Capital. What that means is that it feels the need to run a Crime and Disorder Reduction Partnership managed by an Executive Partnership Group whose work is carried out by the Crime and Disorder Team, the Youth Offending Team, the Drugs and Alcohol Strategy Team, and the Manchester Multi Agency Gangs Strategy. And it still has the highest crime rate of any conurbation in the UK.

Maybe cultural determinism isn't working. Could it be that it's as goofily optimistic as the architectural determinism of the welfare state – a sort of nationalised noblesse oblige in which all the workers got a tied cottage. Good intentions – the belief that better housing creates better people.

The presumption that humankind is improvable let alone perfectible by material means is as laughable as that which blithely prescribes doses of participatory street theatre and mime laboratories for everyone. There has always existed a minority of the race whose instinct for self-preservation is so unevolved that it will trash its dwelling and piss in the lift. Another of Heseltine's prime-cut brainwaves – I'm sorry to keep coming back to this giant – was that granting tenants the right to buy council housing would correct this flaw. It was a canny vote winner. That short-term consideration was, of course, more important than the deprivation of shelter to future generations.

It is a fact of life that the self-made get out of where they come from. Once upon a time in England – though not Scotland – it was to the outer suburbs and satellites. From Manchester to Hale and Knutsford. From Brum to Henley and Meriden. Today it is to shiny apartments and kooky houses within cities. To developments that are billed as *housing market renewal pathfinders*. Live the dream!

They are as much suburbs as the arterial roadscapes of the 1920s and '30s: these are the twenty-first century's first essays in suburbia, the brownfield suburbs. They are the most widespread achievement, if that's the word, of the regeneration lark. The ideal English home throughout the twentieth century derived – at many removes – from the bucolic cottage. Today if you've just freighted a hundredweight of speciality goods from Morocco or Colombia you head for a lifestyle solution that is exclusivitised by representatives of the security community and by representatives of the Salford Quays guard dog community.

The aspirational synthetic modern apartment was the product of a soufflé economy – on which someone has now opened the oven door. It was based in an unsustainable fantasy. It was a product of design culture. It draws on bar culture, restaurant culture, retail culture. Gallery culture, museum culture, culture culture. It was designed by an architect.

It is not built by a mere builder. It is built by an urban regenerator. That cavity, just there beneath a waistband, that temptingly dark hirsute cleavage with optional blackheads for added appeal is an *urban regenerator*'s bum.

For builders to call themselves regenerators is an unexceptionable fib. It is the urban bit that is questionable. Brownfield burbs may be within cities rather than on their peripheries, but they are not part of those cities. They are not integrated. These places are alien growths within the body rather than on its skin. They turn the last century and a half's association of topography and high income on its head. These new burbs are built testimony to a momentous demographic shift.

Since the industrial revolution Inner City has – but only in the USA and England – been shorthand for poverty, neglect of children, racial strife, sexually transmitted diseases, illiteracy, used needles and losers. Poverty was to be found in inner cities, wealth in outer suburbs.

Regeneration has turned that all round. Within a few decades inner-city Britain will come to resemble inner-city Europe. As Britain becomes effectively re-regionalised the great provincial cities will regain their autonomy, an autonomy that such places as Lyon, Turin and Zaragoza already enjoy. The oblique processes began by Michael Heseltine will indeed achieve the regeneration of inner cities. The epithet inner city will cease to connote social woes.

But poverty, neglect of children, racial strife and so on won't go away. The dispossessed and the desperate will always be with us. They will merely be hidden away in the outer burbs. Victims of an economically enforced migration. This is what adherence to the European, specifically French, model means. It mean riots and burning cars and police horses and power hoses *outside* the ring road. Out of sight – so that's OK, isn't it?

(2008)

THE CURSE OF BILBAO

GROINWICH! POTENTIALISING POTENTIAL. Scrofcastle! Going Places. Smeltbury! Enabling inspiration. Dundee! City of discovery. Bubochester! A runway to creativity. Griminster! We're an energy pathway. Swarfield! Trailblazing connectivity. Newport! City of reinvention. Scurviedale! A town in a hurry. Oxterton! Networking receptivity. Scrapieburn! An opportunity compass. Southampton! Cruise capital of the UK. Longut! Incentivising incentive. Festerford! Meeting tomorrow's challenge yesterday. Felchinborough! A world café. Newcafard-on-Sea! Regenerating regeneration.

The night of 29 May 1985 I was at Palavas les Flots near Montpellier. The hotel room telly was broadcasting herringbone tweed. I hurried along the street, then another. Café after bar was telly-free. I kept looking at my watch. Then, at last, a glimmer through a salted window. I entered the bar – which any other night might not have been the bar from hell – and, before even looking at the cantilevered telly, asked the barman if there was any score. He fixed me with a rheumy, possibly alcoholic, certainly psychopathic eye and replied menacingly: 'Yeh. Thirty-five dead.'

This was when I realised that the European Cup Final at Heysel had not even begun. It didn't get better. The house Alsatian, sprawled across the floor like a crocodile, had its master's mien. The half-dozen or so punters were mutterers: '*Angliches??? 'ooli-gans??? salopards??? meurtriers.*' Etc. There was no question of making an exit. The game began. It ought not to have been played. The accusatory stares continued.

Eventually an atypically conciliatory drinker suggested to his chums that I was not to blame. He addressed me and asked, what is it with the English, what is it with Liverpool? I pointed out that Liverpool was no more typical of England than Marseille, then the

European race-hate capital. But, he insisted, they're doing something about Marseille. (Whatever the something was hadn't been apparent in that graffitied war zone a couple of days previously.) What are they doing about Liverpool? Well, some people would like to give it to Dublin, but Dublin doesn't want it. And then, of course, they've got the Garden Festival. That is what they're doing. My interrogator's incredulity was no greater than my own when I realised what I had said. We laughed.

Football apart, Liverpool in the early '80s was notable for strikes, Militant, riots and the loss of 20 per cent of available jobs and 10 per cent of its population. The then Secretary of State for the Environment, Michael Heseltine, sprung into action. He was a square peg in Margaret Thatcher's second administration – whose motto was *sauve qui peut*. Heseltine, on the other hand, was a closet welfarist, he believed in statist noblesse oblige. He had the bright idea of solving all of Liverpool's ills with a 'Five Month Pageant of Horticultural Excellence and Spectacular Entertainment'. It proved to be the start of a new industry, the greatest of contemporary gravy trains. The Regeneration Industry.

Describe what was known, circa 1958 to 1973, as comprehensive redevelopment as Regeneration and it is stripped of its deserved connotations: crassly ill-conceived motorways through the centre of cities such as Glasgow; the alliance of dorkishly neophiliac municipal authorities and vandal builders which ripped the heart out of countless small towns and turned them into parking lots. Describe architectural determinism as Regeneration and the stick of this-vertical-sump-is-good-for-you is replaced by the carrot of this-is-stylish-living-for-a-new-millennium – because you're worth it. Describe volume building as Regeneration and off-the-peg neo-vernacular hutches mutate into off-the-peg neo-modernistic apartments. It is telling that *Building*, the deftly positioned trade weekly whose special pleading for construction and yet more construction is relentless, has spawned an equally parti pris, equally clever offshoot. Entitled, it goes without saying, *Regenerate*.

Describe publicly funded, parochial vanity as Regeneration and you invest the dodgiest of projects with a feel-good, fuzzily spiritual, communitarian blanket of hectoring self-righteousness. How dare you attack our landmark bridge, our iconic icon, our enterprise vanguard!

Like all other endeavours, Regeneration is subject to fashion. And here again, in the late '80s and early '90s it was Liverpool that gestated the aloes. After the Garden Festival came the renovation of its nineteenth-century dock buildings. The endurance of these awesome structures serves as visible reminders of what once was and now isn't: a city's greatness. The buildings had to be given a role. So the Merseyside Maritime Museum was established, devoted to an industry that no longer exists, situated in a building rendered redundant by that very non-existence – this was a quaint way of looking to the future.

If it was to Liverpool that was due the word, the subsequent practice of Regeneration derives from another post-industrial city on the skids. Bilbao was a rust-bucket, known outside the Basque country only to amateurs of ETA, INLA fund-raisers and the peerlessly thuggish 1980s footballer Andoni Goicoechea. Then along came Frank Gehry's Guggenheim Gallery. And the rest is history. Actually it's not. The rest is plagiarism.

As a gallery the Guggenheim is wanting. Its permanent collection is derisory, its exhibition spaces are sacrificed to the geometrical delinquency of the design. But this is to miss the point. The purpose of the Guggenheim was to put Bilbao on the map. And in that it has succeeded. The building might as well be an abattoir or a casino. It doesn't matter. Its function is to provide the city with a three-dimensional logo that is photogenic and telegenic. Bilbao is now world famous, if only for that lump of architectural sculpture and for Santiago Calatrava's bridge across the Nervion beside it. A hundred kilometres along the coast, at San Sebastian, Rafael Moneo's contemporary and vastly superior Kursaal has attracted no international attention. Well, it wouldn't – it's subtle, austere, beautiful, doesn't scream look at me and wasn't created with a lens in mind.

Bilbao was a traction trigger that broadcast an unequivocal message to cities everywhere, that they too needed a headline-grabbing thing. Eminent blue-sky thinkers across the globe thought the same blue-

sky thought: send for Gehry, or for Calatrava, Foster, Liebeskind –
for anyone with an antipathy to right angles and a lavish software
habit to support. The way forward was no longer with the restoration
and adaptation of what already existed. The future was now with
outlandish design, playpen colours, screeching profiles, engineered
gestures, flashily zoomorphic eyecatchers. They have much in common
with eighteenth-century follies. That's to say they are capable of
delight and surprise just as sham castles, grottoes and hermitages are.
Occasionally, follies were invested with allusive meaning. Occasionally,
they were stylistic rehearsals for grander buildings.

But there was no claim that they were other than autocratic or
aristocratic caprices. The follies of the past decade are publicly
funded, thus they have to pretend to do something. After all, the post-
Protestant, market-led conscience cannot allow architectural marvels
– or mistakes – to be created for architecture's sake, for aesthetic joy.
They must possess a social purpose. They must serve community.
We all know that community is as sacred as low-emission diversity,
vibrant sustainability and holistic leverage.

There is a lesson that has apparently yet to be learnt. Namely, that
what is appropriate to Bilbao will not necessarily work for Scrofcastle
or Felchinborough, Zungebad or Bitteville-la-Pipe. A couple of
years ago I suggested parenthetically to an Amsterdam conference
of European endeavour drivers that this was a self-evident truism.
Astonishingly, I found myself treated like the bluest of blue-sky
thinkers. Of course, it was ever thus, the power of fashion. The Gothic
of St Denis was taken up by dioceses throughout Europe north of
the Loire. The Hanse cities adopted Lübeck's highly mannered urban
decorative devices. But these were instances of willed uniformity, like
the buildings of empires. The most signal parallel to today is that of
many small nations and culturally autonomous regions which, at the
end of the nineteenth century, determined to assert their independent
individualism through architecture. Unhappily, they all chose the same
idiom, art nouveau. So individualistic Riga, individualistic Genoa,
individualistic Nancy, individualistic Ixelles and individualistic Tbilisi
are often indistinguishable from each other in their independent

individualism.

The Lowry Gallery and the Imperial War Museum of the North on Salford Quays, the Royal Armouries Museum at the Clarence Dock in Leeds, the (already former) National Centre for Popular Music in Sheffield, the Sage Centre on the Tyne at Gateshead, the Scottish Exhibition & Conference Centre (aka the Armadillo) on the Clyde in Glasgow, the New Art Gallery at Walsall Canal Basin, the Magna Centre in Rotherham, the National Glass Centre at Sunderland . . . These structures, and there are scores more throughout Britain, may differ in their architectural merit: they lurch from meretricious accessibility to high seriousness. Yet they all conform to the Bilbao model. They serve the same optimistic civic or regional aspiration to Regeneration through (sub)cultural tourism generated by the vessel rather than the content. It is because of them and their like that there are no drugs in Handsworth's streets, no muggings in Toxteth.

In most instances they have too, at a micro-local level, effected a physical clean-up of their surrounding inner-city brownfield sites – which are no use to anyone save those with an entropic taste for street crime, yesterday's condoms and polythene bags flapping to release themselves from rusting rheostats. This clean-up has in turn encouraged the construction of synthetic-modern apartment blocks, soft-modern lofts and climatically inappropriate outdoor cafés in their vicinity (such cafés are de rigueur in architects' proposals).

A new kind of inner city or, at least, inner suburb is rising all around us. Especially if we happen to live in Manchester. As they used not quite to say: what Manchester builds today England will build tomorrow. What Manchester is building is on an epic scale. The highest this, the biggest that. But it is all 'infra-sustainable'. Affordable housing? You bet.

But there is a significant proportion of the population which cannot afford the affordable. And the safety net of social housing has been sold off and sold on. Regeneration is not curing social ills but shifting their loci through economically enforced migration. Poverty, crack, neglect of children, disease and riots will not disappear, but at least they'll be outside the ring road.

The Curse of Bilbao

The war criminal Albert Speer posited what he called, with typical grandiloquence, The Theory of the Value of Ruins; i.e. the hardly novel idea that a civilisation is judged on what remains of its buildings – evidently not so in the case of the 'civilisation' that Speer so abjectly served.

However another war criminal, Tony Blair, may take some succour from the fact that, long after his anilingual romance with George Bush and their recklessly catastrophic adventure are forgotten, many of the buildings erected in 'his' regenerated civilisation will remain, in one state or another. And how apt they are: pretty, eager to be liked, preternaturally youthful, pretending to modernity, empty, dangerously vapid.

(2007)

THE WHEEL EXPERIENCE

THE BUILDERS OF STONEHENGE had the ingenuity to bring stones by sea and river from the Preseli Mountains in north Pembrokeshire. Even to have brought vast sarsens from the Marlborough Downs was a supreme feat. They were architecturally sophisticated enough to understand the perspectival device of entasis. They possessed the nous to align the north-east axis with the summer solstice sun.

But they quite failed to maximise the monument's dramatic potential. Unhappily, they were working before notions of the picturesque and the sublime were commonplace. They had much to learn from the makers of eighteenth-century landscape. Instead the builders of Stonehenge III (very roughly what we see today) followed the practice of using the site of their monument's precursors. This is in a declivity – which is why the stones seem so slight when first seen, especially when first seen by anyone familiar with them from a photograph (which means everyone).

The stones cannot compete with the immensity of the downs which rise around them. They are in the wrong place to satisfy secular eyes used to monuments and follies positioned for visual effect rather than in obeisance to redundant superstitions.

Another complaint: the Beaker folk and their successors quite failed when it came to the sine qua non of a good day out. They forgot to provide a visitor centre! So – no gift shop, café, toilets, interactive screens or scale models. Negligence? Put it down to entrepreneurial sloth or to the great misfortune to have lived in an age when 'accessibility' was not a daily mantra. The poor things didn't realise that a visitor centre is not merely an integral part of the 'experience', it is *the* 'experience', for it mediates the incomprehensible and the arcane. The monument – whether it be cathedral, gallery, museum, country house or industrial archaeological site – is simply an add-on, the pretext for a shopping trip and a fizzy drink.

The Wheel Experience

Stonehenge's stewards at English Heritage should study the Falkirk Wheel. They should act boldly. Instead of burying a token stretch of road and creating a bunker-like visitor centre they should move the monument to a position which will do it aesthetic justice. About a third of a mile west would suffice to lend it the drama it deserves; then they can think about the visitor centre.

This is not a prescription for vandalism. And sacrilege does not come into it: the appropriation of the monument by Druidism is an historical whimsy occasioned by John Aubrey's wrong-headed speculation that that cult was responsible for its construction. There are numerous precedents for moving monuments, amending them, rebuilding them. Great age should be no inhibition.

The Falkirk Wheel is three and a half thousand years younger than the last phase of Stonehenge. Lack of age should not count against it. While our thraldom to antiquity may evidence a sense of wonder and humility it too frequently masks an aesthetic myopia: vintage is regarded as a more important criterion than excellence – old does not equal good. The Wheel is a prodigy of civil engineering.

The name is misleading. It is a utile structure, a counterbalanced boat lift which replaces a series of locks at the junction of the Forth and Clyde Canal and the Union Canal. The only other boat lift in Britain, built 125 years ago, is between the Weaver and the Trent and Mersey Canal at Anderton, north of Nantwich. That at Arques outside St Omer in the Pas-de-Calais was replaced by a lock (of exceptional depth) about thirty years ago. Belgium, a country which still sees the point in working canals, has a number of operational lifts on the Canal du Centre and has lately built a new one.

But none of them combines its functionalism with the sheer sculptural bravura of the Wheel. This is a creation which is opportunistically placed to take advantage of topography: it is not cowed by the rampart of the Antonine Wall beside it. Viewed from the north across fields and straggling suburbs it is a moving sight: an enormous installation in the form of an avian profile; of a forgotten mark abandoned by a race of giant proof-readers; a piece of swarf from one of Falkirk's scrapyards cleaned and inflated a thousandfold . . .

Similes are all one has, but they cannot but diminish the primitive strangeness of the structure which, millennia hence, will no doubt come to be confidently regarded as a temple to an extinct water-god. Which, in a way, it is. It is a thrilling and scary place for anyone who suffers both hydrophobia and vertigo: the higher level of the lift is 115 feet above ground.

The imagery of the visitor centre, which is positioned to mask the least dramatic aspect of the lift, is propitious. This building, which is a large part of the whole, is shaped like an upturned boat, a Brobdingnagian boat, of course. As you walk round it a deft act of legerdemain is prosecuted and clinker gives way to glazing: from the lift itself the building is a segmental headlight.

It would be captious to claim that the Wheel is liable to encourage the use of the canal systems it links as a means of commercial transport. Indeed, there is an implicit admission that it is for pleasure in its kitschless correspondence to a fairground gondola or pirate ship – it's a 'ride'.

Beyond that, it will achieve the Bilbao effect of making Falkirk and Stenhousemuir known as something other than components of the poetic litany of Scottish soccer results. The Wheel is a wonder of the modern world and its architects and engineers (RMJM and Arup) are more than the peers of the Beaker folk.

(2002)

STRATEGY SAFARI

WOOLWICH IS IN London though not of London. It is patently displaced. The old notion that it is a Midlands garrison 150 miles too far south is appealing. But it might, equally, have intruded from the north Kent shore. It has the toughness and precipitous grimness of the Medway towns. That provenance is sustained by North Woolwich, a ferry ride away on the other side of the Thames: till just over a century ago it actually belonged to Kent. North Woolwich today is estuarial Essex on day release Up The Smoke.

The vertical horizon of Canary Wharf doesn't reduce this illusion. This place is marginal, so marginal it's off the page. The two most notable Victorian buildings, St Mark's, Silvertown, and the Gallions Hotel are separated by a mere twenty years yet belong, stylistically, to different worlds. What they have in common now is that they are closed, wired off, heavy with security companies' warnings. There are by-law streets of two-up two-downs, a sugar refinery towering over them, deserts of waste ground, burly weeds and sagging buddleia, big sheds, unreconstructed caffs, disused rail tracks, swarf alps, piled-up Portakabins.

Yet the dereliction is far from comprehensive. The flat landscape is constellated by pockets of gleaming newness: an immense pseudo-moderne cliff of white apartments beside the Thames Barrier; marinas in the former dock basins; countless exercises in 'accessible' synthetic modernism; the douce cylindrical buildings of the University of East London.

The message is clear. This is rich brownfield pasture where Regeneration's Fresh Green Shoots can flourish: not only green but every other modish colour too. Like the Isle of Dogs in the '80s it is an exhortation to entrepreneurship untrammelled by the obligations of infrastructure. It's a kind of Klondike.

It must put a song in the heart of businesspeople heading for City Airport whose single runway occupies an isthmus between two former docks. And as they are lifted high above it all they must look down and muse on the endless opportunities it offers. Well, they should muse thus. A Scottish exec seated beside me a couple of years back summoned the stewardess five minutes into a Glasgow flight and complained that: 'I appear not to have my in-flight magazine.' This was worthy of H. M. Bateman. It also marked him down as a loser. I wonder what bleak, blanketed doorway he occupies today, that businessman who, suicidally, travelled without a laptop, without a spreadsheet, without a self-improvement book.

City Airport can hardly be described as charming. But its diminutive scale makes it much easier to negotiate than such virtual cities as Heathrow or Schiphol. It feels provincial. And despite the thorough mediocrity of its buildings, it enjoys other advantages. This is the airport for anyone who is offended by the sartorial infantilism of Vacationing Britbloke – you know, drawstring trousers that reach mid-calf, shiny polysomething soccer shirt, swollen trainers. Such clothes are notable by their merciful rarity at City Airport, where the uniform is suits for both sexes. Their mobile is grafted to their collar. They bring up new windows on their parlously perched laptop and speak into the phone, then repeat what they've said, then repeat it again with messianic emphasis. One recognises their words as English. Here a preposition, there an adjective, everywhere a plentiful supply of more or less familiar nouns. But there's evidently been a run on verbs, they're just about out of stock and when they are employed it's with a disregard for the distinction between transitive and intransitive.

Further, the language has been subjected to a Germanic makeover. Not in sentence construction (which is in this context a hopelessly vague concept) but in the welter of portmanteau words. I observe this not in a spirit of proscription – the one thing we can be certain of about usage is that it is perennially mutable – but with reportorial wonder. How do people learn to speak thus? And are they monoglot? Can they understand the language that I speak and write?

A suit can browse at City as at any other small airport. Scent, drink, tobacco, shoes, ties, scarves, etc. But the shop called World News does not conform to the airport norm. Sure there is the usual array of papers, magazines, trashy best-sellers. However this is a shop that knows its customers. Its (very) subliterary specialisation comprises shelf upon shelf of business books. I have to admit that this is a genre that was alien to me. No longer. I spent an afternoon leafing through *Winning the Merger Endgame, The Performance Prism, A Genie's Wisdom: a Fable of How a CEO Learned to Be a Marketing Genius, Corporate Longitude, Strategy Safari, Venture Catalyst.*

I previously had no idea that within every suit there is a victim of New Age drivel, a sort of post-hippie dreaming of achieving untold wealth by adherence to this or that meaningless mantra. Business books are written by evangelical illiterates to convert the ambitiously gullible. Taking a helicopter view of these books I'd say that they were as bogus as most religious tracts, as full of empty promises, and as immodest: my way is the true way.

(2003)

CONSTRUCTIVE SURVIVAL

PLACE – ON NO MATTER WHAT SCALE – is one thing. Creation of place is quite another. That creation is accretive and continuous, it occurs across time. It is liable to owe as much to serendipitous juxtapositions and, on the other hand, to malign interventions as it is to wilful design.

Again, the interests of those who 'use' place – that's every one of us – are not necessarily coincident with the interests of those who initially make place, those who set out the armature – upon which subsequent generations will impaste layers determined by utility, by successive technologies, by economic fluctuations, and by fashion. Never underestimate the power of fashion, of collective taste.

It is today fashionable to tear down 1960s buildings, the good as well as the mediocre, just as in the '60s themselves it was fashionable to tear down Victorian buildings – whose loss we now rue. The layers which survive the attentions of the demolition community will be amended by yet further generations or stripped away and then replaced.

Every place we ever make carries the germ of its future archaeological dissection. That is an inevitability. So making places or buildings which are provisional, which shout about the possibility of flexibility, which proclaim their willingness to incorporate change is redundant – and a sort of prospective vanity. They *will* change whether those initial makers like it or not, whether they make allowance for change or not.

There are also future wars, imponderable climatic shifts and unforeseeable accidents to take into account. Place can only be made for the present and even then the reaction of the user, the spectator, can confound or disappoint the maker's expectations.

The only sort of place which allows the maker the satisfaction of knowing that it will mean what it is intended to mean is that which is pedagogic, a mental cage: the religious structure – which includes the arenas of totalitarian regimes and immense monuments that living gods erect to themselves. And these, of course, after their thankfully brief period of utility, will be sacked, burnt, destroyed or employed by a contrary programme of mumbo-jumbo – mosques become churches, churches become mosques, the Third Reich's Kongresshalle in Nuremberg became a pound for wrecked, stolen cars – strength through joyriding.

This user – a word that makes place sound like a drug, and so it is – suffers a discomfiting sensation otherwise unknown to him when he surveys Le Grand Bleu, the law courts in Bordeaux, the Millau viaduct – a swelling of base national pride which is otherwise alien or at least buried. This is obviously not what the Lords Rogers and Foster and plain Mister Alsop intended when they made these structures – which happened, through their sheer potency, to have made places. This, I assure you, was not what I intended either.

I was shocked by the way that my inner aesthete was subjugated by my inner lout who presumably wants to put HP Sauce on cassoulet, wrap himself in the flag of St George and bellow about Ingerlandland. It merely goes to show that place is never pure, never simple, never predictable even when place is new and yet unamended. Our compact with place is corrupted even before we set eyes on it.

It was always corrupted by expectation and anticipation, by the tales of men back from sea, by the promise of towers that reached to the clouds and the big rock candy mountain. It was increasingly corrupted by the democratisation of reproduction. Steel prints, oleographs, Stevengraphs, photographs. By the democratisation of print and mass literacy.

Place is today invariably popularly mediated, the distant is familiar. There is nowhere in the world we can go without having experienced an ocular clue to it. And an aural clue: we have heard the language and the music. And a gustatory, olfactory clue: we have tasted the food, we have sniffed it – in Headingley or Redlands, Haringey or Sparkbrook. Well . . . a version of the food.

Today we all arrive, semi-prepared by the virtual and with unprecedentedly greedy access to the actual. This may not be the fate of our grandchildren.

We are very likely living in an era when the world has achieved maximum shrinkage, is smaller than it will be for a long time to come. The vanguards of the consequent insularity are already apparent, already domestically prescient: we are conscious of the depredations occasioned by promiscuous air travel and car travel; we know that much of the world which we still have the means to reach within hours comprises no-go states. We are not going to be getaway people forever. Escape may not be an option.

We are consequently rediscovering our back yard, we are beginning to cherish it, if only in a *faute de mieux*ish way. The signs are there. They may be tiny but they are telling. The allotment movement is thriving. The English seaside resort is at last being seen as having the potential to be something other than linear trash. There is a now almost conventionalised literature of inward travel, of sentient response to the immediately local, rather fancifully styled psychogeography, which uncovers the latent exoticism of the everyday and the overlooked. This literature does what all literature does but without the encumbrances of plot and character. It invests place with mythology, with symbolism, with a kind of magic.

Its preferred, almost invariable subjects are edgelands where underfed horses freeze between interchanges and reservoirs, sewage outfalls, trails of rusty dereliction: but in its treatment of them it is within an ancient tradition of response to place, of reverence for place as the measure of our short life, and the trigger of nostalgia. Just as the pastoral poets of the eighteenth and early nineteenth centuries were nostalgic – for a wilder land not yet taken over by agriculture, not yet enclosed by the hedges that were perceived as industrial intruders. The London Dickens wrote of was a fantastical London of his childhood. Hardy's Dorset was an invention based as much on local newspaper stories from before he was born as on his extreme sensitivity to the places around him once he had moved back there, to the house he designed: thank god that he designed nothing else.

I have to admit to a fondness for pitted former rolling stock dumped in fields and for abandoned filling stations. But man cannot live by oxidisation alone. It's not a question of *either* atmospheric scrappiness *or* gleaming newbuild. It's a question of *both/and*. It's a question of the quality of the atmospheric scrappiness, the quality of the newbuild. Any place is better than the place which invites no response, which breeds indifference. The greatest offence in the creation of place is to attempt to avoid giving offence.

Without doubt the loudest, most widespread sign of our revived localism, our revived sensibility to place, is the manifestation of an appetite for that place called the inner city – which can be extended to include the inner town. And it is inevitable that our now-pristine inner cities of Corten – which is rebranded rust – of glass, of pigmented concrete, of buildings shaped like pyrites and of bars serving climatically appropriate Mediterranean food will in their turn become the subjects of nostalgic reverie and satirical derision.

It goes without saying that the twentieth century in Britain and in North America – though not really in Scotland and certainly not in much of Europe – was the century of *sub*urbia. The history of twentieth-century urbanism in this country is largely a history of a resistance to urbanism. And that resistance has not yet abated. We are constantly enjoined to believe that the future is sprawl and that we should, if not welcome it, then just lie back and accede to it.

That, very likely, is an old-fashioned future. It is, of course, a projection of the present. There exists a tendency to move beyond the resigned acceptance of sprawl as an unavoidable fact of life towards, first, passive toleration – and thence to a celebratory sanctioning of it.

Now, I'm all for perversity. But this is above and beyond. It's an alarmingly irresponsible route to pursue. The application of libertarian laissez-faire to the construction of new places or the alteration of old ones is grotesquely short-term. We should not believe in the sanctity of the free market when that market affects place. To do so is to legitimise a return to the deregulation of the 1930s and ribbon development, to an era before town and country planning legislation, before the establishment of the currently threatened greenbelt.

It is not cooling towers and wind turbines that desecrate this country's landscape but pseudo-Victorian hutches and neo-vernacular closes stranded in infrastructural limbo. It may be ideologically attractive to a certain caste of mind to grant licence to a volume builders' free-for-all, but it will cause environmental mayhem of the most insipid kind. It guarantees a contaminated future. The lesson of the twentieth century is surely very simple: it is – not to follow its example. But, rather, to consider the legacy of place that we bequeath.

The buildings, the streets, the squares, which can do no more than provide the mere armature of place, are not consumer goods. Yet the most cursory scrutiny of this country's constructional practice suggests that they are in certain quarters regarded thus. The most cursory scrutiny of the magazines which reflect and flatter and special-plead on behalf of the professions and trades involved in construction suggests the same. My scrutiny of both happens to be far from cursory. Here's a list of places: Beverley, Bilbao, Chatham, Christchurch, Hornsey, Hull, Leeds, Letchworth, Liverpool, Manchester, Paris, Salisbury, Sherborne, Southwark, Spurn Head, Stowe, Le Vesinet, Winchester.

This is a very partial list of where I have filmed this year in the hope of making sense – however oblique, however notional – of what makes place, how places make us, why place makers make or made the places they do, or did. It would be rash, not to say impertinent, to extract generalities from this topographical overload. But it would be a professional dereliction were I not to. Damned if I do, damned if I don't.

When I talk about the elemental components of place being regarded as consumer goods what I intend to convey is a truism – obvious to everyone involved in construction but perhaps not so obvious to what was once described to me, with no irony, as the lay public – the truism that this is an industry which is compelled to produce – not for the needs of clients, not for the needs of different gamuts of society, not for the needs of place, but to ensure its own survival. Construction is undertaken for the sake of construction, for self-interest. The industry's duty to itself – if duty is not too dignifying a word – is to keep manufacturing. There has to be a constant flow of product. And that product is qualitatively different to any other on earth. It demands space.

Space – space, anyway, in a small overpopulated country – is evidently a finite resource. Yet there is a reluctance to recognise it as such. We acknowledge that fish and oil are running out but we continue blithely to exhaust the supply of space with piratical abandon.

And that abandon persistently results in *unter*-place, low-grade place. Arriving in London from virtually anywhere in the developed world one is struck, of course, by the meanness and the grimness of the endless burbs – but also by the sheer profligacy of land-use. Overpopulation ought to occasion high-density development. But that is far from the case. Britain's lack of communality is surely linked to its wasteful tradition of low density, to its craving for horizontal spread, to the fact that each one of us requires a broader *cordon sanitaire* around us than the people of other European nations. We are much more sensitive than anyone to the infringement of our personal space in public places. The way we cringe and have the word *sorry* perpetually on our lips are expressions of the desire for separateness, of the desire to *possess* space rather than to share it.

Product – that's to say built structures – is also quantitatively different. It cannot enjoy the diminution in size which afflicts many manufactured goods from cars to computers, because humans are not shrinking in size. Further, there are ever more of us. The demands on space and the necessity for controlled husbandry of it increase in direct proportion to our numbers. And such control will not be exercised so long as the government's advisory quangos are composed of members with a pecuniary interest in construction, so long as the executive of the day listens to well-meaning volume builders whose responsibility is to their shareholders, creditors and, maybe, their employees. It's hardly a case of the lunatics taking over the asylum – but *quis custodiet* and all that.

I appreciate that I sound way behind the times. There no longer exists such a thing as a volume builder, any more than there exists such a thing as comprehensive redevelopment. Today every volume builder is an *urban regenerator*. We have rebranded. A new logo, a new name: Thames Gateway becomes Thames Parkway . . . this is the greatest of our national talents, the one we're world leader in. PR, spin, the peddling of willingly accepted delusion. The Ingerlandland-is-going-to-win-the-World-Cup-syndrome is like a virus afflicting every area of our life.

We should not, however, allow the wall of delusion and hype to obscure what is actually being beneficially achieved – achieved for those not in the business of making place, that is. Nor to obscure the probable ramifications of those achievements. The urban regeneration or comprehensive redevelopment that we are witnessing today is the successor to the piecemeal, ad-hoc process that began in earnest 30 or more years ago, predominantly in London, with the reclamation of certain inner-city slums by the young and arty bourgeoisie – the process that would be widely calumnised as gentrification. Why was it calumnised? Liberal self-hatred? Envy? It was the seed of an urban renaissance. It was both economically rooted – the house price differential was then skewed in the other direction – and culturally reactive to the magnet of the burbs that had lured immediately previous generations.

What we are witnessing today is the second stage of that reclamation – gentrification through newbuild. Often visually impressive if somewhat homogenised newbuild: synthetic modernism or neo-modernism or pastiche modernism or populist modernism or accessible modernism. Whatever you like to call it's clear that there is a widespread taste for it. Especially among the affluent young – perhaps because they know nothing else. Private sector domestic building is achieving the level of quality that only the best public schemes achieved in the 30 years after the Second World War. And it will fare better because the self-interest of owner-occupiers is more entrenched than that of tenants. This is a way of suggesting that fate of place depends upon who uses and occupies place.

We're deluding ourselves – again – if we pretend that income and the wherewithal to buy is not as much a determinant of place as the recipes that social engineering once proffered. A proprietorial ethos counts for as much as pedestrianisation, street furniture, lumps of sculpture, clever arborealism, et cetera. That et cetera includes the services – shops, restaurants and so on which belong to the soufflé economy on which the oven door will one day be opened, the not very far-off day when foreclosure becomes a norm.

When we look at, say, the centre of Leeds or innermost Manchester – and here we have to acknowledge the part played by such promoters of regeneration as Gerry Adams and Martin McGuinness – we see another future – the future of English urbanism. A future which is both exhilarating and profoundly disturbing. It is a future which has mercifully shunned sprawl. Those cities and the other great formerly industrial cities, the virtual city-states of the North and Midlands, are now competing with each other as they did in the mid- to late nineteenth century, the era of mighty civic buildings and municipal pomp. Today it is not a question of who has the grandest town hall – it was always Leeds – but of who builds highest, who has the flashiest landmark, who has the most bridges by Calatrava or someone plagiarising Calatrava.

Those landmarks have certainly worked their magic: it is directly due to them that there are today no more drive-by shootings in Hulme, no more crack dens in Moss Side, no more bomb factories in Beeston. What we are actually witnessing is an abandonment of the North American model and an espousal of the French model. The *embourgeoisement* of the inner city combined with a dereliction in the matter of building social housing to replace that which was so carelessly sold off is effecting an economically enforced demographic shift. Social polarities are not going to disappear. The sites of income-defined ghettos are merely being exchanged. They're swapping with each other. A new hierarchy of place is being created. The *haves* move inwards. The *have-nots* move, or are forced, outwards. There is a significant population who cannot afford the affordable. Privilege is centripetal. Want is centrifugal. It can be summed up like this – in the future, deprivation, crime and riots will be comfortably confined to outside the ring road.

That is the pessimistic, dystopian, despairing prospectus. It need not be the only one.

It is no coincidence that the urbanism that is most widely revered is that of the eighteenth and early nineteenth centuries when there existed an explicit enlightenment belief that place could promote happiness, a belief we have lost. The means by which such an ethos can begin to be regained are not complicated – but they are not vouchsafed either to place makers or to users. They demand governmental will. A will to create a new framework. A will – such as exists in Spain – to positively discriminate against chain stores: they are forbidden to open for as many hours as small retailers who are not subjected to the same rate burden as the behemoths.

The refurbishment of the Brunswick Centre in Bloomsbury is exemplary – architecturally exemplary. Urbanistically it is a disappointment because the quirky shops and galleries have been replaced by the same chains that we see everywhere else on this country's corporately inhabited high streets.

It *is* possible to legislate for quality of life and for equitable prosperity. Four-fifths of inward investment to Britain is to the southeast. The reason is simple. Our third world transport scares off the desire to invest elsewhere. Accessibility prompts renewal: look at Marseille. We missed out on TGV: we are thus in a position to leapfrog a generation of technology and go straight to Maglev which would put Manchester within twenty minutes of Birmingham. Birmingham within half an hour of London: such a rebalancing would take the heat off the south-east. This, of course, sounds like infrastructural utopianism. At a micro-level just widening a pavement – and this country's are notoriously narrow – is a device for returning civility to cities.

A notable proportion of the contenders for the Academy of Urbanism's awards are places which owe their specialness to the very fact that they *are* controlled. They are, all of them, atypical. But what is atypical in one era can, if we learn from it, if we seek to emulate it, become typical in the next. You know, tiny stream burgeons to become great river.

(2006)

TVSSFBM EHKL: PROVIDED IT'S VALID WITHIN THE CONTEXT OF THE SCRIPT AND IS DONE IN THE BEST POSSIBLE TASTE

OR

I'VE LOST A HUNDRED POUNDS AND CAN'T THINK WHERE TO LOOK FOR THEM (2001)

OR

SURREAL FILM

TITLES . . . ROLL OVER PROTRACTED AUDITORY AND VISUAL MONTAGE OF INSTANCES OF BROADCAST USES OF THE WORD 'SURREAL' (E.G. ANDREW MARR, MATTHEW BANNISTER, ETC.) AND INSTANCES OF IT IN PRINT. THE WORD ECHOES, DISTORTS, IS ENDLESSLY REPEATED, IS SPEEDED UP, SLOWED DOWN — AND THE CONTEXT IS, IN EVERY INSTANCE, LOST: WHATEVER IS BEING DESCRIBED IS UNCLEAR. THE WORD BECOMES A MERE SOUND. DOCTORED IMAGES, TOO (E.G. MARR'S EARS GET BIGGER EACH TIME WE RETURN TO HIM. 'SURREAL' IS PICKED OUT IN PRINT IN DIFFERENT COLOURS, THE LETTERS SEPARATE AND RE-FORM). JM CLOSE-UP

Surreal. Totally surreal. Well surreal. Double surreal . . .

What does surreal mean? Does it mean anything? Is it simply a synonym of strange, lazy fancyspeak for weird? What do you — the people — think it means?

THREE VOX POPS. IN THE DISTANT BACKGROUND THERE'S ALWAYS SOMETHING GOING ON. WE DO NOT HEAR THE QUESTION BEING ASKED.
JM IN DIFFERENT DISGUISES
a) OLD WOMAN IN MARKET

Oh yes, yes — I know what it means . . . Is that it then?

b) FOOTBALL SHIRT MUTE, BANDIT MOUSTACHE

[mouths words, shrugs]

c) WHITE RABBIT

Funny peculiar? Funny ha-ha? Spot of both I'd suppose — cider and sambuca with a Bailey's top, my baba.

BACKGROUND: LOSELEY HOUSE. TWO NUNS WITH MASSIVE WIMPLES MUGGING A BOUNCER. THEY GO THROUGH HIS WALLET. JM PTC:

It means bizarre.

TWO BLIND MEN (STICKS, DARK GLASSES, HATS, DOGS) MASTURBATING EACH OTHER — 'MY LAST BREATH'. WE CAN'T REALLY SEE WHAT'S GOING ON. HIPPIE VOICE OFF CAM (JM) NERDY VOICE, HEAVY BREATH:

Well, it's that bizarre feel innit it. Yeah . . . bizarre
[pronounced: be czar]

STUDIO: BIGGINS IN CROWN AND GOWN OCCUPYING ENTIRE FRAME:

I thought you said I could be czarina.

SOUND OF SHOT. INCREDULITY ON BIGGINS'S FACE. BLOOD OUT OF HIS MOUTH. HE FALLS OUT OF FRAME TO REVEAL SIGN THAT READS 'EKATERINBURG POP 301' CROSSED OUT TO SAY 'POP 300'. CORRIDOR — JM OPENS DOOR ON TO SLEEPING SURREALISTS

It *does* mean bizarre. But the tamed end of bizarre. The surrealists of the '20s and '30s took the bizarre out of the wild and incarcerated it in a zoo. They domesticated a feral instinct, expunged the savagery, sought to keep it for themselves and their precious movement — when, of course, the irrational really belongs to all of us, to all ages, if not to all cultures.

BACKGROUND: EARTH MOVERS AT PRECIPITOUS ANGLES ON BUILDING SITE DRIVEN BY MEN WITH DAY-GLO PAINT ON FACES. BULLFIGHTER IN FOREGROUND:
He's two words. Sur — the south. And real — which mean royal: palazzio real, you know. Royal south you say.

JM C/U
Wrong. Wrong etymology.
   But surreal*ism* did come from the south.
   And the artist with whom it is, regrettably, most readily associated — the twentieth century's aptest anagram, Avida Dollars — was nothing if not a fawning royalist.

AT BOTTOM OF SCREEN THE NAME *AVIDA DOLLARS* FORMED FROM LETTERS SHOT THROUGH GLASS. JM REARRANGING LETTERS AS *SALVADOR DALI*. ROOM FULL OF ANTIQUES, VASES, KNICK-KNACKS, ETC. COUPLE IN MASKS — QUEEN AND DUKE. QUEEN WAVING QUEEN WAVE. DUKE HANDS BEHIND BACK. CAMERA TRACKS BACK WITH DALI. HE WALKS BACKWARDS, THE WHOLE TIME BOWING ELABORATELY, AS IN A RESTORATION COMEDY. AS HE MOVES HE BUMPS INTO FURNITURE, BREAKING CHAIRS, VASES, KNOCKING OVER FLOWERS, CAUSING MAYHEM
So fawning indeed that after he was presented to HRH Queen Elizabeth II of England and to the Duke of Edinburgh at L'Oustau de Baumanière in Provence, Dalí's anxiety not to display lese-majesty was such that when he took his leave of the royal couple he declined to turn his back on them . . .

ROOM FROM ABOVE. SCENE OF DESTRUCTION. QUEEN AND DUKE WANDERING AMONG THE RUINS INSPECTING THEM. WE MERELY HEAR THEIR VOICES.
QUEEN: What a splendid example of an acte gratuite, *Philip. How generous of Señor Dalí.*
DUKE: To destroy is to create, old thing. Totally surreal . . . of course, it comes naturally to your dago.
QUEEN: Do you remember that story Uncle Louis's friend from the old country told us about Señor Dalí? Oh what was he called?
DUKE: Herr Max Ernst — he was well surreal.
QUEEN: And what was the name of that common little man who — ?
DUKE: Buñuel. Too surreal by half if you want my -
QUEEN: So distasteful. Señor Dalí was quite right to get him sacked.

ARCHIVE FILM OF MAX ERNST: HE IS SPEAKING IN FRENCH — JM V/O TRANSLATION
Luis Buñuel was called in by his employers and asked if he'd made this impious, atheistic film which Salvador Dalí had disowned. Buñuel said, yes, course. And they sacked him.
   A few days later Buñuel and Dalí run into each other on Fifth Avenue, and Dalí sticks out his mitt which Buñuel refuses to shake. Buñuel says, it was you, wasn't it, that got them to give me the boot — and stuck one on him.

JM TALKING TO LUIS BUÑUEL:
Monsieur Buñuel — you hit Dalí: is that correct?

BUÑUEL GRINS. BACKGROUND: ROW OF ELDERLY WOMEN ON DECKCHAIRS SUCKING CADBURY'S FLAKE BARS. JM C/U IN CORNER OF SCREEN

If surreal is now nothing more than another word for odd then the one thing that can no longer described as surreal is pictorial surrealism which —

IRISHMAN'S VOICE (OFF):

Oy you! What were that you were saying about it coming from the south and being royal. Bollocks. It comes from the west. It's republican.

FIVE MEN IN DIFFERENT SORTS OF TERRORIST CLOTHES. EACH WEARS A BALACLAVA HOOD. THEY HOLD A VARIETY OF SMALL ARMS. THEY SING TO THE TUNE OF 'YMCA'. HEAVY OIRISH BROGUE. THEY PARODY VILLAGE PEOPLE'S DANCE ROUTINE

> *Declan, are you looking for fun?*
> *I said Conal, can you hold a gun?*
> *I said Patrick how would you like a chum?*
> *To help you wound, maim, slay . . .*
>
> *Oh join the INLA,*
> *Oh come and join the INLA.*
> *Give yourself a treat*
> *Help us blow up a street.*
> *There'll be Brits missing brains -*
> *There'll be blood in the drains*
>
> *Kevin, you might face arrest.*
> *Liam, you can do a dirty protest.*
> *Connor, you can smear shit with your hands.*
> *Eoin, you can be Bobby Sands.*
>
> *If you join the INLA . . .* **[FADE]**

SMEARED WALL. FAECAL BROWN. JM PTC HUW WHELDON STYLE

Bobby Sands was a provisional surrealist. He sought the infant within the man. In his most celebrated work he pursued *nostalgie de la boue* with a fervour and with a literality that even the coprophile Dalí eschewed. Dalí's medium was paint. Sands went much further — he adopted a stratagem of Marcel Duchamp's and made an ever-mutating installation that was hidden from public gaze and which he knew would be destroyed after his death — which was also a surrealist act: irrational, romantic, gratuitous, anarchic . . . and truly heroic.

CAMERA PULLS BACK TO REVEAL THAT THE SMEARED WALL IS A CANVAS HANGING IN A ROOM LIKE A VICTORIAN MUNICIPAL GALLERY. BELOW IT IS AN ORNATE LOUIS-STYLE CHAISE LONGUE WHERE A HALF-DRESSED NUN LIES MASTURBATING WITH A DOG BESIDE HER. NUN (REPETITIVELY, TO HERSELF):

Holy Mary mother of god, Holy Mary mother of god, Hol — oh oh Mareee . . .

FLASH FRAMES OF BERNINI'S ST TERESA, ORGASMIC FACES IN EXTREME C/U. JOCKY WILSON > TWANG > MADONNA/ROSARY. LONGER-HELD SHOT OF RAILWAY

SCRIPT 4

TUNNEL ENTRANCE. VOICE PLAYED AT WRONG SPEED, SLOWED DOWN
Even when there is no train the tunnel is still there.
FLASH FRAME OF HITCHCOCK'S PROFILE. CORRIDOR — JM OPENS DOOR TO REVEAL LEWIS CARROLL
To the British, Hitchcock's a manufacturer of thrillers. And Lewis Carroll is a mere children's writer. Continental Europe regards them differently, reveres them as great artists, inventors, artificers, de facto surrealists who delve — literally, as often as not — beneath the surface skein of the observable world.

JM AT GRAVE OF LEWIS CARROLL. SUPINE ACROSS IT. THE NAME IS VISIBLE ON HEADSTONE. GRAVEYARD WITH NUNS IN BACKGROUND LIKE IN MILLAIS PAINTING — THEY ARE PLAYING TAG. CYPRESSES. CAMERA MOVES IN AND OUT FROM GRAVE FOR QUITE A WHILE. ON LAST MOVE OUT TO INCREASE FRAME ALICE COMES INTO VIEW. SHE IS A TINY OLD WOMAN WITH METAL LEG BRACES IN A WHEELCHAIR PUSHED BY MAN IN NUN'S CLOTHING. ALICE (CHILDISH VOICE):
He's dead. I'm not. I am forever.

JM:
So's Lewis Carroll.

ALICE:
Not so forever as me, my dear. The creator lives merely by leave of his creatures, his creations. They buried me here in 1934. I am forever — aren't I?

THERE ARE NUNS WRESTLING ON THE CEMETERY LAWNS IN DISTANCE. CLOSE IN ON THEM AS WIMPLE COMES OFF TO REVEAL DAY-GLO HAIR. JM PTC:
That's the power of myth, of art, of fiction, of imagination — it causes eternity to beckon. It kills off the notion of there being no life after body death.

JM PHOTOGRAPHED LIKE ISAAC BABEL WITH HORSE BEHIND. JM C/U:
Right then, surrealism came from the south and from the west — where else did it come from?

SURREALIST MAP OF THE WORLD
This map, published in 1929 in *Varieties*, shows where the surrealists thought surrealism came from — it is a cartographic homage to the cultures that they believed to be their sources and their inspirations, that they believed they owed a debt to. They had merely plundered them — but that's the European way . . . colonisation through appropriation.

STILLS FROM GUN-DOG BOOK
Thus Labrador,

BIGGINS AS CZARINA
Russia,

LONG-HELD SHOT OF MELTING BAKED ALASKA
Alaska,

ARCHIVE OF HAYLEY MILLS AS POLLYANNA
Polynesia,

Surreal Film

V/O FLANN O'BRIEN
No one knows how I killed Old Philip Mathers . . .
And Ireland are inflated.

ARCHIVE GRASSY KNOLL, DALLAS '63, MAN WITH RIFLE FIRING THROUGH GAP IN PALE FENCE
The USA

C/U PEBBLES AND STONES, BLOOD SEEPING OVER THEM
and Spain don't exist.
DWARF WITH MOBUTU MASK, DWARF WITH DE GAULLE MASK, DWARF WITH THATCHER MASK
Africa, France and Britain have shrunk.
 These are neat enough gags. But they're evidence too of delusion, and vanity.
 The diminution of Africa is especially telling.

ARCHIVE: PATRICE LUMUMBA
African art had been influential on western artists since the turn of the twentieth century: it was a way out of the continuum of the Renaissance.

PILES OF BODIES
The surrealists wanted to be different — so: out with African art
. . . and in with Inuit art, Easter Island art.

BACK TO MAP
This map is wrong.
 It denies surrealism's euro-centricity.
 One history of the twentieth century is the history of bourgeois guilt and of its attempted expiation through stay-at-home plundering. Europe tried to absolve itself of slavery and colonialism by appropriating the idioms of alien cultures.
 Europe is a living museum, the warehouse of the world, the great plunderer — which is why it is the centre of civilisation. And of cultural suicide.

BLUE-FACED MAN WITH WHITE GUITAR STRUMMING IN CHURCH. FAWNING OCTOGENARIAN GROUPIES IN BIKINIS. FULL-FRAME BRIAN JONES
Can blue men sing the whites?

WIPE TO HIPPOPOTAMUS WITH THE HEAD OF A KRAIT
Or are they hippo kraits?

MONTAGE OF CATHOLIC RITUAL — SANTIAGO CATHEDRAL, FANTASTICALLY RICH ROBES, VAST CENSERS, THE HILL OF CROSSES IN LITHUANIA, BLACK VIRGINS, STIGMATA, SHROUD OF TURIN T-SHIRTS, SEVILLE PENITENTS, PHILLIPINE CRUCIFIXIONS
A more accurate map of surrealism looks like this. It is equally a map of Catholic Europe. Which is apt. Surrealism flourished in Catholic countries where Christian mythology is sustained by a gamut of rituals which render the fantastical inseparable from the everyday. The dourness and unimaginative literality of Protestantism produces dourly unimaginative naturalism.
 While surrealism may have been fervidly anti-clerical, it knew its enemy. It purloined clerical garb and ecclesiastical

practices. Surrealism was born of the religious impulse, of a desire for mysteries — but not the old threadbare mysteries, though it certainly adhered to the conviction that there is more to it all than the objective world. It was concerned to build new mythologies of its own. It believed in the primacy of the imagination. It represented what could not be seen. It invented worlds rather than vainly trying to copy what is supposedly before us. It made the figurative plastic and concrete. It demonstrated that anything is possible, that the imagination should not be trammelled by facts, circumstances, usage . . .

FROM ABOVE. THE MAP IS MADE FROM SOME MATERIAL THAT IS EDIBLE. IT IS ON A TABLE THOUGH THAT IS NOT IMMEDIATELY APPARENT. MAP REPRESENTS DISTORTED WORLD. IN THIS MAP THE HIGH POINTS OF SURREALISM ARE PICKED OUT IN COLOUR. FRANCE, SPAIN, ITALY AND BELGIUM ARE THUS BRIGHTLY COLOURED. OTHER COUNTRIES ARE GREYISH, WITH JUST A FEW SITES. IAN PAISLEY FIGURE APPEARS, AS THOUGH HE HAS JUST BURST INTO ROOM, ARMS AROUND THE INLA TROUPE STILL IN MASKS. WE SEE THAT THE MAP IS ON A TABLE.
PAISLEY:
> *Thos uss dosgustin. All credit to the Virgin. What a woman — a midfield dynamo in the proud green and white of Celtic.* **[SINGS]**
>> *Rome, Rome on the hills,*
>> *Where the pope and the cardinals play*
>> *Where seldom is heard a Protestant word*
>> *And the sky is not orange all day . . .*
>
>> *Oh — take me back to Toiber, vust me in rayments . . .*

INLA TROUPE STARTS TO EAT MAP, BEGINNING WITH ULSTER. JM TALKING TO LUIS BUÑUEL (SOUTHERN ACCENT):
> Monsieur Buñuel — *C'est vrai que vous lui avez flanqué une gifle — Dalí, lui?*

BUÑUEL GRINS.
PARC BUTTES CHAUMONT, PARIS. JM THERE, ALICE STILL IN STUDIO:
> *My dear, I hadn't noticed how you've changed.*

JM:
> You'd never met me before five pages or seven minutes ago.

ALICE:
> *All the better to see how you've changed. And my, my, you have lost weight.*

JM:
> Yes I have. **[TEARFUL]** . . . And I don't know where to find it.

MALE VOICE CHOIR [OFF]
> *And he doesn't know where to find it* **[LOOP]**

ALICE STRETCHES OUT CHILDISH HAND. ALICE STILL IN CHAIR:
> *I'll show you. I'll show you where to find it. I'll help you. I know where it'll have gone.*

FLASH FRAME OF SLAUGHTERHOUSE. JM IN PROFILE. CAMERA ZOOMS ALMOST IMPERCEPTIBLY CLOSER TO SIDE OF HEAD TILL EAR FILLS THE SCREEN. AT THIS POINT THE CAMERA STOPS

Like I say — anything is possible — as it is when we sleep. What happens when we REM? The world is re-ordered according to our desires and anxieties. Reason sleeps when we sleep.

Surrealism connotes merely a twenty-or-so-year span of the eternal human urge to replicate the reality that is achieved in dreams and in cerebral afflictions where reason absents itself.

CAMERA CLOSES IN ON EAR, APPEARS TO ENTER EAR, FRANK OFF–CAMERA URGENT WHISPER:
Go on get in there.

ROB O-C:
Not without a condom.

FRANK O-C:
Pick of the animals?

ROB O-C:
That's different — you're on.

CAMERA WHIZZES ALONG AS IN DUFF SCIENCE FILM THROUGH INITIAL DARK BIT GRADUALLY LIGHTENING AS THOUGH THIS MIGHT INDEED BE A EUSTACHIAN TUBE. IT CLIMBS, IT SWOOPS. SYD BARRETT 'OCTOPUS RIDE' — 'HEY HO HERE WE GO UP DOWN TO AND FRO' [LOOP]
That eternal human urge seeks to replicate in art what we experience through trance and narcosis when we will reason to absent itself.

THEN INTO EXTREME CRIMSON — BLOTCHY LIKE BLOOD, VELVET, CLOSE–UP MEAT. TOMMY JAMES AND THE SHONDELLS' 'CRIMSON AND CLOVER' — 'OVER AND OVER' [LOOP AVOIDING THE WORD CRIMSON]
In pre-childhood where we have yet to learn it, to learn anything save the luxury of uterine dependence.

THEN INTO EXTREME VERDURE — TRACK SO THE FRAME IS FILLED WITH GREEN GRASS THEN TO MOB OF LIVE LOBSTERS ON THE GRASS — SLOMO REPLAYS OF THEM FIGHTING
The urge has always been with us — it merely lacked a name till Guillaume Apollinaire came up with surrealism. The Surrealist movement ludicrously attempted the codification of that urge.

ANDRE BRETON — SOVIET UNIFORM — AIRBRUSH PROCESS SHOWN — UNCLE JOE
Its commissar Andre Breton imposed rules where there can be no rules. But then he was no artist and not much of a writer. He was the prototype of the Stalinist cultural impresario. He had a taste for purges and manifestos. He diminished the eternal urge by turning it into another art movement, another school of literature, another ism . . .

CAMERA WANDERS UP INCLINED GRASS TO NUN (HYPER WIMPLE) PRAYING BEFORE BLACK DILDO GROWING OUT OF GRASS
. . . when it is, rather, a state of mind, a sensibility, an impulse, an elemental need to exteriorise or exorcise the interior phenomena we experience when we are freed of reason's straitjacket, when cause and effect seems a construct of the most banal kind . . .

SCRIPT 4

CAMERA MOVES ON TO FIND MAN IN BLINDFOLD, CORSET, STOCKINGS WITH ORANGE IN MOUTH ON KITCHEN TABLE SWINGING FROM NOOSE ATTACHED TO A TREE IN A WOODLAND CLEARING

. . . when Judaeo-Christian morality seems like an elaborately frivolous system of suppression, when clock-time is exposed as a job opportunity for the Swiss, when linear logic has yet to be invented, when the notion of meaning has no meaning.

CAMERA TRACKS ON TO FIND FARFISA ORGAN IN THE WOODLAND CLEARING. AN ORANGE-DYED POODLE BESIDE IT. THE POODLE SPEAKS:

Now — Orange Mulligan's Milligan's Orange.

SIMON MULLIGAN ENTERS — ORANGE FACE WITH WHITE LIPS LIKE B&W MINSTREL, BOWS, PLAYS VIRTUOSO PIECE WHICH SOUNDS LIKE IT'S A RECORD GOING BACKWARDS. AT END HE ADDRESSES CAMERA, SPACILY:

And then I woke up — it had all been just a dram.

CAMERA PULLS BACK TO SHOW THAT HE IS WAIST-DEEP IN WHISKY AND INSIDE A GIANT BOTTLE. BACK TO BUTTES CHAUMONT. JM PTC:

The journey in time is one of the four cardinal devices of fantastical narrative — the others are: the double, the contamination of primary reality by secondary reality, the work within the work.

APE-LIKE FIGURE, HANDS DOWN TO ITS FEET PAINTING A CANVAS ON AN EASEL WITH A SPRAY CAN. CAMERA MOVES ROUND TO SHOW THE SUBJECT THAT HE'S PAINTING — IT IS A VERY BANAL BOARDROOM PORTRAIT OF MAN IN SUIT

The first surrealist was the first of our ancestors to make a representation of not what was outside him — the daily grind of hunter-gathering and making fire from flint — but of what was inside his head, of what had no existence before he made it exist, of what might come to objectively exist in the far off-future — if only because it is willed to exist by this vision. That first artist — who did not, of course, realise that he was an artist — possessed an innocent capacity for wonder that his avatars have striven ever since to recapture.

How do we achieve that state? How do we put ourselves in touch with the capacity to see what we alone can see, what is unique to us? How do we enjoin ourselves to regain that ideal of aesthetic prelapsarianism.

CAMERA WHICH IS NOW ON THE BOARDROOM PORTRAIT MOVES PAST ITS EDGE TO REVEAL BARE STUDIO

And why do we wish to? Neuropsychopathologist Vigo Bream . . .

JM AS BREAM IN LAB COAT WITH ROBERT WINSTON MOUSTACHE

Somda onirologistad, somda arbeitistad, somda makfabricoristad
. . .

PICTURE DISAPPEARS. BUTTON-PUNCH. CAMERA MOVES BACK FROM TV SET IN ROOM WHOSE WALLS ARE PAPERED IN VIOLENT CANDY STRIPE. THERE ARE BIRDS — CANARIES, BUDGIES, PARROTS — EVERYWHERE, NOT IN CAGES
MALE VIEWER VOICE, OFF CAMERA:

He used to be all right. I liked that one when he did the disgusting sex act with the llama in the motor boat.

Surreal Film

FLASH FRAME ARCHIVE SHOT FROM FIRST ABROAD SERIES: JM DRIVING MOTOR
BOAT. FEMALE VIEWER:

What else we got?

BUTTON-PUNCH. MALE VIEWER:

*Rick Stein's Road Kill [READS] — this week Rick prepares squashed
otter and shows us how to scrape a fox off a hard shoulder and
braise it so it tastes sweet as dog.*

SCENES OF ROAD KILL. FEMALE VIEWER TAKES REMOTE:

Give us that.

BUTTON-PUNCH. GARY RHODES OPENING GIANT CAN OF CORNED BEEF WITH KEY.
IT STANDS QUIVERING ON TABLE. RHODES:

*The Aberdeen typhoid outbreak of April 1964 was caused by
contaminated corned beef. Anything becomes fantastical if you look
at it long enough.*

CAMERA ZOOMS IN VERY SLOWLY TILL THE MEAT FILLS THE SCREEN LIKE A
PINK AND CREAM ABSTRACT. THIS SHOT HELD FOR A MINUTE. MALE VIEWER:

All cooking isn't it — when it isn't makeovers.

FEMALE VIEWER:

That's art, that one.

BUTTON-PUNCH. ACTOR DRESSED LIKE LAWRENCE LLEWELLYN BOWEN STANDING
BEHIND SOFA AND CHAIRS. MALE VIEWER:

*He's heterosexual you know. You missed that one the other week
when he dropped his hot lunch on those people's carpet and just
left it steaming there.*

LLB ACTOR PICKS UP JERRYCAN FROM BEHIND SOFA. POURS PETROL ON TO
FURNITURE. STRIKES MATCH. EXPLOSION. HOUSE ON FIRE. BUTTON-PUNCH.
BBC2 LOGO. CONTINUITY ANNOUNCER:

*In just a moment in a special for BBC2's Chocolate Night. Loyd
Grossman goes 'Through The Anal Keyhole' to ask whose is this
rectum?*

BUTTON-PUNCH. PETROL TASTING. JILLY GOOLDEN VOICE:

*Mmmmm! lead coming through on the nose, ooh ooh big saturated
branch-chain hydrocarbons saying hello to daintier effete
propylenes, chunky moleculed butylenes pushing through in a
devastatingly complex finish. It says '70s Elfs.*

BUTTON-PUNCH. SCREENFUL OF MOVING CORPUSCLES. SOUNDS OF HER GRABBING
PAPER. FEMALE VIEWER:

Oh, there's that quiz with wassisface on 24. Give us that.

BUTTON-PUNCH. DAYTIME TV QUIZ SET — LOW RENT, WOBBLY, GLITTERY.
VALHALLA OR BUST: TWO TEAMS OF 'FAMILIES' AT CONSOLES. BIG TV SCREEN
BETWEEN THE CONSOLES. THE FAMILIES COMPOSED OF: ONE PARENT, ONE
CHILD, TEAM A — MOTHER, BOY. TEAM B — FATHER, DAUGHTER. QUIZ HOST:
BLOND-WIGGED BIGGINS IN SS UNIFORM WITH ASSISTANT IN SS UNIFORM —
BONDAGE GOING ON *TOTENKOPF*. SHE BEARS A CIGARETTE GIRL/USHERETTE TRAY
WITH QUESTIONS ON CARDS IN IT. BIGGINS:

Welcome back, mein volk, *to Valhalla or Bust.*

SCRIPT 4

ASSISTANT [ATROCIOUS GERMAN ACCENT]:
> Here is the next question Herr Reichs Quiz Meister.

BIGGINS:
> Ooh — Here is a hard one.
> On your ting-a-lings now my sweet things.
> All right-ee. Which one of these put the T in Britain? Attend to
> your screens mein lieblings.

FLASH FRAMES: GOLF TEE (GIANT GOLF TEE IN PRADA), TEABAGS, TEA CUPS, CHIMPS' TEA PARTY, DESMOND MORRIS, MRS T, T-BONE STEAK. BELL RINGS. TEAM A:
> Were it Mrs Thatcher?

BIGGINS [ROLLS EYES]:
> No . . . Team B. Who put the T in Britain?

TEAM B [CONSULTING EACH OTHER]:
> Were it them chimps on the telly?

BIGGINS [INVITES AUDIENCE]:
> Oh dear. Sniff the whiff

AUDIENCE OFF:
> Of pig ignorance [AUDIENCE ECHO]

BIGGINS:
> It was Typhoo who put the T in Britain. Jah das gut nein?
>   Let us think now about Scunthorpe. Who's going to the Eastern
> Front and who's going to Valhalla . . .
> The question is, if Typhoo put the T in Britain — who put the . . .

JM:
> Thor in Scunthorpe. Who did put the Thor in Sc-

KEITH ALLEN:
> It wasn't the Thor I put in Scunthorpe — know what I mean? [SID
> JAMES LAUGH]

JANET STREET PORTER HOLDING BUOYS ON A BEACH:
> I want to state unequivocally that I did not put the Thor in
> Scunthorpe, I was out picking up buoys that day.

MARIE HELVIN:
> I would have lurved to have put the Thor in Scunthorpe. [KISSES
> LENS]

KATHY LETTE:
> Look squire — anyone, anyone who suggests it was me put the Thor
> in Scunthorpe gets what Odin to them.

FLASH FRAME: MARTY FELDMAN > CLOUDS/THUNDER/LIGHTNING/ (IN THAT ORDER). CHRISTIAN ADAMS COMING OUT OF CLOUDS — HUGE, BOMBASTIC, LOOMING OVER STEELWORKS:
> I am the Thor in Scunthorpe.

LOSELEY HOUSE. HAIRY SEMI-NAKED WILD MAN SWEARING AT STATUE. ALICE IN CHAIR AND JM WATCHING ALICE:
> Oh I never liked him — the Mad Hater.

Surreal Film

JM:

Hatter. Mad hatter.

ALICE:

Oh don't be so silly — the illustrator drew a hatter. So he had to stay a hatter.

JM AND ALICE IN CHAIR STAND ALONGSIDE A STACK OF TINS OF MOCK-TURTLE SOUP WHICH STRETCH TO EDGE OF FRAME.

ALICE:

I knew him well too.
We all find our route to immortality, even if it is in liquid form.

JM:

So *you* put the extra T in hater?

ALICE:

May beet.

FLASH FRAME OF MAYPOLE AND BEETROOTS.

JM:

Did you put the T in Britain then?

ALICE:

Oh I had so many Ts. I had to get rid of them wherever I could.

ALICE AND JM AT EDGE OF POND. SHOT FROM POND SO WE SEE BANK ONLY AND A MERE HINT OF WATER. LILIES, FROGS

The Pool of Tears, the Pool of Tears — it wasn't what Mr Dodgson actually wrote.

THEN FROM ALICE'S POV — A POOL OF EARS — HUMAN EARS ALL FILMED LIKE MONET'S LILIES, FLOATING. ALICE AND JM WALKING IN HARDLY FURNISHED COBWEBBY HOUSE AT DUSK. LONG DESERTED CORRIDORS. WE SEE INTO HALF-LIT ROOMS.

ALICE:

It's all because I kept forgetting to wash my tears out. Cotton buds haven't been invented yet dear.

JM AND ALICE IN BARE, ARTIFICIALLY LIT ROOM WITH CANDY-STRIPED WALLPAPER AND NO FURNITURE SAVE ONE FRAMED MIRROR.

ALICE:

I never leave a door open to this day. I don't want to get a chill because of the drafts.

GIRAFFE'S HEAD COMES ROUND DOOR

You see what I mean. I'll get a chill now. Goodbye.

JM:

Hang on — you said you'd help me find the weight.

ALICE [OFF — SHE HAS VANISHED]:

I have. This is the waiting room.

A MINIATURE STEAM TRAIN — HYTHE AND DYMCHURCH SIZE — GOES PAST THE WINDOW IN FULL DAYLIGHT WITH PEOPLE IN BRIGHT SUMMERY CLOTHES WAVING FLAGS TO CAMERA. AUDIO: THE STARGAZERS, 'CLOSE THE DOOR THEY'RE COMING THROUGH THE WINDOW'. THEN IT'S DUSK AGAIN OUTSIDE. JM TURNS

SCRIPT 4

AWAY FROM WINDOW. THE DOOR HAS DISAPPEARED. HE LOOKS ROUND — THE
WINDOW HAS DISAPPEARED. BOTH DOOR AND WINDOW ARE CANDY-STRIPED WALLS
NOW. THE ROOM HAS TURNED INTO A DE FACTO CELL. JM GOES TO MIRROR. THE
REFLECTION IS OF HIM IN A FORMER FILM

    Is that where the weight went? Is that where the past is?

JM TRIES TO PULL MIRROR BUT GETS WHAT APPEARS LIKE AN ELECTRIC SHOCK
AND REELS BACK ACROSS ROOM. GOES BACK TO MIRROR. NOW THE REFLECTION
IS OF A SKULL WHICH IS SINGING 'LAST TRAIN TO SAN FERNANDO'. THEN THE
SKULL STRETCHES OUT OF THE FRAME. CAMERA ZOOMS INTO JM'S EYE.
HYTHE AND DYMCHURCH RAILWAY AT NIGHT MOVING SILENTLY ACROSS
DUNGENESS. SKELETONS WAVING FLAGS TO CAMERA SINGING 'LAST TRAIN TO
SAN FERNANDO'. CANDY-STRIPED ROOM. JM TALKING TO SOMEONE WHOM WE
CAN'T SEE

    This country relegates Carroll to the status of children's writer
    out of fear of what he released, of what he'd tapped into. He's an
    affront to the canons of common sense. To call him silly, childish,
    nonsensical is a way of rendering him harmless. He's a high master
    of irreason: but it's irreason of his own devising, it's not
    religion which is licenced irreason, sanctioned irreason.

CAMERA DRAWS OUT TO SHOW WIMPLED NUN WITH HAIR ON HER BACK AND
SHOULDERS (LIKE EPAULETTES) SHAVING THE BACK OF A SECOND HALF-NAKED
WIMPLED NUN WITH A CUT-THROAT RAZOR

    Religion is the irreason which is permitted, the way tobacco
    and alcohol are permitted but opiates and hashish are not.
    The irony is that the more irrational a church is, the more it
    insists on the potency of its myths, the greater the appetite
    for extracurricular forms of irreason it creates. It's not by
    chance that Carroll, a cleric, wrote at the time of the church's
    strongest hold on the English imagination.

THE NUN WHO IS BEING SHAVED LUNGES AT JM WITH VETERINARY-SIZE
HYPODERMIC SYRINGE. HE LEANS BACK AGAINST WALL WITH IT STICKING OUT
OF HIS CHEST. IT FILLS WITH BLOOD. C/U ON JM'S FACE.
STUDIO. SCREEN STILL BLACK.
LOP-EARS' VOICE — SOUTHERN EVANGELIST:

    Repeat after me: The Lop-ears is my shelter.

SCREEN LIGHTENS. JM C/U — THE DROPPER OF THE HYPO, STILL IN HIS
CHEST, HAS TINY EXOTIC FISH IN IT. WE CAN SEE THAT HE'S MORE OR
LESS SUPINE ON SOMETHING GREEN. HE LOOKS UP. JM'S POV OF LOP-EARS,
A HUMAN-SIZE RABBIT (BIGGINS) WITH HUGE EARS DOWN TO HIS WAIST — NO
BACKGROUND APPARENT.
JM:

    Uh?

LOP-EARS:

    I said — repeat after me: The Lop-ears is my shelter.

JM:

    The Lop-ears is my shelter.

LOP-EARS:

    I shall not want. Take it away!

ZOOM OUT TO REVEAL LOP-EARS ON STAGE WITH A CROSS BEHIND HIM ON WHICH IS A HUMAN-SIZE CRUCIFIED RABBIT.
JM:

He maketh me to lie down on green pasta.

WALL OF BAYSWATER ROAD-STYLE ART MADE FROM CLOCK PARTS. IN FRONT OF IT A RANK OF MEN WITH JEWELLER'S EYE PIECES ATTACHED TO THEIR EYE FROZEN IN POSES SCRUTINISING. LOP-EARS IN FRONT OF JM, PROCESSING, PULLED ON UNSEEN DOLLY.
JM:

He leadeth me beside the still watchmakers.

DRESSING ROOM WITH LIGHTS ROUND MIRROR. LOP-EARS ADMINISTERING SMELLING SALTS TO YOUNG ACTRESS.
JM:

He restoreth my soubrette.

LAKE OR RIVERSIDE. GROUP OF LAYABOUTS DRESSED LIKE PARISIAN APACHES. LOP-EARS FOLLOWED BY JM PUSHES THROUGH THEM. THEY MAKE WAY RESENTFULLY TO REVEAL LAKE IN WHICH GROUP OF NAIAD OR NYMPH (CF LORD LEIGHTON CLASSICAL FANTASY) POURS WATER FROM AMPHORA. BEYOND THEM ON FAR BANK A SAILOR OUT OF BURRA OR GENET.
JM:

He leadeth me in the patches of riffraff for his naiads' sailors.

STEEP RIVERLESS VALLEY (OUTSIDE COOMBE BISSETT). BRIGHT SUNLIGHT.
JM:

Yea though I walk through the valley — why alter so chilling a phrase? The psalmist was a surrealist — yea though I walk through the valley of the shadow of death I will fear no evil . . .

SHADOW LIKE THAT OF CLOUD IN FRONT OF THE SUN DARKENS THE VALLEY.
JM:

Won't I? Won't I fear no evil? The potency of that phrase obliterates and contradicts all hope: pious optimism has no chance against it. When I was a child I believed that this place really was the valley of the shadow of death, I believed I had found the exact location: I believed that from the moment a cloud occluded the bright summer sun. And when I learnt that in France that such a valley, with no watercourse, is called *une vallée morte*, a dead valley, it was merely confirmation that I had got it right. My infantile terror was prompted by this dead valley alone.

SHEEP IN FIELD NEAR ODSTOCK WOOD

Now, this kindred landscape three miles away was different. It was wholly benign. Had a cloud switched off the sun *here* it would have been no more than meteorological happenstance . . .

WHITE HORSE JUMPING OVER GATE AT NIGHT — REPEATED MIX OR WIPE TO 'WHITE HORSES', WAVE CRESTSON THE SEA, CHARLES TRENET'S 'LA MER' — LOOP OF PHRASE 'LES MOUTONS BLANCS'— MIX OR WIPE TO SHEEP JUMPING OVER GATE

When I lay in bed at night as a child this is where the sheep that I counted to put me to sleep jumped over a gate. I don't know why they should have lived here. I don't know why the French for white horses is *les moutons blancs* — but I do know that what is remarkable is not the difference in figurative animal but the instinct in both languages for zoomorphic simile.

PERPIGNAN STATION. DALI FIGURE (WIDE MOUSTACHE AS BEFORE, KNEELING DOING SALAAMS)

Twenty-five years ago I went to Perpignan specifically to see the railway station that Dalí genuflected in front of every time he got here. I stared at it for — oh — ten minutes. It did nothing for me.

ROOF OF L'UNITÉ D'HABITATION, MARSEILLE. IN BACKGROUND MAN WITH BIG COLLAR LIKE DOG'S NECK RUFF. HE TRIES TO GET HIS HAND OVER IT TO SCRATCH HIS HEAD: CONSISTENTLY IN SAME PLACE IN FRAME IN THE FOLLOWING: METZ STATION OR OTHER; QUARTIERE COPPEDÈ, ROME OR BOFILL'S STUDIO; MARIOLATROUS STATUARY AT LE PUY; AUMISM STATUE AT CASTELLANE (IF STILL EXTANT); AUDIO: DION'S 'LOVE CAME TO ME' (THE FIRST LINE IS 'I LIVE IN DREAMS')

And I know too that we don't elect which places touch us. Sense of place is not analagous to such primary senses as smell and touch. The topographical epiphanies we suffer as children are early evidence of our incipiently sentient capacities — for wonder, for imagination, for investing the exterior world with properties of our own devising. Particular places speak to us, act as triggers of this amalgam of qualities which add up to a nascent aesthetic sense. We make our own version of these places, we possess them. They infect us as we infect them. They are the inanimate catalysts of a signal awakening — so it is unastonishing then that we should eternally return to them in the vain hope of retrieving those sensations of childhood.

And we are grateful when serendipity grants us those sensations in adulthood because they take us back to childhood, to that time of our life when we are most susceptible to mystery. That time of our life — a month then is a day now though 60 seconds is still a minute, a faster minute — that protracted time when we slough off innocence. All that innocence means is: knowing fuck all. And the idea that children know fuck all is a fiction of politicians, churchmen, rotarians, Mr Toni Tartuffe and so on. The vast gulf between surrealism and such demagogues is its impetus to abandon control, to let it roll, let it jelly roll.

SUBTITLE: 'Jelly Roll Morton was so called because' AND THE REST OF THE SUBTITLE IS OFF-SCREEN.
PAVEMENT. JM'S SHOES SURROUNDED BY OUTLINES OF SHOES PAINTED IN DAY-GLO PINK. SQUEAL OF WHEELS COMING TO HALT.
BRUMMIE VOICE OFF:

You the one that's looking for the lost weight?

DRIVING. JM'S EYES IN REARVIEW MIRROR. SITTING IN BACK OF MINICAB. BESIDE DRIVER (WHOM WE DON'T SEE) A DOG ON THE FRONT PASSENGER SEAT. ROAD AHEAD SEEN THROUGH THE WINDSCREEN. IT CHANGES WITH SWIFT WIPES — SO THE SCREEN REALLY IS A SCREEN. DRIVER VOICE OFF, DRONING ON:

It all bust up most acrimonious, like, in Moist Groin. It were much more than a marriage — which is what makes it so painful. But musically we'd grown apart the way people do: I was into totally progressive sounds but for Mutilator and Slasher it were novelty numbers or nothing: 'How Much Is That Doggy In The Window' and that malarky . . . Then there was what I'd call irreconcilable differences over the colour of our coats and the direction we should go in. I wanted to go in the Kidderminster direction but they wanted to go in the Walsall direction. I mean: end of story. Have you ever heard of anyone in Walsall buying . . .

INTERIOR CAR. JM'S POV OF DRIVER WHO IS REVEALED AS A DEPRESSED BLOODHOUND

. . . a hot dog from a couple of lilac poodles in a van called Moist Groin playing novelty numbers? No Mister.

HEATHLAND. DISTANT EXTERIOR OF CAR. IT SLOWS AND STOPS. JM VOICE OFF:
Is this where the weight is?

BLOODHOUND:
No, no, I need a leak.

BLOODHOUND GETS OUT. JM'S POV WITHIN CAR. CAMERA MOVES THROUGH SCREEN TOWARDS SKY. CLOUDS IN FANTASTICAL SHAPES. JM MUSINGLY:
There's only a letter's difference between artistic and autistic.
There's only a letter's difference between cosmic and comic . . .

CAR INTERIOR. SOUND OF FOOTSTEPS AND CAR DOOR OPENING. A GIANT LEEK IS PUSHED INTO CAR. JM BURIED BENEATH IT. BLOODHOUND:
Sorry about that mate.

FIELD WITH MECHANICALLY BALED CYLINDERS OF STRAW LIKE STANDING STONES. JM FROM DISTANCE:
There's only a letter's difference between micturition and *Allium porrum*. No . . . — that's not right is it?

STUDIO. LOP-EARS IS CARRYING COUNTLESS BAGUETTES PAINTED WITH CAMOUFLAGE
It is not. Between leak and leek — that's what yo' was meant to say. Yo' missin' the joke bo'. If yo' was in a realistic film — well, I hate to think . . .
Yo' hungry bo'? Snack? I jus' lurv to get some egg on my soldier. Where's the nearest army camp?

STUDIO. LOP-EARS PICKING UP MINIATURE CAMOUFLAGED SOLDIERS, DUNKING THEM IN EGG YOLK, IN VAST EGG, MUNCHING THEM. JM:
The professionals. One day Kosovo, the next day ovine sacrifice.

NUNS. JM — CHEERY ON LOUIS SOFA — IN FRONT OF THEM. EGG TIMERS, LOTS OF IDENTICAL SMALL ONES, GIANT ONE
The ability to soft-boil an egg is not regarded as a virtue in Spain. The more pervious a culture is to surrealism the less interested it is in the most unimaginative thing that can be done with an egg. It is a post-Protestant talent to hang around for four minutes watching water precipitate gastronomic banality. France — where reason struggles against the surreal impetus — boils an egg for three minutes.

C/U SOFT-BOILED EGG OUT OF ITS SHELL. RAZOR IS DRAWN ACROSS IT — YOLK SEEPS OUT OF IT ON TO HAND

SCRIPT 4

Here are some other ways of cooking an egg.

MARCO PIERRE WHITE FRYING EGGS. HE TALKS THROUGH THE STAGES OF PUTTING IN PAN, ETC. — VERY FORMAL, VERY TECHNICAL. JM WATCHES. INTERCUT WITH: JM WITH MAP DIRECTING GROUP OF PEOPLE DIGGING TOP SOIL ON HILLSIDE TO REVEAL CHALK BENEATH. THEY ARE MAKING A CHALKWORK WHICH EVENTUALLY RESOLVES ITSELF AS A MITRED BISHOP BUGGERING A FULLY ERECT DONKEY. CREDITS. SYD BARRETT'S 'NO MAN'S LAND'

END

5

THERE IS NO WASTELAND

Barbed Wire's Allure

GREENISH BELT

THE BUNGALOW IS non-rolling stock. It is intermittently faced in ginger-nut pebble-dash. The paint on the window frames is flaked. Here a window has been replaced by a sheer sheet of glass set in plastic. The porch is skewed. The roof has been randomly layered with tarpaulin. A collapsing lean-to has a lean-to of its own. Here's a former guard's van surrounded by nettles. A horse wearing a damp blanket dares not drink from a rusty trough. Immobile vehicles of several vintages and states of decay are sinking into oozing mud. A matt caravan is listing. Bales of rusting barbed wire have shreds of polythene flying from them like grimy bunting. The distant roar of a motorway is incessant.

Welcome to The Green Belt which is evidently not literally green.

And yes, let us preserve it. These scapes with their breeze-block piggeries are precious. Corrugated iron – which is to be found in abundance – was, a century ago, the material of the future. Sheds and byres built of it are as deserving of protection as limestone dovecotes or cob walls. Britain's 'edgelands' – the ungainly epithet is planning-speak – are disappearing fast. They are unloved save by topophiliac perverts and amateurs of the ad-hoc, among whom I obviously include myself.

I acquired a taste for these marginal places at a tender age. Close by where I grew up, the western edge of the New Forest was all unmade roads fringed by self-built shacks. They spoke of indigence, certainly, but also of resourcefulness and of a thrilling pioneer spirit born of desperate necessity. They were tree houses on the ground. They have almost all disappeared. The land they occupied was more valuable than the bodged structures themselves. So as their occupants died they were demolished and replaced by dinky dream houses of sturdier but much less ingenious construction.

Such colonies were once commonplace: nearby there was also Palestine and Four Marks. There greatest incidence was in Essex, north Kent and at numerous seaside sites. When some years ago I decided to make a film about them I toured the country from Spurn Head to Tatsfield (the 'London Alps' – really) to Birling Gap. The only remaining extant sizable concentration of these structures, the one I eventually shot, was in the Severn valley between Bewdley and Bridgnorth. The shacks there remained – and still remain – because local custom happily decreed that they should be leasehold: the land they stand on is still owned by farmers.

What militates against the preservation of shacks and edgelands is a perceived lack of aesthetic worth. They are expendable despite the fact that they are valuable repositories of working-class history. But Britain's thraldom is to the ancient, to the sacred, to the picturesque and to Good Queen Bessery. These places are not very old (1910 to 1930, mostly); they are low in the hierarchy of building types; they are seldom pretty; they rarely possess important associations – no blue plaques on shacks. Today we are all Binneys, thankfully. The chances of a work by, say, Nash or Teulon or Mackmurdo being destroyed are slight. But such buildings are atypical. We should apply the same protectionist sensibility to the banal and the humble, to the formerly quotidian stuff which doesn't quite qualify as *architecture*.

This may sound like a plea for some sort of tectonic relativism. Well, so be it. Yet there is another reason to cherish the overlooked. And that is the prospect of what might replace it, what indeed will replace it: the executive home, the science park, the retail mall, the big shed. So long as these infrastuctureless, exurban aberrations are allowed to exert their centrifugal pull our cities and towns will decline. Their density will be lessened. We must learn to value the scrappy and the atmospherically unlovely if not for its own sake then for that of urban survival and civility.

(2010)

LAST RESORT

THE WORD 'LAZARETTO' derives from Lazarus. It signified a place intended to receive and hold, for up to forty (*quarante*) days, lazars, the potentially or actually contagious, the crews (and cargoes) of yellow-flagged ships from the plague ports of Genoa, Marseille, Messina, etc. The lazaretto was the successor to the hulks, the precursor of the sanitorium: fresh air and isolation – those were the things. Two hundred years ago a different gamut of diseases was epidemic: cholera, yellow fever, typhus. The terror that they and their vectors fostered is emphasised by the desolation of the lazaretto's site at Chetney Hill, which is no hill, but an island on the Medway marshes: an artificial and circular island separated from land by a convict-dug canal.

The place was not a success, no matter how that is measured. It was abandoned in 1820, only five years after building stopped. Its full capacity was never used. The materials, which had cost an astonishing £170,000, were sold. Still, the Office of Works and its comptroller, James Wyatt, had chosen a site which was nothing if not apt. Gloom does not begin to describe it. The only visible habitations were hulks, and new vessels arriving with fresh viruses for Blighty. Otherwise there was nothing save marshes, mud, water, purslane, sky, tufty saltings, glasswort, wrack, sheep. But the chromatic and textural variations of sky, mud and water are infinite so it can't have been too bad, can it?

It can. It still is today. It is thrillingly cheerless. This is a landscape of despair. It is also a landscape of giga-sculptures, of abstracted shapes. Who needs Christo when we have the CEGB? Who needs Caro when we have cranes the size and articulation of those at Grain? Who needs Judd when we have such chimneys and 100-foot-high bridge piers and henges of hay? The repertoire – it also includes pylons, silos, hoppers, bulky mills, ships that tower over the earth, horizontal bands of smoke – is not large, but its repetitions are endless. On Sheppey and Grain, you are dogged by utilitarian grandeur, by structural rawness, by industry in its nudity, by mostly unwitting sublimity, by the forms of might. The conjunction of environmentally insouciant buildings and machines with flatlands and wetlands is immensely potent: the former provide gross accents which nature cannot deliver. All that is missing is a halma set of cooling towers. Still, you can't have everything. In lieu of cooling towers, there are the names: Deadman's Island, where those who perished at the lazaretto were buried, ranked east-west; Pigtail Corner; Bedlam's Bottom; Egypt Bay which brought me here in the first place.

The plot of Robert Hamer's 1952 film *The Long Memory*, and of the Howard Clewes novel on which it is based, is convoluted, thrillerish. The film does not approach the brilliance of either *Kind Hearts and Coronets* or *Pink String and Sealing Wax*. But it has the late John Slater as a punch-drunk pugilist shadow-boxing his way into tiny rooms. It photographs the mean terraces of Gravesend with a studied expressionism. At Egypt Bay, John Mills lives in a beached boat, a man who has been wrongly imprisoned and is out for revenge. I cannot think of any other film whose loci I have chased, but *The Long Memory* possesses a rare topographical specificity, and Hamer owned such a beguiling gift for the distillation of place . . .

The point is that the north Kent shore is obstinately disobliging: it persistently refuses to correspond to the notion of the soft South, which is a notion born of torpid generalisation, maybe, but none the less often true. So much of it seems displaced: a slice of Humberside or Teesside where the accent has an East Anglian twang. The land of oasts, orchards, hops, hedges, clinker-build and sweet English plenty collides with something indisputably harsh. Idioms of human enterprise (agricultural, architectural), which are sympathetic to their host land, clash with ones which are defiantly hostile to it. The process begins as far west as Woolwich, which may pretend to belong to London, but whose martial brutality and rollercoaster roads are echoed all the way to the three-named but united conurbation of RochesterChathamGillingham, estuarial towns whose streets dive precipitously down parlous gradients and swoop up the next like regimental lines. The collusion with slopes, escarpments and the land's lie, which characterises gentler places such as Bath or Malvern, is nowhere to be found; but, as I say, the quintessence and the very excitement of this protracted dystopia reside in its tradition of bone-headed vanquishment, of impoliteness, of brutishness.

There are two distinct sorts of primitivism here. The beauty that there is – and I would maintain that there is great beauty – is not the beauty of beautification, let alone of prettification. The functions of land, of structures, of marine enterprise, are never dissembled, never disguised, never euphemised. They are often hidden, but that is different – the approach to Chatham dockyard (whose magnificence is not mitigated by its emasculation and rebirth as a tourist outfit) is through a canyon of brick walls that have a Piranesian capacity to talk to the base of the spine.

A lot of north Kent takes its cue from Chatham: big walls are big here. Old ones topped with chevaux de frise; new ones topped with barbed wire; spanking up-to-date ones with razor wire; ad hoc ones built of crates and sleepers, and crudely painted with invitations to end up as a dog's lunch. The main gate of Chatham dockyard sets a mood, if not a style, for much that was to follow it. This is Primivitism (1). The main gate, finished in 1720, is a work of acutely self-conscious art, theatrical, vaguely medieval, certainly inspired by Vanbrugh (who inspired no one save naval, military and prison architects). It has the upside-down air of Battersea power station. Grain power station, the largest in Europe when it was built in the mid-'70s, is another essay in artful primitivism, fearsomely dramatic and deliberately symbolic of power; that symbolism is part of its function. It is an awesome piece of machinery. (Architects: Farmer and Dark.) The predominantly late-Georgian and Victorian dockyard buildings at Chatham and Sheerness belong to the tradition of engineer's architecture, whose laureate is Sir James Richards; this is Primitivism (2). These stupendous sheds, up to 350 yards long, include the Sheerness Boatstore, which is among the earliest iron-framed buildings anywhere; it, too, is long, and exceptionally graceful. I write from the memory of having seen it as the Vlissingen ferry pulled out of Sheerness some years ago; I was unable to see it the other day because I was nabbed by a dockyard jobsworth.

Beyond primitivism, there is the simply primitive. A distinction must be made: it is Picasso on the one hand, Lascaux on the other. The one is wrought by artists, the primitive by . . . well, by non-artists; industrial non-artists and domestic non-artists. Some of them work on a grand scale. At Swanscombe, there are pyramids. They may be made of roadstone, gravel, earth, but they are pyramids, and there are tipper trucks crawling up them, loaded like Brobdingnagian ants; you do not get that at Giza. Had the nearby chalk pit been dug by Romans or Saxons or some other oldsters, it would be venerated, and its amphitheatrical majesty lauded, but it's post-war (Second World War not Gallic), so it isn't. The striations of chalk with the inevitable green icing on top and clunch below are quite as fascinating as any limited spectrum can be. Follow this shore with open eyes and renew your historico-aesthetic sensibility. As in any area where the land has been wantonly scarred, there are chunks of dreariness; they, however, are the chunks that might be anywhere – arterial road naffery, waste plots where the buddleia suffers brewer's droop, light industrial plants of the '50s and '60s. Much, however, reeks of nothing but itself. There is much that is peculiar to this neck of the estuary: the sudden stunning views of the Dartford bridge, ditto of Canvey and its refineries. Even the bales of swarf gleam in a special way – ascribe that to the quality of light.

And there is nowhere else like Rosherville, Jeremiah Rosher's new town of the 1850s, which had a Thameside pleasure garden, an ape of Cremorne where the punters ate so many oysters that their shells made nacreous pyramids (again) that were visible from Tilbury. That, anyway, is what Hippolyte Taine reported. Rosherville is now recalled in the name of Rosherville food market; in the streets between the Italianate villas, Sikh kids play cricket under the Rosherville sun.

Across the river from the ferry at Gravesend is the prodigious oddity called Bata: the model village designed for workers of that (then) Czechoslovak shoe company by Czechoslovak modernist architects in the '30s. Even the disposition of the trees looks central European. The last time I went there, the Bata hotel was still operating. That is Essex, though; that is not Kent. Kent: home of the British bungalow. The town of Birchington has much to answer for: here is the *fons et origo* of this building type. Are bungalows primitive, or are they primitivist?

Birchington, the point on the foreshore where the sub-architectural virus from Bengal first hit, may be Buchan's 'Bradgate', the site of the thirty-nine steps. He certainly worked on the book at Broadstairs. It does not matter one way or the other. But mark this: the first bungalows to be built in England were those at Birchington, on the cliff, and from their gardens led (and still lead) stepped tunnels to the shore.

Moreover, these were buildings associated with aesthetes. In Buchan's crass cosmos, that meant spies – witness, too, his 'Biggleswick', a crude analogue of Letchworth Garden City, in *Mr Standfast*. The bungalows at Birchington are all greenery-yallery, decorated with sgraffito cherubs. Rosetti died here. There is a road named for him. They got something right. The bungaloid successors – did they get it right?

We can all agree that Britain's coast has been trashed. Yet has it ever been so sedulously trashed as on Sheppey? Leysdown on the north-east corner of that isle is terminal England, the last resort, the nadir of bungalow development. It makes Hayling look like Biarritz. Even its kitsch is unconvincing. The bricolage and inspiredly untutored inventiveness of most early twentieth century plotlands is missing. On the shore, half a mile from the village's centre, there are asbestos and wooden hutments, endlessly buffeted by the howling easterly. Bent bushes try to smother them. The groynes in front of them are like the bones of prehistoric creatures. On the road beneath the sea wall, a car is parked facing the marsh: oblivious to the birds and the swaying grass and the distant chimneys, the man in the driver's seat concentrates on the *Sunday Sport*.

Back in Leysdown, in a car park behind a boarded-up amusement arcade, four men are transferring the contents of an articulated lorry into two Transits and a rusty hatchback. They do not like me, I can tell this from their faces. Now, what's that shooting through the air? That is a series of jellied eel bones. A jellied eel family has synchronised its buccal expulsion of bones. There are many jellied eel families; there are also extended problem families, rottweiler families, double doberman families. There are the White Men freelance entrepreneurs who are very proud of their Y-reg Escorts, and drive them at an average of sixty miles an hour anywhere. That's the White Man's right. The English can be terrifying. They like their fun, though. The air of Leysdown is filled by the amplified, beery croak of a bingo caller. Someone is having fun. And there will be more fun tonight when Big Al's Radio Stars Disco Roadshow gets gigging. Right now Big Al's deafening yellow van is parked among maisonettes on an unadopted road. Two huge dolls dressed as bridesmaids stare at it from a lavishly netted window.

Northern Sheppey is dense with caravans, immobile homes, metal habitation hutches. Glinting white tombs, they appear in the distance the way the ranked graveyards of Flanders and Lorraine do. Every caravan park has a pretty name – Lazy Days, Song of the Sea, Sea Cliff, Willow Trees; and every name is just a bit of a fib. These caravan parks are not blots. They possess a despairing summeriness. They are also the most ordered items in an area which really is falling apart.

Man's depredations have, as well as creating a terrible beauty, given much land to the sea – the digging of silica clay from the Medway and the Swale has caused those rivers to spread. And nature has done its part too. Wind and water have taken their toll on friable rock. The coast is eroded. Blockhouses and bungalows have tumbled into the sea. Notices warn of the dangers of walking the cliffs, lest the land slip from under your feet. All places are in a state of permanent mutation, but here the process is accelerated, overt.

The newest wheeze is to bury toxic waste. That figures. A history of mistreatment engenders a future of mistreatment. The by-product of this unstoppable force is an Area of Outstanding Unnatural Beauty.

Oh, and there's much, much more: at Harty Ferry, on the south of Sheppey, stout wooden fruit crates are piled high and as long as a three-storey terrace; Kemsley, beside the Swale, is a neo-Georgian model village of the '20s with, at its centre, a doll's house of a pub and, beside it, a concert hall; not far off, at Ridham dock, is a field of caravans which resolve themselves as you approach into rectilinear bales of paper; a mile or so north of Sheerness, where the lanes of the Thames and the Medway join, there juts from the sea the masts of the liberty ship *Richard Montgomery*, which struck a sandbank and split in 1944. Its cargo included 3,000 tons of bombs, 1,200 tons of TNT, etc. Were it to go up, it would grossly accelerate the process referred to above.

(1991)

TERMINAL

WHEN DID MY FONDNESS for railway stations become suspect? In the generation and a half since Ray Davies composed the most charming of London songs, termini have been stripped of their innocence. To own up to an appreciation of, say, King's Cross is akin to expressing a wish to work in a children's home, to become an ordinand, to train as a Scoutmaster. Yet stations draw me to them. I blame . . .

Well, where does one start? Salvador Dalí doesn't help. He used to genuflect in front of Perpignan station. I recalled this in Barcelona one late autumn when I was too young to know better, and set out for that town across the border by way of Puigcerda, Saillagousse and le Train Jaune. I disembarked at Perpignan, walked out of the station, turned, gaped. And nothing happened, nothing. It entirely failed to move me. But I was so green I booked into l'Hôtel des Moustiques et Hyperpuces which overlooked it. Still nothing happened. Of course it didn't. I stood in the forecourse and realised something: that one man's madeleine is another man's potential tragic traffic casualty statistic. So it was an epiphany of sorts, after all. A lesser epiphany, one that afflicted my reasoning brain, not the small of my back. My reasoning brain told me that I had stations of my own.

St Pancras. I had first seen it one all-blue summer's day in my teens. It seemed to belong to a beneficent fairy tale. The bricks – I was evidently ignorant of their Midlands provenance – glowed, the structure was fantastical. I'm afraid that it was Scott's pseudo-Flemish pastiche rather than Barlow's engineering that appealed to me then, and it still engages me to this day. This is, I realise, aesthetically incorrect. But we don't elect what excites us.

Sloane Square. This District and Circle Line stationstop had (has?) an improbable bar on the westbound side. Why? What was it doing there? It felt like it was custom-built for spies/drops/illicit meetings.

The reason I don't know whether it still exists is that it feels as though it might be fated. The two people I knew who 'discovered' it are both dead. David MacTavish (the singer in the largely forgotten psychedelic band Tintern Abbey) and Tariq Yunus (actor, writer and supreme card sharp).

Paddington. There is a social gulf between the places served by the west of England and Welsh route and those served by Waterloo. Oxford, Bristol, Bath, the Cotswolds and north Wiltshire are apparently inhabited by a rather different tribe to that which roams in the Surrey heathlands, Basingstoke, Southampton, Bournemouth. At Paddington, posh accents are far from extinct and received pronunciation is commonplace. So too are Barbours, British Warms, loud corduroys, hefty checks, culture tourists, books.

Waterloo. This was my entry to London. Not my station of choice. But I repeat: we have no more say in this matter than we do in that of our parents – whose domicile is liable to be a determining factor. So until my mother's death just over a decade ago, this was the station that I was best acquainted with. And the sites beside the track from Salisbury to Waterloo are etched deep in my memory. Bleak downland, extinct watercress beds, a mill that manufactured banknote paper, the once derelict, now repaired Basingstoke Canal, houseboats on the River Wey Navigation, golf course after golf course, the pines and rhododendrons of Brookwood Cemetery, Brooklands' embankment, the pretty mosque at Woking, the orphanage beside it, a nearby wedding photographer's pebble-dashed corner premises, about which Cecil Beaton sniffily remarked in David Bailey's film about him, 'There but for the grace of God . . .'

Cyril Connolly wrote with acuity of the point that exists on every line into a London terminus where he would be beset by an inchoate terror of returning to the metropolis. I wonder if all provincials experience this in their early London years. I certainly did. It was the jolly floral clock in front of the art deco palace of Sutton's Seeds near the Kingston bypass that signalled a sure lowering of my spirits. The last time I returned to London on that line I could not summon even an echo of those sentiments. So something changed for the better.

And, though it may be shocking to say it, so too has the train service changed for the better: faster, cleaner and on time.

As for Waterloo itself? The cinema which had begun life as a 'news theatre' has gone. So has the machine which enabled you to pointlessly stamp your name on a piece of metal. And the DIY recording booth is no more. But the sad truth is that none of these facilities was really much good. Further, I'll hazard that the existence of countless undifferentiated stalls and shops selling the same espresso, ham, fruit, pastries, 'freshly squeezed' juice, is a marked material improvement on the days when the choice was a Wimpy or nothing.

However, retail's exploitation of public and semi-public concourses is not without cost. Although lip service is endlessly paid to the notion of public space's worth, this country seems bemused by the point of it and is lax about implementing prescriptions which might inhibit our inalienable right to buy, to buy, to buy. One unwritten law of the free market is that every space possesses potential as a marketplace, and that to fail to realise that potential is an offence against mammon. Public space for the simple sake of public space is, I fear, a European idea too far. Its provision encourages ambling, flânerie, hanging out, wasting time, dreams of escape – dreams that were once triggered at Waterloo by posters promising 'The Strong Country'. This was not some blackshirt fantasy but a long-running advertisement campaign by Strong's Brewery of Romsey near Southampton, which featured the work of the landscape painter Rowland Hilder. Of course, the intention was to sell wallop. But it was achieved doucely and with what were not then called environmental benefits. Indeed, the environment had yet to be invented.

(2004)

THE OTHER GROVE

AND AT THIS POINT I make a detour into Newton Road, a secluded street of somewhat Cheltonian classical-survival villas among which stands a singular oddity, Denys Lasdun's first building: a strangely graceful house completed in 1938. But it's hardly a landmark. It lacks the cachet of Highpoint or the De La Warr Pavilion. Maybe this is because Lasdun survived the architectural hiatus of the Second World War and its long aftermath in a way that the majority of first-generation modernists did not. (Younger readers should be instructed that not only, for obvious reasons, were few buildings made during the war, but that there existed for almost a decade after it the proscriptive tyranny of 'building licences': thus British architecture 1939–1954 is a blankish page.) Then again, maybe no. 32 Newton Road is off the map because it doesn't accord with received notions of that tentative British modernism. First: it is symmetrical. Second: its emphasis is vertical. But Lasdun's work was ever thus. He made up his own rules and was out of step with his time. Indeed, in retrospect much of his stuff seems positively eccentric. This diversion into tranquillity and the solace of art was taken to escape Westbourne Grove, a frenetic street of such diversity and scuzzy richness that a little goes a long way.

In Fresco, a bearded young Buffalo Bill in a floor-length fuchsia velvet coat and cowboy boots goofed a stoned grin as he ordered a smoothie using every available ingredient. Across the road a newsagent was selling *The Kerryman*, *The Limerick Leader*, *Western People*, *The Munster Express*, *The Longford Leader*, *The Nationalist*, *The Kilkenny People*, *The Irish Post*, *The Connacht Tribune*, *The Clare Champion* and so on, and on. The electrical shop next door had a luminous tortoise in the window. What is a luminous tortoise for?

This street is as layered as a millefeuille. It is also singularly anachronistic. Its appeal derives from its reluctance to slough its multiple pasts. Forty years ago it was the high street of Rachmanland, and the neighbouring roads, Newton included, were all gas-meter lives, transience and damp lino. Today there's not a slum to be seen. Domestic property is as expensive as anywhere in London. Multiple occupation is no longer a norm; 'original features' are – you can clock them at dusk. Of course, they aren't really original but re-enactments of the cornices and roses and fireplaces which were removed in the years of destitution.

Somehow the eastern end of Westbourne Grove has failed to get the message about the area's demographic and economic ascent. Many of the shops and services suggest that this is still bedsit land, that it's forever 3 a.m., and that Neil Young is droning from the room across the hall. A marked shabbiness abounds. The clunky all-night vending machine which dispensed dodgy pies has gone, and so too has the International Film Theatre, an upholstered ashtray where I saw, inter alia, Buñuel's marvellous *El*. Yet there remains an ample supply of empty launderettes, iffy supermarkets, sparsely furnished letting agencies, unreconstructed Indian restaurants, beer halls, booths offering rock-bottom price international phone calls, money exchanges, cheap carpet shops and heavily defended minicab offices.

Gentrification's progress has been halting at best. There is certainly a whiff of heavily laid on local colour about the businesses, but would you want to buy anything from them? Impasto ambience and retail utility do not necessarily go together. This is probably a street made for that puzzlingly overrated recreation 'street life' – at least it would be were the pavements not so narrow.

(2004)

RAW

'HE'S NO OIL PAINTING.' In which case, is he a watercolour, an acrylic, a gouache, a pastel? The implication is clear: oil is the supreme medium, thus a work executed in it is necessarily better than one wrought in those other, baser, chromatic agents. I exaggerate (of course), but not that much. Despite Girtin, Cotman, Crome and Burra there still exists the notion of a hierarchy of materials, an inapposite hierarchy when you consider how pervious English light and sky are to representation by watercolour.

In the case of buildings an even more preposterous situation arises. Here there are two forms of hierarchy at work. The first is that which so persistently infects and unbalances the greatest architectural-historical enterprise of the past half century, Nikolaus Pevsner's *The Buildings of England*: it is the unquestioning assumption that a sacred building is somehow superior to a secular one. Pevsner was far from alone in decreeing that a building's worth is determined by its usage rather than by its intrinsically architectural qualities. He was merely adhering to a form of thought born in the nineteenth century and which was accepted as a given by his generation.

And it's not dead yet. Witness the touristic popularity of cathedrals; though, of course, age abets that particular popularity. Oldness plus holiness is a most potent combination. It sells tickets. In this hierarchy, department stores, bridges, airports, factories, multi-storey carparks and so on are consigned to the lowest tiers. They are not god's house, they are not even man's house. They may be works of imaginative energy and soaring accomplishment but they are still meant to know their place. Social mobility is our norm but building types remain resolutely stuck. They are set in stone. And make sure it is stone, not some 'lesser' material.

Here is the other pernicious hierarchy which we apply to the built

environment, the hierarchy of materials. It is more complicated than
the hierarchy of building types because it has always been subject to
regional variation. Is Portland stone intrinsically the ne plus ultra of
our quarries' yield – or does the cachet attached to it in some measure
derive from that isthmus's former inaccessibility and its limestone's
consequent expense: it's a point worth pondering because Weymouth,
a couple of miles distant, is predominantly stucco in apparent
rejection of the local material. There are countless examples of such
eschewals. To build with stone in brick country, and vice versa, was to
demonstrate wealth and freedom from the vernacular.

The use of 'redbrick' as a snobbish derogation of late Victorian
universities and their alumni went missing a couple of decades back.
But industrial red bricks themselves, Accrington Bloods and their like,
are forever with us and not much more liked than they ever were.
Their glaring harshness and their inability to age gracefully (or to age
at all) are qualities which put them beyond the pale: and the same goes
for terracotta and faïence. But none of these materials owns the same
problems as concrete.

You know the griff. Concrete is a serial offender, a sullen, grey,
stained recidivist. Building material? Catalyst of social mayhem, more
like. It is forbidden to utter the word without one or more of the
suffixes 'wasteland', 'jungle', 'monstrosity'. Concrete is at the very
bottom of the pile. Or would be were it not so often 100 metres above
ground, affronting everyone else's eyes.

There's clearly something wrong with my eyes. With my taste, too,
no doubt. I'll happily admit to enjoying much brutalism – which name
derives not from brutality but from béton brut, raw concrete. The
derivation of the actual sculptural, plastic forms that characterised its
use in the '60s is less happy. It is not difficult to discern a direct link
back from, say, Rodney Gordon and Owen Luder's wonderful Tricorn
Centre in Portsmouth to the Organization Todt's West Wall and the
defences it built in Occupied France and Guernsey. But it was evidently
not that covert provenance which turned this country against concrete
but, rather, the sheer lack of ingratiation of the buildings made from
it: might, muscle, perversity and bloody-minded ostentation are in

opposition to a public, collective taste which even to this day values prettiness and the picturesque.

A gamut of what sound like occluded lodges – the Reinforced Concrete Council, the Ready-Mixed Concrete Bureau, the Structural Precast Association, the Basement Development Group, The Concrete Society – have mounted an RIBA exhibition, wrongly named Hardcore, in an effort to rehabilitate its product by demonstrating its versatility and its mimetic qualities.

Here, then, is concrete that looks like Lurex, fruit-studded nougat, marble, fun-fur which has lost its pile, marshmallow. Anything, indeed, other than the rough, tough stuff of forty years ago: the quaint modernist shibboleth of truth-to-materials is quite abjured. We are reminded, instead, of Roman concrete and of early Victorian essays such as a house in Kent which is the epitome of the chocolate-box cottage.

We've been here before. Hardcore – which should have been Softcore – proves that there is nothing so intractable that it can't be made over to appeal to a base appetite for eezee-on-the-eye frivolity. At last, the people's sand and cement, accessible aggregate for all, precast slabs for the kiddiz. I had never previously conceived of having the opportunity to describe concrete as populist, as fun. I'll swear your lips moved when you read that sentence.

(2002)

SHEDS: A–M

ALLOTMENTS: THE ALLOTMENT capital of Britain is England's second city. The BDAC (Birmingham and District Allotment Council) oversees more than a hundred sites. Certain of them stretch to the horizon. They are racially harmonious sites. Brummies of differing ethnic origins – Poles, Gujeratis, Welshmen, Spaniards – are united by their love of plots and sheds. The future socialist MP Jesse Collings, whose slogan was 'Three Acres and a Cow', founded the city's Allotments Extension Association in 1882. The diversity of veg is startling: okra, aubergines, show-marrows, chrome yellow courgettes, runner beans, etc.

Blokes: Blokes are not necessarily bloke-ish or boorish. Still, it's blokes rather than chaps who have sheds.

Corrugated iron: A century ago corrugated iron was the building material of the future. Really. And it might have come to pass. All that it requires is a tosh of paint. The former Empire is blessed with churches that were prefabricated in Britain; the garrison church at Blackdown near Aldershot is the best preserved I know. 'It used to be painted red and orange, and looked exactly like a toy church.' (Ian Nairn, *Buildings of England: Surrey*.) There is also a scout hut in Kilburn Park Road, and there are scores of corrugated iron sheds. Get painting, gents!

Dinky Toys: My chocolate-brown and navy-blue model Bedford lorry was bought in Chumleigh, in Devon, in 1952. Down in the valley, beneath the ruins of Eggesford House near the Fox and Hounds was a railway and a railwayman's shed which housed a linesman's trolley which was propelled by pushing a lever up and down. I, illicitly, took the vehicle from the shed and rode it up and down the quiet Barnstable track. These were the best of times.

Evesham: My uncle Hank – who never found the right girl – went 'home' to his mother, my grandmother, in Evesham every weekend, to his mother and to his virgin sister, my literally maiden aunt. I am sentimentally attached to Evesham because of, inter alia, the shed. There was a dry smell of knife-grinding Carborundum and of oil and of paper: Hank, the town-clerk of Burton-on-Trent, had kept every issue of *Picture Post* (which is the *fons et origo* of all newspapers' magazines today). The shed was wood, steep-pitched, narrow, crammed. Hank was an amateur of sheds – provided they were owned by people of the social class he aspired to. He once took me to the house at Stow-on-the-Wold owned by a member of the Riley family; Riley was, until it was swallowed by BMC, a famously sporty marque of car. The Riley who lived at Stow had a stone barn as his shed.

Engineering, Downton: In the first half of the '60s BMC Minis driven by Paddy Hopkirk et al. constantly won the Monte Carlo Rally. These were not ordinary Minis. They were not even Mini Coopers. They were Minis that had been 'tweaked' by the entirely eccentric organisation called Downton Engineering. Downton is a large village between Salisbury and Fordingbridge. The company had begun in a shed and had expanded into premises on the north side of the village, on an aspirant industrial estate. It was owned by the Hon. Daniel Richmond, a delightful man who was married to Somerset Maugham's niece Bunty; who fished Avon salmon with my father; who made a pass at me when I was sixteen and pretty. I turned him down. We remained friends; my father was merely amused. I still have a chromed piston from one of his winners that he gave me. This is a rare instance of shed gone global.

Eccentricity: It goes without saying that sheds not only promote eccentricity but are deliberately constructed to that end.

Flat hats: Were once the headgear of choice of all shed users and were routinely worn both in and out doors. In Northern Ireland they were, and perhaps still are, known as 'dunchers', whose etymology is puzzling. Worn when smoking Park Drive or Weights.

Greengage: Is a sort of plum. When I was a child there was a fruit-and-veg shop in Salisbury named for its owner Clement Gage. He was known to my father and me (if no one else) as Green Gage. In those days there were hothouse plums, grown locally on the Downton Road by the exotically named Hedley Combes. His glasshouses and sheds stretched for a quarter of a mile beside Edgar and Cath's house down to Britford Lane – which was a wonderland of sheds sited between the trees of former fruitholdings. My reckless friend Adrian would cycle among them like a daredevil forever skidding on rotten fallers and wasp-infested plums. I was more timid, and besides I didn't want to fall off because there were adders hereabouts. Mrs Braithwaite had found one swimming in her pond.

Les Hortillonages: On the edge of Amiens along the north bank of the Somme is a remarkable area of mostly former market gardens irrigated by narrow leets known as watergangs. Few of the plots are still commercially exploited. They are now pleasure gardens. Each is equipped with at least a hut. Some of the huts have grown to the size of bungalows. The people of Amiens recreate themselves there at weekends and on summer evenings. The place possesses an unmistakable gaiety. The river is thick with scullers and canoeists. Flâneurs ambling on the towpath have every chance of being mown down by megaphone-wielding rowing coaches on racing bikes. Old boys sit on their stoops smoking curly meerschaum pipes. Families tuck into gargantuan meals surrounded by gnomes and plaster windmills. 120 kilometres due north at St Omer the scene is virtually reproduced on le Marais Audomerois; here, though, a majority of the plots is still used to grow vegetables which are transported in flat-bottomed vessels like punts; the sheds are duly less decorative. Again, in the outskirts of Ghent there are sheds in immaculately bijou gardens surrounded by watercourses. At these Flemish sites the shed is not a male preserve.

Isle of Wight: The range of sheds on the island is impressive. They range from the distinctive boatsheds at Bembridge to a plotland development near Newtown to beach huts which require annual painting and may not be slept in.

Jerusalem, New: Seldom lives up to the name. Has been built of tarpaulin, former railway rolling stock (which was once available from Cohen's of Kettering), the fuselage of gliders, old buses, etc. Invariably ends in tears.

Kinema in the Woods: Woodhall Spa is a Lincolnshire watering place which had its heyday between about 1890 and 1920. It was built on sandy heathland and its abundant sylviculture is determined by that terrain: pines, rhododendrons, birches, broom – it is akin to Arcachon with villas sheltered by the trees. The Dambusters of 617 Squadron were stationed nearby and are commemorated in the hotel whose bar they frequented along with Guy Gibson's labrador, Asterisk: his name may not be spoken. That hotel like most of the rest of the village is red brick with fanciful half-timbering. The most interesting building, however, is different. The Kinema in the Woods was built as a genteel sports pavilion and was extended when it was converted to its present use in the '20s. It has, to this day, the air of an inflated shed. It is wooden, eccentrically proportioned, with a bizarre ground plan. There is no other cinema like it in the UK.

London Alps: Tatsfield is one of the many shack colonies which grew up on the North Downs in the aftermath of the First World War. It was advertised, in earnest, as being situated in the London Alps. Former soldiers were able to obtain grants to start smallholdings. The Bethersden company called Colt, which sold prefabricated wooden chicken coops, soon discovered that its structures were being used as domestic dwellings and branched out into supplying houses. Shack colonies – or plotlands – have disappeared for two main reasons. Successive planning laws have militated against them. The land on which sheds and shacks were built has increased in value in the way that the properties themselves have not. So where there were ad-hoc structures built of what could be begged or recycled or found for free there are now dream bungalows. The largest concentrations of extant shacks are on the eastern end of Sheppey, beside the Severn between Bewdley and Bridgnorth, at Jaywick Sands near Clacton (where the streets are named after forgotten marques of British car: Lanchester, Alvis, etc.).

Men's secrets: A garden shed was, perhaps still is, a man's private domaine, a man's cave, a grounded tree house, a haven. Whatever went on in garden sheds was covert. And it was meant to stay that way; hence the elaborate locks with which shed owners protected their little world from the prying eyes of their family. The hobbies pursued in sheds were legion. Collecting (beer labels, matchboxes, cigarette cards, pebbles, etc., etc.). Tinkering – the Midlands, especially, had a tradition, killed off by the introduction of the MoT test, of men who built cars called 'specials', bodged with a Birmingham screwdriver from components of other cars. The roads of the Black Country were graced on Sundays by these wonderfully Heath Robinson contraptions. Radio: until the introduction of the transistor radio hams would construct their own receivers. Astronomy: this requires a skylight. Light relief: the father of a childhood friend had a shed piled with boxes crammed with copies of *Health and Efficiency* – a naturist magazine, daring in its day – which published photos of airbrushed nudes throwing beach balls.

(2002)

[SCRIPT #5]
ISLE OF RUST (OFF-KILTER, PART 3, 2009)

PART ONE

EMETIC PICTURE POSTCARD PHOTOGRAPHY — DANGLING BRANCHES, ULTRA-CUTE, OVER-COLOURED. STIRLING — CASTLE AND ENVIRONS. HERITAGE SCOTLAND AT ITS MOST OBVIOUSLY PRETTY. PEOPLED BY TOURISTS OR, PREFERABLY, GROTESQUE PRETEND TOURISTS — JM PLAYS ALL OF THESE. CASTLE CAR PARK. COACHES. TOUR GROUPS. [CM: FIGURES IN HOODS. MAYBE SYMBOLIC REPRESENTATIONS — EG, AMBULANCE SIREN AND FLASH OF VEHICLE. BLOOD-SPLATTERED CLOTHES. LORRY RADIATOR. SYBILLANT ENCOUNTER IN TOILET. WORK BOOTS. CHEF'S CLEAVER, ETC.] POSTCARDS HOME. TO DES MOINES. ALICE SPRINGS. JO'BURG. QUEBEC (IN FRENCH). V/O:

> Why do they do it? Why do seen-it-all A&E personnel from Auckland, granite-faced truckers from Council Bluffs, cartilage-splattered abattoir operatives from Manitoba, brutal diesels from Durban, butch sous-chefs from Wagga Wagga, hard-cottaging marine mechanics from the Tigre Delta, leather-skinned coiffeuses from Port Moresby, cynical lawyers from Boston . . .

CHURCH, GRAVEYARD. STIRLING GRAVEYARD OFF LOWERCASTLE HILL/GOWAN HILL (NOT MAIN STIRLING CEMETERY)

> Why *do* they do it — why do they come to Scotland in search of their ancestors who are no more than names? They pore over all but illegible faded copperplate in parish registers with no regard for the damage inflicted on they eyes.
>
> They scrape the lichen from headstones.

DOUNE CASTLE FROM NEAR DEANSTON DISTILLERY: RIVER TEITH AND BRIDGE. IDYLLIC PICTURESQUE COUNTRY. PTC (INDICATES SHEEP WITH NOD)

> They walk beside burns where someone with rickets who happened two centuries ago to share 100 per cent of their mother's surname and 1.575 per cent of their DNA might — just might — have walked in inadequate footwear past that sheep's ancestor.

QUICK CUT TO SHEEP — AND BACK AGAIN. PTC CONTINUES

> What is the matter with them? What is this obsession with *roots*? And why is Scotland world leader in the roots caper?

THE TARTAN GLOBE — NEWFOUNDLAND, CANADA, NEW CALEDONIA, AUSTRALIA, ROME, LONDON, SOUTH AFRICA, CALCUTTA, ETC., MOSTLY PTC

> Why Scotland? Because Scotland's chief export — like the Basque provinces', like the Auvergne's, like the Campania's — has always been people: brain, muscle, cannon-fodder: powerful Scots send humbler Scots to die in distant countries. People leave in search of a living wage, a better life, the chance of material improvement. It's largely a matter of economic survival; partly, too, of the desire to seek a bigger stage. Scotland has seldom been able to provide for all of its people. So they make a hard-nosed decision to leave.

DUNBLANE: CATHEDRAL WITH MUSEUM IN BACKGROUND

> And their descendants make a soft-nosed decision to come back, coo at the quaintness and delude ourselves that they are somehow

Isle of Rust

connected to this place . . .

. . . The stuff in their veins is blood group Mac. When they bleed they bleed tartan blood.

DUNBLANE: LEIGHTON LIBRARY

Scotland's tourist industry naturally encourages this genealogical enquiry. Indeed Scottish tourism is founded on the obsession with burrowing for old bones. This year is the year of something called Homecoming, a summer-long orgy of golf, whisky, tartan, cabers, kilts, pipes: a parallel world of cliché, pure corn, fabrication — and income.

The internet has been a great boon to fans of pornography and amateur genealogists.

One pursuit will make you go blind . . .

STILL. MONTAGE. MR JACQUI SMITH IN FRONT OF GIANT KLEENEX BOX.
LEGEND: 'NO MESSING' IN OFF-WHITE SCRIPT WHICH LOOKS FONDANT, DRIPPING

. . . the other will merely boost sales of Superstrength Kleenex Mansize New Labour Expenses Wipes.

ABSTRACTED FLESH. 'DO IT TO ME ONE MORE TIME' BY CAPTAIN AND TENILLE

The principle of pornography is incremental.

SHEEP WITH PIXELATED FACE

There must always be more — more participants, more contortions, more sheep, more sheep in burkas.

MONTAGE: OLD PHOTOS MULTIPLYING; MUMMIES (SAVOCA, SICILY); NEANDERTHAL MAN.

The same applies to genealogy. The past has a past has a past has a past. It's not enough to stop when you've discovered your great-great-great-grandmother's birth certificate — you have to find her mother and her father and her father's father's father's christening mug with the dent from the Big Bang . . .

It's a sort of addiction.

DOUNE CASTLE

Best never to get started — for at the very heart of this dismal enterprise lies a form of recessive communitarianism, of ancestral tribalism, of a literal backwardness. We demean ourselves if we define ourselves by the community that we supposedly belong to — whether that alleged community is determined by skin colour, religion, political persuasion, football club, or any other criterion.

STIRLING: MAR'S WARK

This coarse pigeon-holing is like apartheid. It is the enemy of miscegenation, social mobility, assimilation: it exacerbates differences, indeed it manufactures differences.

LODGING OPPOSITE MAR'S WARK

It's an instrument of inequity. And in tracking down our ancestors we are simply subscribing to another way of defining ourselves by group or clan. We are creating a new file, a new enclosure, a new separateness. We're also kidding ourselves.

SCRIPT 5

STIRLING: OLD JAIL HOUSE

Telling ourselves a lie of the old days. We have the good fortune
to speak a language which — unlike Arabic or Sicilian — has a
future tense.

STIRLING: ST NINIANS CHURCH. DANGLE. PTC

My maternal great-grandparents were both born and brought up
in the Stirling suburb of St Ninians. My mother spent childhood
holidays with their extended families in the nearby burghs of
Dunblane and Bridge of Allan . . .

BRIDGE OF ALLAN: KENILWORTH ROAD, CHALTON ROAD

. . . whose grand villas they could never possibly have aspired
to.

BEAUTIFUL LANDSCAPE SHOT IN FINTRY HILLS SOUTH OF DOUNE — MUST BE GREEN, DOUCE

She and her sister had to sleep in a button-bed, a niche next to
the oven. It was an arrangement akin to that used by Kazaks and
Turkmenis who placed palette-like constructions over ovens. This
ad-hoc bucolic alcove would come to be formalised as a built-in
feature of many Glasgow tenements.

OLD STIRLING HIGH SCHOOL, SPITTAL STREET (NOW A HOTEL)

Baird, McInnes, MacAllister, Taylor . . . I know the names of my
forebears but they do not touch me.

Stirling High School is by the architect James Maclaren who
died in his mid-thirties, leaving a mere handful of exquisite
works. It is infinitely more interesting to me than tiresome tales
of the Taylor brothers' elaborate ruses to prevent their parents
discovering their chronic truancy from the school.

SEVERAL DIFFERENT VIEWS OF THE WALLACE MONUMENT — ALL FROM DISTANCE, NO C/U

The Wallace Monument's archivist is unable to confirm that my
great-great-grandfather was its first keeper. That, however, was
what my mother told me. Keeper means caretaker, janitor. Hardly,
then, a boast that anyone would bother to make up, so I guess it's
true — if unproven.

That frail familial link does not dissuade me from the
conviction that this astounding structure commemorates a murderous
medieval terrorist. It is the supreme architectural emblem of
Scotland's dodgily fabricated past. Brute force in the service of
gross sentimentality and mawkish victimhood.

DUNBLANE CATHEDRAL FROM ACROSS RIVER. DANGLE. PTC

The lavishly false heredities that Homecomers eagerly weave for
themselves are as predictable and as bogus as tartan and Highland
games. They perpetuate the myth of a land as sloppily counterfeit
as Merry England, a land that hardly was.

A dead land, a millstone — all those Bairds, MacInnes, Taylors
stretching back to elemental mulch.

There are other Scotlands — vital, *living*, untarnished by
Walter Scottishness: places of the strangest energy and greatest
delight. Time for a rebirth . . .

JM WALKS AWAY HOLDING DANGLE.
FALKIRK WHEEL (THOUGH WE DON'T REALISE IT) JM C/U — BACKGROUND
IMPOSSIBLE TO DISCERN

> Places that have no ancestral claim on us, where the yolk of a
> ready-made collective history is absent. Which we can talk of in
> the future perfect rather than the past historic.

AT THIS POINT REVEAL THAT JM IS IN A BOAT ASCENDING THE WHEEL — MORE
EFFECTIVE IF WE DON'T SEE THE WHEEL IN ITS ENTIRETY

> Places that we can elect to *go towards* rather than *come from*.
> Places which are potential — where anything is possible, where
> everything is waiting to happen.

SCREEN GOES TO BLINDING WHITE ACCOMPANIED BY WHITE NOISE. THEN
SCREECHING SEAGULLS.

PART TWO

LEWIS/HARRIS. NATURAL MAGNIFICENCE — THE OBVERSE OF CUTE. EPIC
CLOUDSCAPES. LOCHS. LUNAR ROCKSCAPES. THE SEA. MOUNTAINS. C/U
LICHENS, STONES, SAND. IN THIS PART ABSOLUTELY NO SIGNS OF HUMAN
HABITATION. NO HUMANS. PLACE LOOKS DESERTED. CHROMATICALLY AUSTERE
GOING TOWARDS MONOCHROME. MUSIC: HENZE SINFONIAS NO.1 + NO.2.
RAVEL TRIO. EXPLOIT ALL OF THE LOCATIONS BY SHOOTING IN DIFFERENT
DIRECTIONS, IN LATER PARTS WE SHALL SHOW HABITATIONS, ABANDONED
BUILDINGS AND VEHICLES ALL OF WHICH ARE TO BE FOUND EVERYWHERE. IN
THIS PART THEY NEED TO BE KEPT OUT OF FRAME. IN SUBSEQUENT PARTS WE
REPEAT THESE SHOTS BUT AMEND THEM WITH A FRAME THAT IS MARGINALLY
WIDER OR SLIGHTLY ALTERED TO SHOW WHAT WAS OMITTED HERE. SO THE SHOTS
NEED TO BE DISTINCTIVE ENOUGH TO BE REMEMBERED 30 MINUTES LATER.
WEST COAST OF LEWIS. A BOAT ON A SEA LOCH. INITIALLY ONLY SEA IS
VISIBLE.JM IS DRESSED IN DIFFERENT CLOTHES — TWEED JACKET, ETC.
(THOUGH THE BOAT SHOULD SEEM THE SAME — AS THOUGH HE HAS TRAVELLED
WHOLE WAY ON IT). BOAT POV LAND APPROACHING

> There are places that we choose to attach ourselves to . . . Or
> which choose us. Whose *genius loci* is so potent that it incites
> us to love them. The spirit of the place renders us incapable of
> resistance.

LOOKING SOUTH FROM A858, MAP 8 REF NB 300 298. (NO HABITATIONS VISIBLE)

> That spirit is not spiritual. It is, rather, a combination of
> geological, meteorological and man-made circumstances. It's the
> enveloping presence, *the essence* which incites sensations of
> transcendence, awe and bliss. These sensations do not depend on
> the exhortations and boasts of some intercessionary deity.

PARODY OF CSI-STYLE SHOT OF WORM

> . . . of a wrathful worm-god slithering through the eustachian
> tube to demand oblations for having worked to create this . . .
> this paradise.

MAP 13 REF NB 070 343. CANYON (WITHOUT SHOWING ROAD)

> This is the world's work of countless millions of years against
> which we measure ourselves and are found wanting. Passers-through
> and all that. A blink in the eye of aeons.

MAP 13 REF NB OIO 308. LOOKING E & SE

Here, at the north-westernmost periphery of Europe is what feels
like a presage of the future . . . the distant future . . . the
furthest future . . . After which there'll be no future at all.
This is the Island of Rust — known, too, as Lewis and Harris.
It is a blueprint, a working model of the day which will have no
tomorrow.

MAP 14 REF NB 374 144

It is an island of countless mysteries, enduring mysteries.

MAP 13 REF NB 152 388

The light is singular . . . constantly changing . . . The island
is porous — like petrified sponge . . . Its landscape is not
*land*scape. It is liquid . . . Its waterscape is not *water*scape.
It is solid . . . There are areas of glacial erosion . . . and of
sedimentary transportation.

MAP 8 REF NB 520 665. BUTT OF LEWIS

The metamorphic gneiss [pron: <u>nace</u>] is 3,000 million years old.
It is the result of several deformations. Nace, then, maight be
said to be laike the Morningsaide accent. It is the oldest rock
in Scotland — if this is Scotland, which I doubt. Obviously it is
administratively Scotland but otherwise . . .

MAP 14 REF NB 401 183. BOULDERS

During the last ice age the alternate pressures of freezing and
thawing, expansion and contraction, combined with erosion to
turn what had been monolithic into a moraine of millions of loose
boulders.
In Hindu mythology similar rockscapes were formed by monkey gods
— bad-tempered and, evidently, very strong brothers frenziedly
hurling boulders at each other.
　　Not here. There are no fratricidal monkey gods on the Island of
Rust. Oh no!
　　Calvinistic Presbyterian mythology — which claims to be
something other than mythology, but isn't — decrees that God, the
one God, created this Calvinistic Presbyterian island — boulders,
lochs and all — in six days.
　　Fair enough, we all know the lad's got a great engine on him —
he's 100 per cent work-rate, a celestial Scholsey.
　　But . . . what if God's day is anyone else's couple of million
years. What if his conception of clock-time and calendars does not
correspond to the mundane artifices invented by emperors, popes and
Swiss watchmakers?

**BEACH AND MACHAIR. MAP 13 REF 083 365. MALE SINGER. NEO-CELTIC
OUTFIT. SONGS OF PRAISE-STYLE. SONG IN GAELIC WITH THESE SUBTITLES
(Translation required)**

> *On the seventh day he went down to the sea.*
> *He made water shimmering blue like his truelove's eyes.*
> *He made sand fine and yellow as his truelove's locks.*
> *He made machair so soft, his truelove lay down to sleep*
> *forever.*

BEACH. PTC
Which may not have been part of the grand plan. No doubt he too
hurled boulders in his wrathful grief.
The quaint misattribution of this island's provenance, the
denial of fact in favour of creation myth, should not bother us.
The wonders of evolution are infinitely more subtle and fascinating
than the coarse fabrications proposed by those repositories of
ignorance called sacred texts — forgivable 2,000 years ago as
a means of trying to understand our physical surroundings, but
inexcusable today.

MACHAIR MAP 13 REF NB 105 358. (MANY OTHER EXAMPLES)
Machair — pretty much peculiar to the Atlantic coast — is a work
of millions of years of sand and crushed shells being blown and
swept on to acidic peat. This creates a thin but exceptionally
fertile soil which is further enriched by guano of ground-nesting
birds — oystercatchers, sandpipers, terns, lapwings, redshanks,
corncrakes, peewits — and by grazing cattle's dung.
Does this mean that cattle are coprophagic?
Let us dwell, rather, on orchids, clovers, daisies.

DUNES. MAP 13 REF NB 105 358. (MANY OTHER EXAMPLES)
And on the dunes between the sea and the machair where marram
grass grows — a sort of armature, a set of reinforcing rods, that
inhibits the shift of sand.

PEAT BOG. VIRTUALLY ANYWHERE ON NORTHERN LEWIS
Peat occurs in exceptionally wet temperate or cold climates where
vegetable matter, notably nonflowering plants such as sphagnum
moss, is prevented from totally decomposing by an excess of water.
Certain species of sphagnum have spore capsules which shrink to
the point where the air within them is so compressed that they
explode — noisily. This may partially account for the belief, the
literal belief, that peat moors are inhabited by boisterous bog
sprites which infects so much twilit Celtic folklore.

**NEAR DARKNESS. HARDLY DISCERNIBLE SPRITES OF HUGELY DIFFERING SIZES
WITH BROWN/GREEN FACES AND BODIES COVERED IN MOSS. JERKING AS IF WITH
ST VITUS' DANCE. SOME LUMBERING AND FAR FROM ELFIN. SOME DELICATE.
NON-HUMAN YET NON-ANIMAL**
Peat used to roast barley famously lends the whisky made from
that malt an operating-theatre bouquet. When taken in sufficient
quantity that liquor can doubtless convince the most obdurately
rational mind that green creatures dance by night on the desolate
moors.
These creatures — some amiable, some malevolent, some the
spirits of the dead, some fallen from heaven — were thought to
inhabit unseen places beneath the earth and under pools which
would be revealed as the world ended.
Peat is dug and dried for slow-burning fuel. It is host to many
heathers.

EXTREME DETAIL. FRAME FULL OF HEATHER > C/U MOSS IN POND
Because of a lack of oxygen and an abundance of acids it preserves

animal bodies and human bodies — the bodies of those who have stumbled into its bogs and those who have been murdered or ritually executed.

Almost 2,000 bodies have so far been discovered in such diverse places as South Uist, Cheshire, Jutland, Varberg in Sweden, County Meath, Groningen in the northern Netherlands.

But none has been found yet far on the Isle of Rust. The likely reason for this is that the topography is inimical to the industrialised peat cutting which has revealed most of the bodies so far. But we can go on hoping.

The ligature or noose will often be preserved too. The process is closer to pickling than to mummification. And it is rapid. Though most of the bog bodies that have been discovered are more than a thousand years old, a Second World War soldier has been exhumed in a perfect state of preservation.

CAMERA SUDDENLY PLUNGES INTO WATER. THE WATER IS BROWN BUT CLEAR

These bogs are full of future potential. It is not inconceivable that devices which can detect the limbs of the dead beneath metres of peat are in an advanced state of development.

So we can meet our ancestors by design rather than by the chance of a plough or harrow. And if we cannot meet them in the flesh we can at least meet them in the leather. It beats reading a name on a birth certificate.

THE CAMERA ALIGHTS UPON A BOG BODY FOETALLY CURLED WITH GAROTTE ROUND NECK. MOST PHOTOS SHOW BODIES WHICH HAVE SHRUNK AND WHICH LOOK AS THOUGH THEY ARE MADE OF LEATHER. (PROP REQUIRED — OR WILLING PERSON)

There are golden eagles in the mountains. Soaring no doubt.

And buzzards drifting on thermals looking to transform a harmless mammal into lunch.

White-tailed sea eagles — whose wingspan is over eight feet, two and a half metres — patrol the coast. They live on fish with carrion on the side. There is little competition for prey with golden eagles who eat rabbit, hare, grouse — which makes them unpopular with gamekeepers and hunters.

C/U OF THESE SPECIES. FACES PIXELATED IN SOME CASES, BLACK STRIPE IN OTHERS TO PREVENT IDENTIFICATION. THE CLAWS AND LOWER LEGS OF A BIRD WITH A TARTAN BLANKET OVER ITS HEAD, WINGS AND BODY WITH THE LEGEND 'HELPING WITH ENQUIRIES'

There are ospreys . . . hen harriers which are white-bellied buzzards . . . peregrine falcons . . . kestrels. . . shags. . . fulmars who can eject an oil reeking of putrefying fish from their stomach to fend off attackers: this oil swiftly solidifies to a sort of wax which can be incapacitating. Beside them hoodie crows who merely pluck out the eyes of lambs seem paragons of moderation.

It should be noted that all of these species are recidivist sociopaths, murderous delinquents even into their maturity — they are thus bad role models for any youngsters watching.

They may inhabit paradise but they make it hell for their prey.

DEER

There are deer everywhere. They don't fly — save in France, where a kite is called a flying stag, *un cerf volant*.

There are hare, rabbits, otters, bottle-nosed dolphins, seals, basking sharks that are allegedly harmless though it's hardly worth the risk of finding out, brown trout — the ones with the white flesh which taste so much better than gaudy rainbow trout, salmon, cannibalistic crustaceans.

BARVAS MOOR. MAP 8 REF NB 242 390

There are slow worms — which are legless lizards. But no snakes.

So despite the ancient belief that Barvas Moor was the Garden of Eden it obviously wasn't. Unless, that is, the snake — as maligned as Judas, but as necessary to the story's structure - moved on once it had performed its role.

No doubt some of the many other claimants to Edenhood — the source of the Tigris, a spring in Jerusalem and so on — are more climatically plausible and, of course, they are in the Middle East, font of all monotheistic mumbo-jumbo. But are any of them as primal, as primitive, as truly paradisiac as this elemental moor whose Gaelic name is Barabhas — like that of the outlaw or terrorist whom the mob had necessarily to wish pardoned in order that it be Jesus who died.

SUDDEN DARKNESS

It is more Calvary than Eden. Abattoir rather than cradle.

TERRIBLE NOISES OF CATTLE SQUEALING AS THEY DIE

PART THREE

LEWIS/HARRIS. FIRST TIME THAT WE SEE SIGNS OF (ANCIENT) HUMAN INTERVENTIONS ON THE ISLAND. PHOTOGRAPHY SHOULD STILL AIM FOR MAGNIFICENCE, SUBLIMITY. NO HOUSES, NO ROADS, NO PYLONS, NO CARS. ALL STANDING STONES SHOT IN DIFFERENT LIGHTS, DIFFERENT CLOUDS, DIFFERENT SCALES.

SCARISTA (SINGLE STONE BY BEACH) WITH SEA IN BACKGROUND. MAP 18 REF NG 020 939

This is the north-western extremity of Europe.

It prompts wonder.

It prompted wonder and bewilderment to our distant, preliterate forebears.

SUBTITLE

> *Oy atpqayrf epmfrt smf nreozfrtqrmy*
> *yp pit, smf fodysmy atrzoyrtsye*
> *yp pit. Gptnrstd?*

CLACH MHICLEOID (TALL SINGLE STONE DRAMATIC POSITION ON PROMONTORY). MAP 18 REF NG 041 973

And it prompted too the urge to investigate and measure what was around them.

STONES BESIDE BRIDGE TO GREAT BERNERA. MAP 13 REF NB 163 343

They had no recourse to the supernatural.

SCRIPT 5

AIRIGH NE BEINNE BIGE (SMALL STANDING STONE). MAP 8 REF NB 222 356
They calibrated their place in their surroundings by marking *the actual:*

CLACH AN TRUISHEAL STANDING STONE (THE TALLEST IN SCOTLAND). MAP 8 REF NB 375 537
where the sun sets at the vernal equinox;

CALLANISH 2 & 3 ADJACENT STONE CIRCLES. MAP 8 REF NB 211 330
where it rises at the summer solstice; the positions of constellations, the vagaries of the moon.

CALLANISH 1 MAP 13 REF NB 213 330 (SHOOT IN DIFFERENT LIGHTS, IN SEVERAL STAGES. THE NEARBY HOUSES MUST BE KEPT OUT OF FRAME)
This was practical. This was the first germ of what would become science. Curiosity sated by investigation rather than by crude hypotheses.

These sites may have been calendars of a sort. Unlike the much more extensive menhirs at Carnac in southern Brittany it's improbable that they were instruments to foretell earth tremors — no area of Britain is less susceptible to seismic activity.

It's likely that they were also places of celebration . . .

. . . of the sun which provided power and warmth, greater warmth 5,000 years ago than it provides here today

. . . of the moon which determined tides and illumined the interminable winter nights.

Celebration isn't the same as worship.

The presence of ritual — which appeals to a certain cast of mind with a fondness for occluded drama and ancient conspiracies — cannot be confirmed. Nor for that matter denied, though it is worth noting that while these stones are 5,000 years old, Wicca, Druidism and Paganism are little more than 300 years old and much ancient ritual derives from polytechnics in the 1970s and '80s.

FEMALE SINGER IN NEO-CELTIC GARB ON SHORE, WATER LAPPING NEAR FEET. SONGS OF PRAISE STYLE. SONG IN GAELIC (WITH SOME ENGLISH WORDS FOR WHICH THERE IS NO GAELIC EQUIVALENT) SUBTITLES:

> *Theoretical models of Wicca inclusion/exclusion,*
> *Workshopping Paganism's assessment strategies,*
> *Multiple-choice date in Druidical knowledge sourcing.*
> *These I learnt in my third year in the Celtic Studies*
> *Department.*

CALLANISH JM MULTIPLIED. LOCKED-OFF
It was about that time, too, that Celtic folk rock established itself here and in Brittany, Ireland and Galicia — a cleverly marketed kind of pop music based on nationalistic sentimentalism, on a sense of victimhood, on the Celtic world's secessionist destiny and on an antipathy to the primacy of the English, French and Castilian languages. The entire package is as rooted in historical certainty as tartan and Ossian's epic. Which doesn't mean that it is not real. Bogusness has its own reality.

The nature of these sites' funerary roles and their uses as commemorative cemeteries is moot.

Isle of Rust

But . . . the disposition of boulders by evolution is visually random. We obviously witness it on this island. We see it too on the Breton coast near Carnac and in the sarsens on Marlborough Downs near Avebury — sarsens that were often mistaken for sheep. Which must have caused some embarrassing injuries.

The presumption that henges and menhirs and dolmens are sacred is based more in the interpretative and speculative biases of the early modern age which first concerned itself with such phenomena than in irrefutable evidence. The ascription of fancifully conceived primitive belief systems to our ancestors is wrongheaded.

The contention that the religious instinct was paramount in all peoples and in all epochs is taken for granted: which is a good reason to question it.

Why should ancient man have been any different from us — some of us are credulous, more are not. In societies where there is no cultural coercion or political obligation we believe or not according to our conscience and intellect.

It is patronising and ahistorical to assume — as neo-Celtic musicians appear to — that the ancient Gaelic world was monocultural. Some sites may have been sacred, others secular.

What *is* indisputable about these sites is that they tell us mankind is a maker. *That* is what is paramount. They do not tell us why they were made. And we shall never know for sure.

Given a few pebbles and an idle moment many of us will position them in lines, grade them according to size, form elementary shapes with them. This could very likely be diagnosed as symptomatic of obsessive compulsive disorder. Which it was not called all those millennia ago.

Mankind obsessively devises patterns. Maybe these sites have no meaning. They are patterns made for their own sake or for the sake of controlling their environment. Maybe they are pre-Euclydian exercises in creating a non-representational kind of art, specifically what we know as land art. An art based in the collection and regimentation of local materials which happen, in the case of gneiss or granite, to be durable. There may have been thousands of kindred sites constructed in areas of friable material — chalk or sandstone or gritstone or wood. Sites which have not come down to us because they rapidly eroded or rotted leaving no trace. Our knowledge is dependent on geological chance.

DRAMATIC CLIFFS, CRASHING WAVES
The most circumstantially persuasive argument for these neolithic sites being sacred rather than the result of collective OCD is their very location.

Which is topographically extreme, climatically extreme.

GAUNT LANDSCAPE EMPHASISING ISOLATION
It is in such *exaggerated* places that religions are born and it's where they flourish: deserts, steppes, tundra, mountains, salt flats, aridity, permafrost — these are propitious conditions for the shift into hallucinatory irreason that is dignified by the name of faith. The more extreme a landscape or climate the greater its effects on its inhabitants.

Temperate climes are, happily, less effective cradles of what David Hume called 'sick men's dreams'. They promote the greatest of human virtues, cynicism — in its true sense. The UK, France and Denmark are the least observant countries in western Europe. For the moment.

RODEL CHURCH, SOUTH HARRIS. MAP 18 REF NB 048 832
The Isle of Rust is entirely atypical of the UK as it is today. There are 50 churches — one for every 400 people — in a population of 22,000. Over a third of those people attend a church every Sunday. A vastly larger proportion than anywhere else in Britain. Though it must be noted, too, that two thirds of the population is not observant.

This church at Rodel, founded by a gentleman called Hunchbacked Alasdair, is the most impressive medieval building on Rust. But it would be nothing special elsewhere in north Britain. It is the exception that proves the rule of the island's churches' unfailing banality.

TYPICAL CHURCHES, CHAPELS, PRAYER BOXES AND THEIR OVERFLOWING CAR PARKS (CAR PARKS WILL OBVIOUSLY HAVE TO BE FILMED ON SUNDAY WHEN THEY ARE CROWDED AND CARS ARE ALSO LEFT ON VERGES). WE CAN HEAR GAELIC PSALMS WHETHER OR NOT THERE IS A SERVICE TAKING PLACE. THERE ARE SERVICES THROUGHOUT SUNDAY IN BOTH ENGLISH AND GAELIC.
GRABHAIR 12.00 GAELIC, 18.00 ENGLISH (REMOTE CHURCH WITH LARGE AND FULL CAR PARK). MAP 14 REF NB 375 156
And unfailing dreariness. The idea that a church might be visually enjoyable is alien to Calvinism. *Enjoyment?!*

Indeed while Presbyterian sects are forever diverging over doctrinal nuances invisible to the lay observer they are united in their contempt for the pursuit of beauty. The jezebel called architecture is a papist strumpet. Icons are sacrilegious. Music is the devil's work. Ritual is theatre — which is an abomination. To delight the eye is an offence against modesty. These paltry insipid buildings are slights to mankind: they are incitements to lead a life in fear of life. They are the very obverse of the great stone circles which are hymns to mankind.

CROSSBOST 12.00 GAELIC, 18.00 ENGLISH. MAP 14 REF NB 387 246
Unlike Anglicanism Calvinism demands commitment. It is a matter of conviction and unconditional acquiescence. It affects the way its adherents lead their constrained life.

Even though it is today a minority pathology — just — it still attempts to create a cultural tyranny and to impose its morals and wacky brand of tolerance on the majority — who can think for itself.

KINLOCH CHURCH 12.00 GAELIC, 18.00 ENGLISH. MAP 14 REF NB 322 221
The most evident manifestation of the sacred's contamination of the secular is on Sundays. The Lord's day. The Sabbath.

Even though, despite the clamorous minority's objections, a Sunday ferry now sails, the island closes down. The majority of the vehicles to be seen are those of church-goers. Work of any sort is disapproved of, but so is leisure — don't tend your garden, don't go for a walk, don't listen to music, don't go fishing, don't swim. Don't . . . don't . . . don't . . . Proscribe this. Prohibit that.

STORNOWAY. DAY. MAY NEED TO SHOOT EARLY MORNING. VARIETY OF DESERTED STREETS. CLOSED SHOPS AND PUBS. NO TRAFFIC. NO PEDESTRIANS. SILENCE — HOLD FOR LONGER THAN MIGHT BE EXPECTED

The cause may be delusional credulousness — but it results in the greatest of modern luxuries. Silence. Calm. Something close to serenity.

It causes us to realise that there is a secular case for sabbatarianism. For, anyway, a day that is different — not necessarily Sunday — a day when there is nowhere to lout about at lunchtime, to chant in the afternoon, to get into a stramash, to buy pointless things in an atmosphere where your ears hurt because of drivelling pop-noise and your nose hurts because of burger stench.

It is, of course, conceivable that the Sabbath is uncomplainingly observed because much of the population is in no condition to do anything but slump after the previous evening's nordic exertions.

FRANTIC LOUD HEAVY METAL. STORNOWAY. NIGHT — OR PARTIAL DARKNESS. ROWDY STREETS. BARS

Our race's most widespread means of escapism are religious faith and insensate intoxication. For a totally satisfactory result combine the two in the manner of dervishes, berserkers, shamans.

SERIOUS LEGLESSNESS. PEOPLE THROWING UP. POOLS OF VOMIT

Oh! They certainly know how to escape themselves.

A kip in the pews and a copious supply of Irn-Bru is just the ticket after a gallon of heavy plus chasers.

Until the beginning of this decade all alcohol was imported.

ANDREW RIBBENS WITH BOTTLE OF BERSERKER. FORMAL, ARTIFICIAL - LIKE VICTORIAN PORTRAIT. (HEBRIDEAN BREWERY, BELLS ROAD, STORNOWAY)

In 2001 Andrew Ribbens established the Hebridean Brewery . . .

MARCO TAYBURN WITH BOTTLE OF WHISKY. ARMCHAIR ON BEACH. (ABHAINN DEARG DISTILLERY, CARNISH) MAP 13 REF NB 029 324

And last year Marco Tayburn set up the Aveen Jarraeck Distillery. The Red River Distillery. Apt if belated, for whisky is the only Gaelic word most of the world knows and this island has Scotland's largest concentration of Gaelic speakers —

GAELIC SIGNS VERY C/U SO THEY DON'T READ

If, that is, one discounts Glasgow's Gaelic-speaking ghetto of: Gaelic language media dimensioning operatives, Gaelic language propagation strategists, Gaelic language participation leaders, Gaelic language education vanguard light managers, Gaelic language ambit teachers' teachers, Gaelic language vision lobbyists, Gaelic language sit-on-my-face-book inscribers, Gaelic language horizontal sausaging matrix implementers, Gaelic language translatability facilitators, Gaelic language fellation development officers, Gaelic language signage advisors and Gaelic language supremacists.

The self-important sociolinguistic industry is, of course, risible but the language it supports most certainly isn't. That only 60,000 people speak it is no reason to allow it to wither on the tongue. The fact that it might be more *practical* to allow the English language to vanquish it is of no moment. Practicality is an infirm criterion.

Think of Gaelic as a threatened system of cultural ecology or compare it to a species nearing extinction or a spirituous liquor which might disappear because the recipe was lost. As in a way it was. There is, of course, a *generalised* Gaelic culture as bogus as the plaid and pipes of the Highlands — bardic and bearded with a brand of folk rock and a typeface that links Galicia, Brittany, western Ireland and the Hebrides in a cisatlantic union of minoritarian defiance.

C/U GUGA OR BUDGIES

But there is much that is peculiar to Rust.

Guga, for instance. Gannets. Ten men make the hazardous forty-mile journey each August to the uninhabited island of Sula Sgeir. They spend a fortnight trapping and salting around 2,000 chicks. This was once alimentary necessity. The meat apparently tastes like chicken fed on fish meal. I shall give it a miss, like I gave fermented trout a miss.

WHALING STATION RUINS. MAP 14 NB 125 044

I wish I'd given whale a miss. But I was only three — at a picnic on a Dorset beach. It would have been discourteous to refuse the lumps of grey gelatinous meat beneath an undercooked pastry crust. It was like chewing an eraser.

JAMIE FLYNN SONG (GAELIC INSTRUMENTS AS WELL AS LYRICS — BUT RECOGNISABLE AS 'SAILING'. TRANSLATION REQUIRED) SUBTITLES

> *I am whaling*
> *I am whaling*
> *Very soaked, Mum*
> *To the skin, Mum*
> *Harpoon's sharpened*
> *Did Cowdenbeath win?*
> *I am whaling*

This whaling station was operational from 1904 till a few years after the Second World War.

WHALE'S JAW BONE BESIDE MAIN ROAD. MAP 8 REF NB 291 476

Whales were processed for ambergris — a constituent of perfumes — for oil and wax — which were used in soap and candles, and margarine, which explains a lot. The meat — cheap, plentiful, unpleasant — was a staple part of islanders' diet.

FISH FARM, CREELS AND PENS, BESIDE BRIDGE TO GREAT BERNERA. MAP 13 REF NB 167 342

A lot of food is still produced here. But much of it is beyond the inhabitants' means — scallops, lobster, crayfish and so on. But not salmon which is so intensively farmed that its price has plummeted along with its quality.

Isle of Rust

The lochs are factories.

**MARINE HARVEST, MIABHAIG PIER, KIT, TRAPS, SHED, ETC. MAP 13 REF NB
091 343**

And like factories they manifest what might be called an
industrial style — though that's pushing it a bit. Sub-utilitarian
is more like it.

EXTERIOR OF BUILDING, STORNOWAY FISH SMOKERS, SHELL STREET, STORNOWAY

Herrings and salmon are kippered, salmon is cold smoked and cured
as gravadlax.

**BUCKET OF BLOOD. [OR: BUTCHER MAKING BLACK PUDDING; PROBABLY BEST
TO SHOOT AT CHARLES MACLEOD'S PREMISES, ROPEWORK PARK, STORNOWAY -
PREMISES NEED TO BE DRAB OR INDUSTRIAL)]**

Macleod and Macleod, Charles Macleod, W. J. Macdonald, A. and R.
Morrison, A. France. All of these butchers make the best, the
truly original, the most authentic Stornoway black pudding — *marag
dhubh*. It is the peer of the best blutwurst, boudin noir and
morcilla. It can sit at the top table and hold its own.

CATTLE IN FIELD

Highland cattle. Crowdie cheese. Beef.

**SHEEP ON SALT MARSH, TAOBH TUATH, SOUTH HARRIS. MAP 18 REF NA 995
905. HIGH PITCHED BLEATING**

Sheep. Lamb. Hogget. Wether. Tweed.

PART THREE AND A HALF

**PHOTOGRAPHY. JOHN BRETT STYLE HEIGHTENED NATURALISM.
STORNOWAY, QUAY STREET. HARBOUR AND LEWS CASTLE BEYOND**

That toy fortress was built by Sir James Matheson in 1848. It's
in a style that a plutocrat might have chosen twenty-five years
earlier, long before Victoria's accession.

LEWS CASTLE

Which is not surprising because Matheson had been out east for
all that time. He was a trader. That is to say a drug dealer —
on a heroic scale that the Medellin cartel can only look on with
envious wonder.

This god-fearing teetotaller bought the entire island with a
fortune made from transporting opium from India to China and from
promoting addictive dependence in tens of thousands of people in
southern China.

It was on behalf of Matheson and his fellow dealers when they
were expelled by the Chinese government that Britain waged the
ignominious opium wars. Matheson should have been imprisoned.
Instead he received a baronetcy and, like many criminals before
and since, became a member of parliament whereupon he proposed
that the grim lump of rock called North Rona, sixty kilometres
north of here, should be turned into a high-security jail.

He built his castle on the site of the island's only distillery
— thus encouraging the use of illicit stills.

LEWS CASTLE PARK — SINGLE-LANE ROAD GOING WEST THROUGH RHODODENDRONS,

SHRUBS, ROCKY WOODLAND. IT HAS THE LOOK OF VICTORIAN HYPER-REALIST PAINTING. THIS SHOULD BE CAPTURED. MAP 8 REF NB 362 328

While the castle is, despite its size, rather trivial, the remarkable grounds lead us to understand that zoos and aviaries and parks and arboreta — despite their association with the Enlightenment's propagation of knowledge — were founded by men trying to play at God. Trying to mimic what was still widely considered to be his creation. They belong, sentimentally, to an age before reason. Or at least to an age before humility and reason told man that he was no more capable of controlling the world than a god had been of having created it.

RUINED CROFTS OF MID-NINETEENTH CENTURY. LOW WALLS, ALMOST ILLEGIBLE AS HOUSES. WE DO NOT WANT ONES WHICH HAVE BEEN PATCHED UP WITH CORRUGATED IRON. POSSIBLE: FUAIGH MOR (ISLAND OFF GT BERNERA - REQUIRES BOAT), MAP 13 REF NB 130 350 AND CALBOST, MAP 14 REF NB 409 169

In Matheson's native northern Highlands the clearances of crofters had been undertaken by landowners and clans chiefs in what they believed — or at least deluded themselves — was a spirit of philanthropic *noblesse oblige*: the so-called Improvements were as much a project of Enlightenment pedagogy as of agrarian capitalism.

CAVES ON FUAIGH MOR (AS ABOVE) OR ELSEWHERE

People who could barely read or write and who subsisted in insanitary bothies, burrows and caves were forcibly removed from the land — the land which hardly yielded them a living.

SHEEP — THREATENING EN MASSE, ALMOST SWARMING. CAMERA AT SHEEP'S EYE LEVEL. SHEEP EVERYWHERE INCLUDING: ARNISH MOOR, MAP 8 REF NB 367 284 AND ROAD GOING NORTH FROM MAP 14 REF NB 381 178

Some were provided with houses, jobs and education in the new grid-plan fisher-towns of the Sutherland and Caithness coasts, some were transported across the Atlantic. The moors they had inhabited were made over to sheep.

By the mid-nineteenth century all philanthropic pretence had been abandoned.

Matheson proceeded to treat the islanders — his islanders — with the same sensitivity that he had displayed in Canton, though they were not provided with the balmy solace of narcosis.

He did not build a fisher-town. In the early years of the 1850s almost 2,000 inhabitants were forcibly expatriated, mostly to Canada, their passage paid by Matheson. The suppression of whisky was calculated. The suppression of Gaelic was a clumsy consequence of the involuntary diaspora.

LONG STRAIGHT ROAD. NO HOUSES VISIBLE. RATHAD A'PHENTLAND LOOKING WEST. MAP 8 REF NB 355 340

What Matheson did build was roads — they provided work during the famine years of the 1840s. He intended that they should be used to transport peat so that it might be transformed into tar: the scheme was a nonstarter.

On the mainland such roads came to be known as destitution roads because they were built by the destitute and it was along

them the indigent marched away from destitution to their uncertain
future on the other side of the world.

The name is prescient.

Today they still often lead *to* destitution.

PART FOUR

**THE ISLAND STARTS TO LOOK LESS IDYLLIC. KITSCHY HOUSES, STORNOWAY.
EAST END OF MEMORIAL AVENUE AND WILLOWGLEN ROAD, A858, (100 M W. OF
PERCEVAL ROAD JUNCTION)**

Whatever wealth is generated is mostly well concealed. It is very
seldom manifest in the built environment. And when it is . . .

**DUALCHAS HOUSE, BREINIS (ON WEST COAST, SOUTH OF UIG), MAP 13 REF NA
993 267**

There are some works of architectural merit on the island. Well,
two. And they seem impertinently pretentious measured against the
startlingly abject norm.

**BANAL INHABITED HOUSES — ALL OF THEM GREY HARL. (NOT WRECKS OR
SHACKS). FORMALLY SHOT, STRAIGHT ON - AS THOUGH THE HOUSES ARE
POSING. SHOT TO EMPHASISE SIMILARITY AND INSIPIDITY. WE SHALL NEED
AT LEAST A DOZEN TO COVER V/O AND PTC. ANYWHERE, E.G. NORTH LEWIS
— THE TOWNSHIPS ALONG THE A 857 BARABHAS TO BUTT OF LEWIS ROAD AND
LEVERBURGH, MAP 18 REF NB 015 860, AND TAOBH TUATH, MAP 18 REF NA 980
900, AND CALLANISH — HOUSES SHOT FROM BETWEEN THE STONES**

This is perhaps the only place in the world whose townships and
villagescapes, urbanism and landscapes are wholly infected by the
Calvinist mentality . . .

that is, by a blindness to prettification . . .

an ignorance of any recognisable notion of beauty . . .

by an aesthetic bereavement so absolute that it is a sort of
insouciant anti-aesthetic.

In a way the everyday buildings are the very contrary of
Matheson's. They suggest that to compete with this most magnificent
terrain would be both hubristic and certain failure. So they
simply didn't bother to compete.

The unsurpassable strangeness of the island resides in the
chasmic gulf between the naturally evolved and the negligently
created, between scarp and scrap, between the sublime and the sub-
standard.

**RUN RIG FIELD SYSTEM. THERE ARE MANY EXAMPLES, E.G. 0.75 KM SW OF
ORASAIGH, MAP 14 REF NB 365 115, AND LIURBOST, MAP 14 REF NB 378 252,
AND 1 KM W. OF BAILE AILEIN, MAP 14 REF NB 275 206, AND GEARRAIDH
BHAIRD, MAP 14 REF NB 365 194, AND CALBOST, MAP 14 REF NB 416 169,
AND LEUMRABHAGH SHOT FROM POV, MAP 14 REF NB 378 113**

Not that the natural is invariably natural: it generally isn't.

The landscape most formed by man here is not that of menhirs
and dolmens. In fact *man* is moot in this context. The cable-
knitted hillsides are the work of the sheep that man introduced.
They sit alongside the field systems called run rig or lazy beds.
This corrugated pattern of drainage furrows — the runs; and of
fertile ridges — the rigs, is mostly in desuetude . . . but it is

so widespread that it is one of the defining characteristics of the landscape.

These systems, like stone circles, possess a kind of abstract integrity that often attaches to the purely practical.

PART FIVE

WRECKS, RUINS, BROKEN VEHICLES, ABANDONED COACHES, RUSTING BULLDOZERS, UNSEAWORTHY BOATS, ETC. THERE IS A WEALTH OF MATERIAL. BUT SHORT OF RESORTING TO SHOW AND TELL THERE IS NOT A LOT TO SAY. SO ALTHOUGH THIS SECTION IS VERBALLY BRIEF IT CAN BE TAKEN SLOWLY, ESPECIALLY GIVEN THE VISUAL STRENGTH OF IT ALL. PHOTOGRAPHY — VERY COLOURFUL, SATURATED. ESSENTIAL TO COMPOSE FRAMES THAT SHOW CORRUGATED IRON, MACHINERY, SCRAP, ETC., IN THEIR SURROUNDINGS. EMPHASISE THE CONTRAST BETWEEN NATURAL GRANDEUR AND SCRAP SQUALOR. MAP 14 REF NG 155 928, GREOSABHAGH. SCRAPYARD SPILLING ON TO ROAD

The steadings and smallholdings, too, are purely practical.

But where the field systems are ordered the buildings are extraordinarily chaotic.

BAILE AILEIN ON A859, MAP 14 REF NB 300 210, CARAVANS, SHACKS, BACK OF ARTIC > 2KM W AT JUNCTION OF A 859 AND B8060, MAP 14 REF NB 269 201

The island's shackscape is the apogee of the northern Hebridean anti-aesthetic.

MAP 14 REF NG 132 915, CAOLAS. FINE CARAVAN AND SHACK ENSEMBLE

Members of the National Trust and kindred bastions of insipid taste will doubtless fail to recognise the fantastical beauty of what is here — they should be pitied and soothed with a Glamis tea towel.

MAP 14 REF NB 415 185, MARBHIG. CORRUGATED IRON SHACK RUSTING

These scapes are beautiful in the way that a lupus is beautiful,

MAP 14 REF NB 390 245. RUSTING HARROW AND, NEARBY, ABANDONED TRACTOR AND ASSORTED WRECKED THINGS

or mould on fruit,

A858 BETWEEN MAP 8 REF NB 330 285 AND REF NB 288 299. THERE ARE TO THE SOUTH OF THE ROAD AT LEAST FIVE ROOFLESS HOUSES WITH FINE LOCH/ LANDSCAPES BEYOND THEM

or decaying meat,

or scar tissue,

or amputations,

or diseases of the skin,

or anatomical freaks. They exert the same pull.

MAP 8 REF NB 359 265. CORRUGATED IRON SHACKS ON BOTH SIDES OF UN-NUMBERED ROAD TO LIURBOST, ALSO A PAINTED BUS. NUMEROUS OTHER BOATS IN BACK GARDENS, SHACKS AND RUINS ALONG THIS ROAD WHICH CONTINUES TO CROSBOST (SEE FIELD SYSTEMS ABOVE). ALSO SHED BUILT FROM BOAT HULLS

This is the great outdoors as it might have been conceived by hothouse poets of the darkest indoors, by virtuosi of polymorphous perversion: Francis Bacon and Arthur Tress, Joel Peter Witkin and Felicien Rops.

Isle of Rust

It's everything that planning can never achieve. Art born of artlessness and carelessness — as though man has hurled temporary structures like the old gods hurled boulders.

MAP 18 REF NG 058 848, 0.5 KM S. OF LINGREABHAGH. BURNT-OUT BLUE VAN
It is a thrilling environment made by a beguiling lack of respect for that most sacred of cows, 'the environment' — worthy, green, dull and tiresomely backward-looking. Put a petrol head in a pedestrian precinct and you'll get a fascinating crash.

MOINTEACH AIRINIS, SPECTACULARLY SCRAPPY AGRI KIT WITH SAILS OF WINDFARM PEEPING OVER HILLTOP BEHIND. TRACK OFF A859 1.5 KM AFTER TURN OFF TO B897, MAP 8 REF NB 378 297
And once the entropic template was established it was repeated with endless amendments and in countless permutations.

MAP 14 REF NB 252 175. BIZARRE BULBOUS SCULPTURE LIKE AN INFLATED ONION WRAPPED IN NYLON NETTING. BROKEN-UP CATAMARAN
There is nothing that cannot be used as an ad-hoc folly, as a felicity beacon, as an iconic icon.
    And there is apparently no one on the island who is immune to the appeal of the found object and the oxidising object.

MAP 14 REF NG 158 963. MIABHAIG. IMPLODED BLUE VAN
Whoever would have thought that the last remaining bastion of fundamentalist Calvinism would become the site of a scrap cult.
    That the most seductive collective expression of Gaelic vernacular culture would involve immobile bulldozers, holed boats, Allegros buried in dunes . . .

MAP 14 REF NG 117 908. GEOCRAB. HOUSE BECOMING INDISTINGUISHABLE FROM ROCKS
Whoever would have thought . . .
    We should think again. This island nation, this rather enchanted island is a law unto itself. *De jure* it belongs to Scotland. *De facto* it belongs to itself — it's a tartan-free zone.

STORNOWAY OUTSKIRTS. A859. COLLAPSED CORRUGATED IRON BUILDING OPPOSITE WESTERN LODGE OF LEWS CASTLE. SHOOT C/U. MAP 8 REF NB 404 326
Far from being stuck in the Scottish past it presages the not very distant future when continents and nations and regions and towns and villages split . . .

MAP 14 REF NB 401 244. SYSTEM OF SHEEP PENS AND FENCES. A PRAM WITH CRATE ATTACHED — PRESUMABLY USED FOR TRANSPORTING SEAFOOD
    . . . into ever smaller fragments according to their language, their patois, their schismatic religion, their wacko culture — and their peripherality.

MAP 14 REF NG 112 905. BEACRABHAIC. WRECKED HOUSES, AND A CARAVAN TIED DOWN WITH GUY ROPES IN LUNAR SURROUNDINGS
Other extremities of this continent spell out their situation: Finistera, Finisterre, Land's End.
    Rust may stay stumm about it but this is the end of the world . . . The old world. And the beginning of a new one.

MAP 8 REF NB 234 321. CALANISH. PURPLE HOUSE — NEW AGE?

It does indeed foster a minority culture which has the strength of its self-consciousness.

Its language is a means of communication but it is equally *a cause* spoken by very few people.

MAP 14 REF NB 335 198. NEAR TABOST. FINE STRIPED CARAVAN AT ODD ANGLE

It is abnormally religiously observant. Other nations and nation states will catch up. It will come to seem normal. There can, regrettably, be little doubt that this century will witness a religious revival like that of the second quarter of the 19th century.

GROANING GRINDING NOISES TO SUGGEST TECTONIC PLATES. JM WALKING THROUGH BOG - HE ARRIVES TO REVEAL BLUE TRANSIT AT END OF PTC 1 KM N OF LEUMRABHAGH, MAP 14 REF NB 373 127. SHOOT TRANSIT FROM ROAD LOOKING S TOWARDS LOCHS AND SEA

It suffers exemplary conditions to succour the twenty-first century's gradual isolation and subsequent atomisation. This is the wi-fi wilderness — when the wi-fi is not climatically disrupted. Like poor crofters and weavers before them its inhabitants are discovering the hell that is home-working. Offices, labs, studios, canteens are social places. There are no canteens in the wilderness. There are no snoutcasts gathered outside a door shivering into their gaspers.

NUTS/FHM COVER WITH SHEEP IN LINGERIE ON THE COVER

How long before we succumb to the woolly temptresses on the moor who shamelessly go naked even on the Sabbath?

How long can we survive being trapped in a virtual cage? That is a bit of a mystery. There are, however, much greater mysteries to dwell on.

How do you get a Ford Transit into the middle of a peat bog?

UPSIDE-DOWN CAR IN PEAT. S. OF A858 1.3 KM AFTER TURNING OFF A859, MAP 8 REF NB 343 273

I'm not expecting an answer to that. Nor to . . .

How do you get a Mini Metro into the middle of a peat bog *and then turn it upside-down*?

These vehicles will gradually be enveloped by sphagnum moss. Several millennia hence, in a world ignorant of internal combustion, they will be discovered in a state of at least partial preservation and revered as sacred objects from a distant, pre-apocalyptic age — a paradisiac age which ended with a bleat and a rusted carburettor.

FULL FRAME C/U SHEEP. NOISE OF CAR REFUSING TO START. JM RUNNING (LIMPING) ALONG BEACH. TOWARDS CAMERA. AT LAST MOMENT IT BECOMES APPARENT THAT IT IS SHOT IN MIRROR. HIS BACK COMES INTO FRAME.

END

6

SELF-PORTRAIT WITH PLACE

Forebears' Shades, Literal Haunts

FATHERLAND

I CAN SEE the orange night-light. I can hear my father's voice.

'Lord Rabbit came out of his rabbit hole . . .'

'In the quarry . . .'

'In the quarry. He had an important day ahead of . . .'

'What was he wearing?'

'Because it was cold and frosty he was wrapped in his British Warm
. . .'

'Not his blue coat?' I preferred Lord Rabbit in his blue coat, which
I imagined to have silver buttons, discreet frogging and tails which
flew behind him as he ran.

'He went brrrr because it was so cold and he rubbed his paws
together. Then he bounded off across Bredon, racing across the fields,
leaping over the walls . . .'

My father was of a generation whose ideal, expectation and
actuality was a job for life and a wife for life. I am of a different
generation. And when I had children of my own I told them stories
of Zuma, who was polymorphous, endlessly mutable – now a girl,
now a fabulous beast, now domesticated, now savage, and who lived,
variously, in cities, jungles, caves, palaces. Thus do we define ourselves
through the bedtime characters we invent. Seventeen years after my
death my children, even supposing they're so inclined, will seek in vain
for Zuma's topographies. Seventeen years after my father's death Lord
Rabbit's itineraries are readily traceable, for while he entertained me
and lulled me he was also recalling, with formidable precision and
with the clarity of twenty/twenty hindsight, the place which he had
loved most as a child and which had, I suspect, been a source of solace
after his father died when he was ten.

There is a village called Bredon and another called Bredon's
Norton. But to my father, and so to me, Bredon always meant, always

will mean, Bredon Hill, the largest of the Cotswolds' north-western satellites, which is geologically akin to that range – hence the former quarries which yielded the limestone to build the villages on the Cotswold side of the hill but not on the Vale of Evesham side.

We tend to believe that vernacular architecture is determined by the availability and proximity of this or that building material. Here, to the north of Bredon, it would seem that a quasi-aesthetic decision was made to ignore the local stone in favour of brick and, earlier, of half-timbering. The villages do not, then, possess the homogeneity which characterises and deadens those of the Cotswolds. Of course, the walls that Lord Rabbit soared over were dry-stone ones which – and such is the border nature of Bredon – are hardly found on its north side.

Bredon, as A. E. Housman's exegetists are obliged to point out for the sake of the lad's prosody, is pronounced Breedon. He wrote of it before he realised that Shropshire or, as my Uncle Hank and my Uncle Wangle would have it, Shropshire place names was to be his subject. This avuncular insistence was ascribable to Housman having been a Worcestershire man who by his own admission hardly knew the places he wrote of. My uncles wanted to reclaim him for Worcestershire. Wanted to? They did. My father loved Bredon certainly; it was where he felt himself to be Bevis or Huckleberry Finn. But he loved it this side of idolatry. My uncles were different. They may have been literally atheistic, but this did not inhibit their profession of a quasi-religiose attachment to Worcestershire and to certain though by no means all of its landscapes. Worcestershire was for them ur-England and Bredon was their *colline inspirée*, as the arch nationalist Maurice Barrès called Sion-Vaudémont near Nancy. It was a sort of sacred site – sacred because they believed it to be so rather than, like Barrès's hill, because it had a history of ancient mysteries, sectarian righteousness and pilgrimages (other than of Brummies at Whitsun, picnicking). Their obsessive attachment was not merely dubiously spiritual and intellectually questionable – though it was thought-out and consistently systematised with the specious confidence which is common to all faiths. It was also physical. The mortification of the flesh involved in attaining the top of Bredon was just as important to them. Sure, it's only 300 feet at its highest point, but the trudge up took two hours and was only undertaken on days that were very hot, very cold or very foggy – or is my memory as flawed as the weather it insistently conjures up.

My uncles' Bredon was pseudo martial as well as pseudo holy: they were Aertex-clad striders; I was a nancyish little wimp running to keep up with them as they lustily inhaled gusts of Housman's air which was also Elgar's air – not that they had much fondness for Elgar, but he too was a Worcestershire man and that was enough to secure his entry to their respective pantheons of Sibelius, Grieg, Nielsen and Rex Warner (Wangle) and Beethoven, Chesterton, Joyce (Hank) – these apart from such shared enthusiasms as Lawrence and John Cooper Powys.

My father rarely joined in these route marches while making it clear that they were good for me and that, no, I was not to be excused them. Sometimes Wangle's sickly wife Ann would start them but she'd invariably turn back: Wangle eventually did for her when he took her camping in the Cairngorms in order to assist her recovery from open-heart surgery. (Imagine Roger Sloman and Alison Steadman in Mike Leigh's *Nuts in May*, take it on a bit and you've got the picture: Sloman even looks like Wangle, who drove a black Morris Minor with a red soft-top which he never put up.)

By the time I was about nine Bredon had a multiplicity of meanings for me. Home of Lord Rabbit who, by then, I was rid of in favour of the sanitised psychopath William Bonney – the last thing a boy of that age, in the full throes of cap-gun machismo, is willing to acknowledge is the anthropomorphic rabbit of his bedtime stories of long ago, when he was a little child. Site of my father's childhood adventures, the way that Harnham Hill on the edge of Salisbury was mine. Though even then I recognised the discrepancy. Harnham Hill was suburban, mostly tamed. Up there I was never going to meet gypsies who'd share hedgehog cooked in clay (to remove the spikes) with me as the itinerant pickers around Bredon had with my father. Place of mild torment, avuncular mockery, discomfort, fatigue, humiliation. Repository of secrets – which would one day be vouchsafed to me. I'm still waiting. I guess that's because one key to them is a willingness or capacity to grant to the inanimate properties that the inanimate does not possess. Places do not have spiritual attributes. A rock is a rock. A tree is a tree. They may, of course, be invested with such attributes by humans but those attributes have no independence any more than a fiction such as God has: of course God exists because humankind invented him/her/it just as I invented Zuma for my children.

Hank and Wangle, however, subscribed to the conviction that there was more to Bredon (and to the Malverns) than geological happenstance and the terrain's consequent sympathy towards a particular gamut of trees, grasses and insects. My uncles were infected by a virulent strain of the pathetic fallacy. They believed – with an unquestioning want of reason – in some inchoate bond between humans and the places they come from. Not the bond which is forged by familiarity, detailed knowledge, affection, custom. Rather a bond which is mutual, in which the hill or river or heath or whatever is an active partner.

The second key to Bredon's secrets was, apparently, Housman, whom I was enjoined to read in my earliest adolescence, which is precisely when the bitter glorification of doomed youth and of young death is most affecting. I found it all horribly poignant. My two childhood heroes, both of them Worcestershire lads, the footballer Duncan Edwards and the racing driver Peter Collins, had died, aged twenty-one and twenty-six, within a few months of each other when I was eleven – Edwards, two weeks after the Munich air crash having survived in an iron lung which I conflated with an iron mask so that I thought of him as The Man In The Iron Lung, and Collins in the German grand prix when he was leading the F1 championship: his Ferrari left the Nurburgring at a corner called Pflanzgarten, the garden of plants. It was the first German word I knew save the few gleaned from War Picture Library.

So by the fortuity of those two deaths Housman did mean something even if it wasn't what my uncles intended. They were older then than Housman had been when he conjured his morbid magic, old enough, that is, to have outgrown it and to have come to realise the frailty and spuriousness of the glamour that is attached to curtailed lives. And there is a world of difference between a twelve-year-old boy prospecting his own romantic end ten years hence and a fussy, desiccated don looking back to the age when death should have taken him but failed to. And I, too, am old enough to have outgrown those silly paeans to loss and to hopelessness and to hanging – but amused contempt for meaning doesn't obviate a susceptibility to the verses' sheer allure; deftness is a form of suasion in itself. I cannot but associate Worcestershire with a soothing fatalism and omnipresent mortality – I know it should be Shropshire, Clunton and Clunbury, Clungunford and Clun; should be but ain't. Shropshire is somewhere I visit as a beguiled stranger. Worcestershire is where I go to be touched by the shades of my forebears, to be haunted by my former self, to measure my march towards eternal coma. I may not set out as a death tourist but that's what it comes to – death goes dogging everywhere.

All the places I went as a child remind me of my dead father and of his three childless siblings, all of them dead, of his childless cousin whose name is perpetuated on his building firm's boards, 'Meades of Evesham', of his friends who have, as the French so discreetly put it, disappeared, many of whom I never knew but whom he exhumed, brought back to life when I had tired of Lord Rabbit: Jumbo Evans, submariner, drowned in the North Atlantic, 1941; Eric Rae who came from Rous Lench and used, at the age of thirteen, to steal his father's car in the dead of night – which meant that they had to push the vehicle till it was out of earshot of the house; it would move ghostly through the lanes, high above the hedges. The roadside shacks selling, according to season, asparagus or strawberries or onions; the old greenhouses; the pubs I'd loiter outside with crisps and pop while my father and uncles drank within – these all now coalesce into a memento mori whether I like it or not.

There is one place that was always thus, even when everyone was still alive – when even Mrs Hopkins, my father's nonagenarian former nanny who hugely impressed me by remembering newspaper accounts of Custer's death at the battle of the Little Big Horn, was still alive. Between Offenham and Cleeve Prior there is a weir on the Avon, and a ford across it. The river runs parallel to and tight beside the road for a hundred yards near the Fish and Anchor. Then the straight road vanishes into Cleeve Hill. This, I believed when I was tiny, was where the world ended: if I climb that hill I will die. I have dreamt recurrently of pub, river, road and hill. Were there ever orchards on that hill? Yes, because I've put them there in lieu of the scrubby blackthorn bushes. And it's there, among pagan groves of apple trees, that Claudius comes to minister to Hamlet's father's ear. It's there that the fruit pickers in David Rudkin's *Afore Night Come* stage their ritual killing in foggy homage to their immemorial predecessors. Me, I wouldn't dream of writing fictively of Worcestershire – there's nothing left to transform, it's already an idea rather than a place, already a state of my mind.

(1998)

PINT-SIZED

I SPENT THE 1950s in pub car-parks. In summer I kicked gravel, walked parlously on wall tops, scaled trees. In winter which is what it usually was I sat in my father's car watching my breath condense, miming driving (Collins, Hawthorn), scrutinising the underside of Issigonis's dashboards, reading maps. Irrespective of season I ate crisps, drank fizzy lemonade, dreamt but not so much that I forget the day in '56 when my father discovered that Messrs Smith of Cricklewood had increased the price of a packet of crisps from 3d to 4d; still, there was no diminution of his generosity.

Irrespective of season, what I did mostly was wait. I waited longest outside pubs near Evesham, pubs beside the Avon, pubs beneath Bredon Hill, pubs with fruitholdings and hop gardens all around them, pubs with their own foot ferries, pubs of immemorial beams and bricks.

I waited longest there because my mother, antipathetic to my father's family, rarely accompanied us to his home town, to the Edwardian past that was my grandmother's perpetual present, to drink with his strange brothers my uncles, to eat grey beef at three. The stink of the gust of the bars whose doors I steeled myself to put my head round, trying to catch a familial eye, was invariable. Grown-up life smelt bad, stale, fungal, fusty. Grown-ups breathed pea-soupers, motes floated in their exhalations, the many-denier opacity of the air combined with its stench to make me humble back to the car, unnoticed and crispless.

My retreat was, I knew, from beer, from beer's quarry, from beer's sites, from the intense jocularity it fostered. Sure, there are gin palaces and cider bothies but that English peculiarity and institution, the public house, is founded in beer. Beverage and building are mutually dependent. To dislike one was to dislike the other. To dislike either was, in my uncles' eyes at any rate, to abjure a rite of entry into adulthood.

Pint-sized

The Boy's First Pint was about as close as middle-class, middle-century, middle England got to the bar mitzvah. Anglicanism may have been strong on piety, on body-and-blood baloney, on tiresome porkies about Mary's immaculacy and the resurrection, but it was typically shy about hormonal mayhem, about initiation into adulthood. Anglican First Communion comprised taking the worst that bakery and viticulture could offer from a paedophile's papery hands in the grim fastness of a cathedral; it was a ceremony that signified nothing. The First Pint, on the other hand, was a main line to an older, deeper England, to John Barleycorn, to the very land in which had grown the cereal, to the stout yeomanry that had cultivated it.

That, anyway, was the gist of the avuncular spiel. These two solitaries were not in thrall to the Anglican god but to a home-made pantheism in which *veldtschoen*, Richard Jefferies's Bevis, Housman, tweed, arts and crafts, Vaughan Williams, cleft sticks, *Stalky & Co.*, *The Countryman*, Social Credit, Aertex, Chesterton, xenophobia, the Malverns and baked hedgehog all played a part. Beer was sacramental.

I wondered: it wasn't merely that I couldn't stand the taste of the stuff. I was not keen to buy into this maypole mysticism: I didn't want to join. Besides, beer was more than sacrament to my father's elder brother; it was, albeit indirectly, his livelihood. He was town clerk of Burton upon Trent, brewery to the nation, brewery to the Baltic, brewery to the Empire and ne plus ultra of the single industry town.

Burton was beer. It was ale stores, cooperages, malt houses, a high incidence of sclerosis of the liver, Bass's private narrow-gauge railway, Bass's almshouses, Bass's many churches. It was all brick, stylistically sober, monumental. And H. Turner Meades presided over its self-destruction. I guess he knew no better. His vision of England was profoundly anti-urban (which beggars the question: what was he doing in such a job?) and he was generationally conventional in his distaste for Victorian architecture.

He and the councillors he despised and the brewers he sucked up to would have seen no virtues in hundred-year-old industrial buildings. Especially not in the white heat of the Keg Era: that sort of beer, no nicer and no nastier than the preceding stuff, I thought then was the brewing industry's contribution to '60s neophilia. This was the beer of the future. Soon the world would be all monorails and robots, and non-iron suits with no lapels, and colonies in space. And we'd toast our progress in Red Barrel and Party Sixes and Pipkins.

When I say we, I evidently exclude myself because the First Pint was still undrunk, and likely to remain so. And so long as it did, I knew very well that, pace developments to my balls and my voice, I should remain stranded at the portals of English manhood. I rather suspect that I'm still there; it wasn't until I found myself in Belfort at the age of fifteen that I tasted beer I actually wanted to drink.

Belfort is not in England. It is in France's non-cricket-playing republic. Beer there was free of quasi-eucharistic mystique; it was not a bibulous salute to the religiose nationalism of Barrès to a foreign teenager, it was just a pleasant bevvy, one which was not habitually drunk in quantity, and certainly not for oblivion. That state could be otherwise achieved. Beer and its derivatives are vastly impractical agents of intoxication, and dangerous ones, too, more dangerous than distilled liquors which are, infamously, more heavily taxed, more dangerous than hashish and unadulterated opiates.

The English have been made to suffer more in the inevitable and universal quest for intoxication than have the peoples of less proscriptively governed countries. What might appear to be volumetric masochism is state-prompted sadism. Addiction to beer is indicated by various vitamin B deficiencies, pseudo-dropsical traits such as hanging jowls and guest symptom wet beriberi. The point is this: spirits and, to an evidently lesser extent, wine effect intoxication in quantities which do not fool the body into believing that it is being fed. The copious amounts of beer required to achieve this end do just that.

My specialist adviser on this topic, Professor Guy Neild of Middlesex Hospital, warns that beer is so wanting in proteins that 'drinking enough to get pissed is like eating sugar all day'. He adds that vast quantities of alcohol are harmless provided the patient pursues a decent diet. Beer discourages that pursuit. Or, rather, the way that the English and the rest of northern Europe drink it discourages simultaneous guzzling not least because food, other than thirst promoters such as crisps, inhibits bibulous appetite.

It might be argued that beer is the enemy of food and that a country in which beer has primacy is bound to suffer culinary impoverishment; Belgium is the lone exception. And the gastronomic burgeoning in this country over the past couple of decades has coincided with, has perhaps been pushed by, the ever-increasing availability of affordable wines. But beer is culturally potent, its example is unforgotten; the English now drink wine as though it was beer, without food, standing up, at parties. Same mistake, different bev. Imported bev, too – a fact that might be used to emphasise the fragile base on which this fresh gastronomy is founded.

Sure, there is English wine and, yes, the Romans, unhampered by the knowledge vouchsafed to celebrity weather girls, did cultivate the grape as far north as Lincoln. This is not, though, a vinous country; it produces only white wine. And a vineyard in Surrey – you are thinking, rightly, of the one just north of Dorking on the A24 – is as marvellously freakish, as admirably inapposite as, say, the mosque at Woking which the beer victim Ian Nairn (fourteen pints per lunchtime, latterly) described as 'extraordinarily dignified'. Both mosque and vineyard are untempered by location. The mosque is by an unknown Sunderland architect, W. I. Chambers; the vineyard is by its proprietors (who want to sell). The vineyard is an intruder, a hop garden isn't.

The Englishness I eschewed, the Englishness of my uncles, may be an unembraced faith but the integrity of the landscape which underpins that faith is sort of sacred. The ordered, ranked, fructuous lines of Worcestershire and the Weald touch me in a way that my native Downs don't. Those downs are good for mutton and as canvases on which to represent the optimistic ideal of manhood (the Cerne Giant), hippofetishism (everywhere), Anzac bereavement (Fovant). They don't connect. Oast houses and hop kilns do. It doesn't matter that they are all now houses, domestic dwellings.

As I grow older, I am more and more inclined to some sort of concordance with the blight I was born to: against my will I'm sucked into a terminally English appreciation of the terminally English medium of watercolour which is as weak as beer, an analogue of it, and the means by which our scapes and architecture are best represented. Depictions of Wealden oast houses by Rowland Hilder strike some pre-aesthetic part of the brain; I find myself dorsally thrilled by places whose like Henry James called 'deepest England', places whose abundance and arboreal richness informs Michael Powell's *A Canterbury Tale*, the country round Goudhurst, the majestic intimacy of Fawsley, where Warwickshire, Northamptonshire and Oxfordshire collide; the Marches seen from Clee Hill. To drink anything other than beer in these places would be a solecism, an act of ingestive treachery, dead wrong.

Beer is unescapable, the way nationhood is, the way the poor are. Beer assails us – there is no building type so common as the public house, apart, that is, from the private dwelling. There are getting on for 100,000 of them in this country, a figure which stands as some sort of indictment of the collective taste. And for every exuberant gross-out like the Philharmonic between Liverpool's cathedrals there are hundreds of dismal city pubs. And for every forgotten treasure like the room with two barrels and an out-of-date KP ad beside the Severn at Ashleworth, the Boat Inn, there are hundreds of dumps in the sticks hosted by bonhomous pint-raisers. Here, let's raise a glass to olde Englande, a glass of crème de menthe with a sambuca top.

For it's olde Englande to which pubs belong, to which they increasingly belong. The apogee of pub architecture was, roughly, 1880 to 1910; I'm not referring to inn architecture or hotel architecture, but specifically to boozers, purpose-built to intoxicate and, incidentally, to delight.

Imagine the gulf between a mean terraced house and these palaces, erected by brewers for their people, their subjects. Bevelled glass, mahogany, abundant carving, marble, velvet – these are the materials of brassy sybaritism. The great pubs were wonderlands, fantastical playgrounds, stages of easy exoticism: their decorative richness assisted intoxication, they transported their customers. They offered more than merely liquid escapism.

And that persisted during the only subsequent era of pub architecture that is of much moment. The roadhouse belongs to motoring's first age, when cars were a pleasure rather than a chore, when the now quaint idea of driving to a beer hangar beside an arterial road was reckoned a good one. These predominantly outer-suburban boozers were rigged out in period styles from the great dressing-up box of architecture's past.

Now and again you'll come upon an essay in the international modern style – the Comet at Hatfield or the Prospect Inn at Minster-in-Thanet – but usually it was full-blown Tudor or Wars of the Roses or Dutch Renaissance or Quality Street Regency. The place to see this stuff in its blowzy ripeness is Birmingham where two great breweries, Mitchell & Butler and Ansells, competed to produce ever larger, ever more theatrical pubs which were, if you like, hoardings for themselves – they had to be big enough to catch the eye of the passing motorist.

The stylistic revivalism of these places is light-hearted; you can imagine Errol Flynn swinging across them. There is no intention to deceive. Pastiche not fakery was the rule. That is no longer the case. As the production of beer is increasingly executed in plants which have no technical or architectural kinship with the sort of building we readily identify as a brewery, so do the places where beer is sold and consumed advance with increasing alacrity into the past. A past which is realised with dour solemnity.

All the theme pubs of the '70s and '80s – the Colditz pubs and the car rally pubs and the crazy cottage pubs – have disappeared. Brewers large and small have only one theme today, and that is a simulacrum of yesterday. It's as though they know that bibulous solace is only to be found in some illusory 'heritage'. It's not much of a prospect – wrong word, that.

(1994)

FRAMED

WE CANNOT ESCAPE our earliest domestic topographies. They adhere down the years as a sort of measure. The slum is as potent in this regard as the secret garden. The escapee from the slum will later recall it with rosy partiality – witness urban northerners and cockneys, passim. This sort of nostalgia is independent of a yearning to actually return to that home. Things are different when you grow up in more propitious circumstances.

Sure, the house I lived in till I was fifteen may have been merely two-up, two-down, rented and part of a terrace. But this was a thatched terrace – a rarity even in Wiltshire – and its walls were cob and it was near the several rivers that conjoin and diverge to the south and west of Salisbury Cathedral Close. And when I was fifteen my parents moved, in the week of the Cuban missile crisis, to the only house they would ever own, one they had built at the confluence of two of those rivers, the Avon and the Nadder, the latter so straight over about 200 yards that it might have been canalised.

Standing here Constable had painted the cathedral and his friend Fisher's house, though this sort of landscape and cowscape are more readily associable with Cuyp. Between the rivers there were lush floated meadows, that is meadows subjected to rudimentary hydraulic engineering, incised with straight watercourses called leets, the flow in which had been controlled by means of primitive sluices operated by men called drowners: only one drowner remained, the octogenarian Mr Thick who was crippled by his trade disease of arthritis.

Even before they moved house my parents had a rowing boat on the river, moored beside an unused brick abattoir of circa 1890 which still retained the stalls in which kine had awaited their death. I led the life of Bevis, rowing myself round Alligator Island and up to the mill where the white water was my very own Niagara and where the still water was on four perplexing levels and the leet system might have been modelled on a scheme by Escher or – such was the small-scale might of the aqueducts which carried the courses over each other – Piranesi.

Down Cow Lane beyond the abattoir was Hawk Bowns's bodged farm, all corrugated iron and rusting buses. My father and I built an eel trap out of an axle, bike wheels and chicken wire. Through spring I would haul it from the water with the creatures wriggling within. Through summer I would swim among the lithe green weeds where prognathous pike awaited their prey. They scared me in a way that the sloughing snakes in the leets never did.

Now, I regarded these rivers and meadows as my adventure playground (the construction was yet uncoined), my qualified paradise, my private domain. There I could be Livingstone, Drake, Earp, a spy, a Chindit, an escapee. This was how I inhabited these places, how I possessed them. But I was all the while apprised that the leets' first purpose was to enrich the pasture to feed the cows to make the milk. The rivers drained the uplands of the Plain and the downs. The great spire which rose above the chestnuts was not a mere eyecatcher, a vertical accent of unparalleled height and signal sublimity, but an exhortation to obeisance and worship.

I knew this because I lived inside this terrain as much as I lived inside my head: the compact of make-believe it occasioned did not preclude comprehension of its primary reality and utility – it's a question of both/and rather than either/or. My nostalgia for this roofless pleasure-dome is tempered by the knowledge that I no longer have an appetite for slithering on my tummy from ditch to willow hiding from German troops with tracker dogs. But say I hadn't lived there, say I had only seen it from the outside insouciant of the seasons, of the perpetual mutations, say I had lived all my life in a city or – statistically more likely – a more ordinary suburb, one untouched by bucolicism . . .

All through the summer months tourists to Salisbury would stray down the hidden road where my parents' new house was seeking Constable's pitch. They would already have borrowed his eyes to gaze at the cathedral from Long Bridge to the north-west, from the lake (a pond, actually) in the grounds of the Cathedral School. What they sought, whether or not they articulated it to themselves, was the mediation of the picturesque – which etymologically signifies no more than in the style of a picture or apt as a picture's subject.

The paramountcy of the picturesque is a national bane. It has us all in its thrall. It militates against an understanding of rurality. What does a picture do to a landscape? It reduces it to two dimensions. It amends, transforms and idealises it. It frames it: a 'scene' is rendered finite. It tells you about the taste and preoccupations of its author. Shapes, chromatic range, weather, lucence – these are made, they're inventions. A painting or a photograph creates a reality, it doesn't represent what's there: scrutinise, say, Edward Piper's photographs in the Shell Guide to Buckinghamshire – he renders the county as though it's perpetually in the throes of an apocalyptic electric storm. Filters. We know that such work is all artifice but fail to act on the knowledge. I'm no more immune to the picturesque's lure than anyone else: driving across North Yorkshire a few years ago I could not resist the temptation to go to Greta Bridge. Within seconds of arriving I realised my stupidity and didn't stay long enough for retinal actuality to supplant Cotman: it's that painter's vision of Carr's bridge that remains in my mind's eye.

And there you have it. The picturesque's appeal is self-evidently to the eye. It causes us to look, to see – we may even see with our own eyes though we prefer to have a view sanctioned by at least a postcard: the bits in between those which have received such validation don't count, we're blind to them. The British experience of the rural is almost exclusively visual, almost exclusively led. And when gazed upon from a car the country is happily framed by windscreen and windows.

The cult of the picturesque can only thrive in circumstances where the majority of the (sub)urban population suffers an overwhelming ignorance of the country, an ignorance which manifests itself in idealisation and in fear.

There may be no visual connection between the meat in a supermarket and a lamb's limbs but, surely, those who eat the stuff would reveal, under tough questioning, their knowledge of where it comes from. In the same way those who wish the country to be framed, to comprise set-pieces such as Castle Combe, Montacute and their like must know that there is more to it than that. No, perhaps they don't.

(2002)

A BOY'S-EYE VIEW

A CHILDHOOD IN south Wiltshire is a childhood in which the vestiges of many pasts are omnipresent, everyday. Salisbury Cathedral, Stonehenge, Grimsdyke, Bokerley Dyke, countless barrows and tumuli, sarsen-stones, hill forts, Old Sarum, field systems, Great Yews, the Mizmaze. History and prehistory are seldom clearly layered. More often BC and AD are scrambled in a subterranean pudding. Different cultures inhabit and adapt the same sites. The relics of one civilisation are bequeathed by its predecessor. It is hardly surprising that Wiltshire and Dorset formed the cradle of English archaeological investigation.

Now, I was aware from a tender age that it was my duty to revere cow-trampled earthworks and broken flint weapons and dusty shards of pottery: here, after all, was my patrimony as a son of this land. These were artefacts that supposedly linked me, by some baffling means, to grunting hunter-gatherers who suffered the misfortune not to live in the age of the Dansette portable record player. But I remained obtusely unlinked and unmoved.

It was not until I was eleven that the ancient earth's potency and its buried legacy would suddenly make sense. A gardener was grazed by a rose thorn. He contracted tetanus and died. One explanation offered to the inquest was that the ground which nurtured the rose's roots was contaminated by diseased bodies buried long ago. It was chemically haunted. This may have been bad science but it sent a shiver down my spine. That was more like it. The present was infected by the past. The land seemed suddenly to possess a thrilling immediacy.

One of our dominant urges as we achieve sentience is to classify. We begin to distinguish between states of reality, kinds of uncle, car engine noises, dog breeds, inanimate phenomena. Salisbury Cathedral was, obviously, categorically different from most of the area's monuments. It was used. It had as unequivocal a purpose as a garage or a fire station. It was a living building. Most importantly, it was an intact building. And because it was intact I thought it was comprehensible. It wasn't some fragmentary jigsaw puzzle of a site which demanded the mediation of an accompanying lecture in order that its original form, let alone its supposed uses, might be grasped. Those sites belonged to the world of ideas.

All that the cathedral required was a pair of eyes and an incipient capacity for awe. I had that capacity: I can remember staring up at the spire from my pram, an ideal point of observation. From the age of four I spent nine years at school in that most complete anthology of English domestic building types, the Close, in the cathedral's massive presence and long shadow. This was a visual and topographical privilege. And I knew it, not merely because I was told that it was so. The cathedral may have been an object which I constantly witnessed, but familiarity bred wonder. It beguiled me.

I was evidently not yet apprised of Flaubert's dictum – it might equally have been Warhol's – that the more you stare at something the more fantastical it becomes. But I would recognise its truth when I encountered it because it applied to the cathedral. I was its willing, curious, earnest, trusting votary. I was fascinated by the variety and abundance that were accreted in this treasure-trove: lichen, broken tombs, Purbeck 'marble' turned matt with age, occluded astrological statuary, worn misericords, ragged pennants, reeking pre-revolutionary Russian stoves (the only heating). But more, I was constantly overwhelmed by the sheer might, size, dimensions. They proclaimed God's majesty. The immensity spoke of certitudes.

So, too, did the clergy. In that era Church of England clergy tended to be believers, and quiet proselytisers. It was their job. Like less martial Jesuits they knew at what age to get a child, to capture his soul and shape his mind. In the micro-culture of the Close and its several schools there existed a tacit presumption of observant faith as a behavioural norm. It would have been unthinkable to separate the magnificence of the cathedral's structure from its liturgical and propagandist roles.

Years later, the first time I returned to film it, it occurred to me that there was a correspondence between the tiny impressionable child I had been and the medieval shepherds, reed-cutters, lime-burners and hurdle-makers whom the cathedral was built to ensnare. For me, as for them, learning was the property of the Church. For me as for them the cathedral was the mightiest building they had ever seen (that they would ever see). In my unquestioning infancy the Close was like the entire world had been for them all their short life.

Unlike them I could escape, for Anglicanism's well-meaning hegemony was confined to the ambit of the cathedral. And, besides, time was running out for it. Not least because during the years of Vatican II the Church of England, slavishly following Rome, came out doubting, and began to dismantle the apparatus of irrational mysteries which defined it. It expressed its self-secularising openness by espousing theatre-in-the-round in new churches and excising wicked (and still unfashionable) Victorian incursions such as Sir George Gilbert Scott's screen in Salisbury Cathedral.

Thitherto I had thought of the cathedral as immutable. Its stasis had been part of its appeal, together with its anachronism, its decrepitude, its other worldliness. Now it began to change, exponentially, as though to make up for that stasis. A stone yard was established to the south of the cloisters, on the site of what was aptly known as the Jungle, in sheds which would disgrace an eyesore industrial estate. It became the first English cathedral to levy an entry fee. An entrepreneurial dean proposed demolishing a stretch of the Close's medieval wall and building an ecologically injurious road across water meadows to allow more coaches access to the cathedral's immediate precinct.

A bizarre vertical experiment in cleaning was conducted on a two-metre wide stripe of the entire West Front, then abandoned. The fabric of the structure is still being replaced, stone by friable stone: the rigour of the craft and the science cannot be questioned, yet it demands only the faintest speculative aptitude to conclude that by the time Salisbury Cathedral celebrates the 800th anniversary of its consecration in 2058 it will be an on-site replica of Salisbury Cathedral.

For a devotee of decay, entropy and ruins this is an unalluring prospect – but maybe that very devotion is just sour grapes founded in the mortal knowledge, gleaned from tombs and tetanus in the chalk and the death cult called Christianity, that we do not share the capacity we grant to inanimate structures for eternal renewal. We die, they continue. And a zealous forgery of Salisbury Cathedral is surely preferable to no Salisbury Cathedral? – intact cloisters to the ruinous cloisters depicted by Turner?

When he painted and when Constable painted, voraciously, there was no question that the only worthwhile purpose of such a building was as an adjunct to landscape, an eye-catcher, a sublime force of near-nature which happened to be extravagantly spiked. It was thus treated even by the dean and chapter who allowed James Wyatt, still well known by the older clergy in my childhood as 'the Destroyer', to demolish the freestanding bell tower and to remove hundreds of headstones for the sake of appearance. The picturesque's lure triumphed. Views vanquished ideas. Architecture had overcome medieval superstitions.

A delightful old boy – who may have been the precentor – once took us nine-year-olds to the sites where Constable had put his easel. He was at pains to emphasise the painter's debt to his friend Fisher, Archdeacon John Fisher, to whom he had first come on visiting Salisbury. It was as though he was trying to render Constable's art sacred by association. I, of course, acceded. Then.

(2005)

FIRST LOVE

I RECALL IN the mid-'60s, in my mid-teens, remarking to the chatelaine of a grand Sussex house how much I liked the place, how fond I was of high Victorian building. The steely old bird went ballistic: 'This house is not Victorian – it is Elizabethan.' Which was partially true, but even then I had a taste for courteously expressed offence. And I knew very well that Victorian was a dirty word to her generation, to my parents' generation (b. *c.* 1910–15) and to those around it. To call her house Victorian was akin to disputing her legitimacy.

It's an attitude which persists. Preparing *Victoria Died in 1901 and Is Still Alive Today,* I wanted to film the Victorian statuary on the west front of Salisbury Cathedral. I was told when I rang that the west front had just been stone-cleaned and that the Dean and Chapter were not willing to grant permission to the BBC to film it simply to traduce it as Victorian. The cathedral's marketing line – and the man I spoke to was doggedly insistent – is that this is a medieval structure.

In which case, I asked, why do you have that stone yard and a brigade of carvers just outside the cloisters if not constantly to renew the building? Many medieval buildings are stone-by-stone copies of themselves, and the process of that copying began and was most sedulously pursued during Victoria's reign. And if it is medieval why has it been stone-cleaned to render it new?

Our marketing . . .

Well, of course: marketing must always take precedence over architectural-historical truth, any historical truth. And marketing a Victorian cathedral is evidently rather different to marketing a medieval one. There is nothing like age to attract mass tourism. Oldness is all, real oldness (whatever that means). In this country oldness and 'historical association' (Queen Bess/Charles Dickens/ Judge Jeffreys slept here) will always triumph over aesthetic worth.

I eventually filmed Lichfield Cathedral's west front in lieu of Salisbury's. It is an even more comprehensive example of George Gilbert Scott's winning capacity to make the Middle Ages what they would have been had they had the sense and nous to be the high Victorian age. But it's not this fakery that I find so compelling. It is, rather, the gap between the fakery and what it purports to be.

Medieval masons and craftsmen were not the beneficiaries of industrial means of production and industrial means of transport. The architects who worked at the high point of Victoria's reign, roughly 1857–73, the long '6os, created something quite as mannered, tough and loathed as the architects of the twentieth century's long '6os: they created buildings which were without any kinship to the land where they stood, which were monuments to humankind's ingenuity and not to an enslavement to the debilitating limitations of local materials, local traditions (mostly occasioned by the properties of those materials). Buildings such as S. S. Teulon's Elvetham and William Butterfield's Keble College and F. R. Pilkington's work in Edinburgh and Kelso are a match for Rodney Gordon's or James Stirling's stuff of a hundred years later. They possess the same sculptural perversity and lack of ingratiation; they revel in their own artifice, in their archaeological inaccuracy. And yet . . . they are beholden to multiple precursors – whom they wilfully misread. It is this tension between exemplary model and self-advertising idiosyncrasy, between generalised past and specific present which so excited me, which still excites me.

These buildings abhor the strenuous fiction called naturalism. They are loud, rhetorical, unreticently clever, self-proclaiming, ill-mannered, hallucinatory, grotesque. It might be said that they are, thus, atypically British. But for all our dread of colour and our toe-curling obeisance to the dictates of good taste and devotion to that basest of thought-systems, common sense, there is no cause to believe that low-key, self-effacing apologetic art was a norm for more than a weird interlude, a long hiatus in the mainstream of British endeavour, a hiatus which began with the late Victorian reaction to Teulon and Pilkington, to Dickens and Lewis Carroll.

The architects and artists born in the 1850s and '6os, whose first works appeared in the '80s, were the people who devised the aesthetic example that this country succumbed to through two-thirds of the twentieth century. The 1890s are much more characterised by cuteness, sweetness and plainness, by the work of inspired little Englanders such as Voysey and Lutyens, than they are by worldly decadents. And there was a commensurate moral shift: had Wilde sued for libel twenty years earlier it is improbable that the love which dares not speak its name (and cannot, because, as Tynan had it, its mouth is full) would have caused his downfall.

The young Lutyens was, alas, the greatest English architect since Vanbrugh. I was thirteen when I first saw Marsh Court; I suffered a rush, a sense of wonder, a sort of inexplicable happiness. It was my first experience of sheer aesthetic bliss. Had, of course, I chanced at that age upon the relic of Vanbrugh's Eastbury (and I might have done – it's the same distance the other side of Salisbury) or Juvarra's Stupinigi or Le Corbusier's chapel at Ronchamp (which I first saw two years later) or the roof of his l'Unité in Marseille or Cothay Manor, then one of those might have been the key that unlocked an insatiable, lifelong appetite for buildings. But however much I might wish it otherwise it is that stylistically mutating, materially constant (clunch, i.e. deep-bed chalk) house above the Test, designed in the last year of Victoria's reign, that did it for me.

It's not the case that I was previously insensitive to buildings or topography: I went to school in Salisbury Cathedral Close from the age of five till I was thirteen, and it astonishes me that anyone who enjoyed such a privilege, such a persistent subjection not merely to the cathedral (awesome, yet unlovable) but, more, to the peerless anthology of domestic architecture around it, should not be tectonically obsessed, or at least tectonically sentient. I suppose it goes without saying that this environment was extra-curricular. It was not talked about. We were not enjoined to appreciate it – thankfully. The important things were religion and sacred music. I am largely indifferent to the latter, and as for the idea that a cathedral is a 'functioning' building . . . I loathe the smell of Salisbury Cathedral. It is the smell of pious superiority, of clerical snobbery, of preposterous liturgy, of institutionalised delusion, of pompous asceticism. It is the smell of a lie, a much graver lie than the one about the date of the west front's statuary. Religious belief is a form of mental illness, a wonderfully rich and endlessly complex form which never ceases to beguile this onlooker: I regard believers with the amused pity and horrified fascination that eighteenth-century visitors to Bedlam felt for the inmates. Salisbury Cathedral scared me. It contaminated the Close around it. The perfect houses seemed tainted: they seemed to be rendered sacred by their purpose, which was to provide shelter and ready-made hubris for the officers of the church. They appeared welded to their former function.

That initial encounter with Lutyens thrilled me not least because Marsh Court was so patently secular. Here was great architecture untouched by a programme of superstition. I didn't realise that it was great architecture whose ultimate effect was the partial ruination of this country. I suffixed Lutyens with 'alas' because his brilliance was so seductive that the generations which followed him could not but attempt to copy him and his not much less brilliant coevals. And they copied him so ineptly. They got the mannerisms right, and the eagerness to please, and the winsome tweeness, but they quite forgot the talent. Vanbrugh's aim was to put on what he called 'a masculine show'. Lutyens's early work is feminine. That of his literally thousands of followers is effeminate, which is different. And it is land-hungry, spendthrift of a resource which in this country is in short supply. Britain may have been the first country to industrialise. But Britain today is much more the product of the late-Victorian reaction to industrialisation's grip on cities, a reaction which was an evasion of the problem: let's get out as soon as we can. It's a process which, incredibly, continues to this day. A despicable collusion persists between laissez-faire planners, vandals called volume builders, and successive governments who know that the only civil liberty the subjects of this proscribed country are interested in is the right to own an aesthetically bereft 'castle'.

Just as Salisbury Cathedral contaminated its environs so did the patrons and proselytisers of the Gothic revival attempt to turn the secular sacred with pointed arches and yesterday's tracery. They sought to bless with a gamut of architectural devices, a gesture as potently quixotic as saying grace before a meal. It was not until I was able to dissociate structure from meaning that I began to overcome my antipathy to the Gothic revival, to see it not as a fearsome instrument of suppression and irreason but as an inventively absurd symptom of a pathology, and a monument to mania and altered states: I did think of calling the Victoria film *God, Pox, Laudanum*. It was, after all, that trinity of afflictions – the supreme delusion, tertiary syphilis and liberal use of opium – which combined to make high Victorian medievalism glorious.

(2000)

FIRST SHACK

AT WEST HARNHAM, which I called 'Dog West Harnham', I at least understood the sources of my pram terrors. There was water: black, impenetrable and algae-edged when I looked left; white, churning and vaporous when I looked right. There was the water's roar of joy as it was given murderous life by the fall through the sluice. Then there was the Alsatian, the leashless beast from the former mill beside the sluice; it was, I knew, a chromatically adjusted wolf, and its roar was as big as the water's. West Harnham accounts in the most fundamentally causative way for my lifelong cynophobia – I loathe any dog that stands higher than my (adult) shins – and for my qualified hydrophobia.

But East Harnham is different. The source of infantile terror there has turned out to be a pleasurable and endlessly pursued preoccupation of adulthood. Harnham Hill is a tract of agriculturally useless land to the south of Salisbury, which a bishop of that diocese gave to the 'people' of the city in the middle of the last century, there to recreate themselves too literally, I fear: the bushes are cobwebbed with underwear and torn tights.

The outskirts of all towns possess such places: places where urbanism has not triumphed but which are not yet full members of the Green and Pleasant; places where there are allotments and smallholdings and fences reinforced with bedsteads. But Harnham Hill is so steep it allows no elision. On one side of the cinder path beneath it rises a tufty chalk cliff (great for maniacal boy sledgers), on the other are the gardens of Victorian houses – well, there were: they are all built over now. Thirty-five years ago there was, at the end of one of these gardens, the thing that terrified me more than the Old Mill's Alsatian, more than the white water, more than their conjoined roar. When I reached this point along the path on my Raleigh trike I would happily identify (wrongly) this or that flower for my father.

'What's that, John-John?' he would ask. And I would come up with a dutiful 'cow parsley'. I would come up with anything in order not to have to look the other way, towards the thing at the end of the garden. Not long before he died, I asked my father if he remembered it; by then he had got used to my preoccupation, but still could not work it out and nor could he remember what I will always remember as My First Shack.

What I averted my eyes from was a railway carriage, converted in, say, 1903 into a summerhouse and subsequently neglected, to the point where it was all dust and creepers. The maroon paint was peeling, this was a metal piebald; it was as though an inanimate, inorganic object had subjected itself to a form of adaptive coloration. Later, when my terror had abated, I would cycle back there on a mere two wheels now and gaze at it. Later still I would cycle further, to the perambulations (boundaries) of the New Forest where, in places called Noman'sland and Bohemia, shacks had been erected on common ground in the early years of this century. And there was another place called Ogden's, and one called Lover. It seemed that the Wild West was on my doorstep. Corrugated iron (a hundred years ago this was the material of the future), asbestos, tree trunks, telegraph poles . . .

I wondered who lived in these feats of architectural bricolage – a railway carriage here, a glider body there. I put two and two together. My mother, who had taught in the port of Southampton, told me about children sewn into rabbit furs for the duration of winter, of others who wore newspapers beneath their threadbare clothes. I put two and two together and probably got it wrong.

Or did I? High above Hay-on-Wye, at dawn, a wildman waving a 12-bore hurtled from his shack towards the car from which I was surveying it. Beside a black reservoir in Staffordshire, a young man exhibited a profane vocabulary that seemed to belong to a hundred years ago. More usually, though, I'm met with kindly incredulity, hospitable curiosity and by a puzzled sense of pride; there are, after all, not perhaps as many shack-spotters as there are visitors to grand houses. What is it that makes me seek out the dispersed but not disparate analogues of that first shack under Harnham Hill? What sends me up

Clee Hill in the Marches, down single track lanes near Bridport, across the moonscape of Dungeness, along the coast near Middlesbrough? Why is Bewdley in Worcestershire a Mecca-like magnet?

In the environs of this pretty Georgian town, world-famous as the birthplace of Stanley Baldwin, is the greatest extant shack colony in Britain. It is here, beside the Severn, that you can get the most powerful whiff of something otherwise entirely lost in these islands – the pioneer spirit. Here are structures no architect has had a hand in, here is a place uncontaminated by planners' wisdom. I guess my initial terror turned to enthusiasm when I realised that shacks were within the capabilities of everyone – they held the same appeal as tree houses. But they are more important than that. At Bewdley one witnesses nothing less than the triumph of untutored invention, of primitive individualism. At Bewdley you learn that everything is mutable, that there is nothing which cannot be recycled. In my lifetime most of Britain's shacks have disappeared. Officious local authorities have accounted for some, landslides and climatic infelicity for others. But the majority have gone because the land they were built on was more valuable than the structures themselves. Where once there was brightly painted wood there is now, more often than not, York stone and double glazing. The Severn valley is different because there farmers leased rather than sold land to those who wanted to erect what were intended as weekend houses.

Shacks mostly date from a brief period when easy mobility (a proper railway system, the early years of mass motoring) coincided with minimal planning laws. Bewdley served Brum and the Black Country, places where things were made, where cars were constructed in lean-to workshops from parts of other cars, where bodging was a way of life, where the night sky was red with the light of a thousand forges. It is not notoriously like that any more. The West Midlands had its heart ripped out, it was rationalised by the very planners who have sought time and again to bulldoze the Bewdley shacks. They appear not to realise that these wonderful structures are the last survivals of the creative urge that earned for Brum the sobriquet 'Workshop of the World'. But Bewdley is not a museum, emphatically not; it is a living community, a village of the utmost strangeness and delight, of a non-conformity (and this is truly rare) which is unselfconscious. It might be rash to style it an unwitting social experiment, but it is more genuinely and successfully that than are the most damaging exercises in 'housing' of the architectural dark age from which we are only just emerging. What we find beside the Severn is folk architecture, a demotic idiom which is the built equivalent of slang: it has that vitality, that urgency, that humour.

An environment which is self-determined to this degree is unusual in Britain. Home-made homes are not part of the mainstream domestic tradition. Of course, there are streets in north London where every house has been built by an architect for himself but that's a different phenomenon; and so is officially sanctioned self-build. Bewdley's are marginal dwellings. They may not be outside the law but they're not quite in it either. They represent a heartening failure on the part of their makers and their inhabitants to go by the book, to subscribe to the quiescent notion of the all-shaping environment. Here they got out and shaped it before it shaped them, and they shaped it with wit and charm and guile and on a generally empty purse. This has fostered a common resilience in the otherwise varied inhabitants. Local pride can be an ugly thing, but here it's born of an intimate knowledge of the sheer struggle it took to create the place (the same families have tended to live here, generation upon generation). This pride extends to the invariable euphemism of 'bungalow' for shack. It's a harmless enough fib, a sort of defence against those whose use of the word 'shack' would be necessarily pejorative.

I obviously don't intend it thus. There's no reason why a shack should not be as grand – as imaginatively grand – as a castle. We're too inclined to respect age and size and noble associations. None of the shacks at Bewdley is more than seventy-five years old, none is large, and I very much doubt that any of them has ever been in the hands of a titled family. But together they add up to a site of collective surrealism, a magic village, a dream of Eden on a shoestring. They, and their begetters, have pluck precisely because they lack so much else.

(1990)

THE FOREST, MY FOREST

THERE NEVER WAS a time when I didn't search for fungi. From earliest childhood I would be woken in the dark and piled breakfastless into my father's rickety car. Dawn would barely be breaking when we crossed the vaporous floated meadows of the Ebble and began the wheezing ascent to the downs, where certain fields – 'our fields' – would be constelled white like a driving range. An hour or so later we would return home to the edge of Salisbury, baskets brimming, appetite whetted, to a breakfast of field mushrooms and horse mushrooms, bacon, eggs. Of my father's hunter-gatherer pursuits this is the only one I enjoyed: beating pheasants was boring; shivering beside March leets while he fished for the first run of salmon was boring and cold.

And my enjoyment of mushrooming was far from entire. It was mitigated by the chalky open terrain where, save mushrooms, there grew only grass, thistles, blackthorns and rusting water-butts. The pleasure was in the gathering. Not in the landscape. In Welsh and French, cwm and combe signify a valley. In Wiltshire English, coomb signifies a riverless valley. Such valleys frightened me: I identified one, in particular, as the Valley of the Shadow of Death (clunkily renamed 'the darkest valley', which doesn't do the trick).

The terrain that delighted me began, however, only a few miles away. Grassless, coombless, sylvan and – when I was a child – fungusless. I don't mean that the New Forest was bereft of fungi, but, despite my mother having gathered 'penny buns' there when she was a child, I was brought up in ignorance of the esculent properties of 'toadstools'. That typically English ignorance, born of mycophobia, was eventually allayed in my teens by the tenant peasant at a friend's château in Entre Deux Mers. He obligingly taught me to identify the few species worth gathering and the few that are harmful: most are neither gastronomically interesting nor poisonous.

There is, unsurprisingly, no English word for the slangy *cepeux*. It means abounding in ceps. The New Forest does just that. This new knowledge reinvigorated My Forest, granted my childhood playground a change of use suitable to adolescence and adulthood. Searching for fungus is the very opposite of playing golf. It is a good walk improved. It is the greatest of autumnal pleasures. And if you gather nothing, well so be it – for you will have leaf-peeped and spotted jays and smelt wood smoke and touched velvet moss and pondered the source of droppings. Are these spraints or are they fewmets? The latter is more likely, for they are the faecal matter of deer, which are everywhere. A few Octobers ago, in a thicket at Bramshaw, I came face to face with a magnificent stag with the build of a bull. It was, thankfully, more alarmed than I was – it bolted. But otters? They certainly live within the confines of the Otter, Owl and Wildlife Park. It is characteristic of the new New Forest that there should be signs warning drivers that these cuddly lovable creatures may be crossing the road – i.e. proclaiming their presence. (Vipers have yet to enjoy kindred advertisements.) Parts of the old New Forest constituted an illusory wilderness, the best that suburbanised southern England could manage. Now, that illusion has not been entirely ruptured, but it is certainly more fragile than it once was.

Oh, the mushroom hunter is as well served as ever. A few days ago we picked four kilogrammes of ceps (predominantly *Boletus edulis*, *Boletus versipellis* and *Leccinum quercinum*) in an hour or so. And had we wished to murder or hallucinate there were copious quantities of death cap and fly agaric: the problem with this latter is that toxicity does not correspond to the size of the fruit body.

The Forest's fungi remain immutable, to date. But it is not beyond the capability of humankind to amend a mycorrhyzal ecology or at least to attempt to subjugate it. Why? Because the exercise of control is a potent drug. Those who doubt it should scrutinise the new New Forest, the National Park in waiting, the smallest and most densely inhabited of such parks. Almost 40,000 people live within its boundaries (or 'perambulations'). Its southern fringe is bordered by the relentless spread of outer Bournemouth, its eastern by industrial villages serving power stations and refineries. The will to grant the Forest statutory protection from the unchecked, aesthetically base development that has despoiled so much of England seems unexceptionable. Yet the Forest has till lately rubbed along rather well, as if to demonstrate that an administrative and social anachronism can flourish. The most visible evidence is, of course, the commoners' ponies and cattle, which enjoy rights of pasturage, and the pigs, which in the pannage season have right of mast, that's to say all the acorns and beechnuts they can snout out.

These and various kindred practices, which have largely fallen into desuetude, belong not simply to the Forest of the nineteenth century when the Court of Verderers was established, but to the Forest which endured till less than two decades ago. That Forest was dominated by rural indigence. Smallholders lived in shacks. Devotees of bucolic squalor were in their element. Corrugated iron was plentiful. The ragged fields were put to work. Men with devastatingly low brows stared out at strangers. Every hedge contained the rusting gearbox of a fifty-year-old Fordson tractor. Every thatched roof had mange.

And today it's an outer Surrey. The heaths are responsibly managed. Trees are pollarded. Ferns are cropped as though they are on a municipal roundabout. Crocodiles of walkers hung with maps for safety and wrapped in purpose-built clothes for warmth venture from one car park to the next: the illusion of wilderness still holds for them, then.

(2004)

ON POMPEY

PORTSMOUTH IS A CITY of 200,000 people. Pompey is a novel of 200,000 words. And that numerical fortuity may be the entire sum of the affinity between my invention and the 'real' place whose nickname I have appropriated.

Portsmouth did not suffer, in the early '70s, a venereally vectored plague; nor was there ever a 200-foot-tall cross on Portsdown Hill; no brewery ever used the slogan 'Blighty's Best Does Pompey Proud'; if there really had been a dream house on the dunes at Hayling, what would have become of the (ever popular) miniature golf links? And although the grotesque revenge exacted by a chippie on a whore did actually happen, it didn't happen in Portsmouth, but near Basra, and in '44, and the man wasn't a chippie but one of my father's troops.

I'm pretty sure that no cultist anywhere, not in this century anyway, blinded herself staring at the sun under the misapprehension that sun and Son of God are not only homophonic but one and the same; it certainly hasn't occurred on the beach at Southsea. There isn't a single tower block at Baffins to hurl a pig from in order to show that it can fly, is thus fowl, and so not subject to dietary proscriptions. The prevailing wind is not known as the Spithead Bite. Packs of feral dogs, unwanted presents taken away by people called Puppymen, do not roam Hayling Island. The ex-servicemen's club called the Done Our Bit is in north London, not on the south coast. And so on.

Why, then, invade this innocent island city? Why pick on this place to sully? Why not coin a new name for it? Why not invent an entire decor for the nightmare to unfold against – the characters and incidents are imaginative constructs (or filchings), so why not the stage? Why transport the freaks and geeks, the monsters from my back brain, here? What has Portsmouth ever done to me . . . I don't know a single one of the 200,000.

On Pompey

Before I conceived the germ of the shadow of the idea of Pompey, I had been there twice: a childhood visit to the Victory; a ferry to St Malo to avoid the Jubilee weekend in '77. Just the first trip, as a twelve-year-old, would, however, have been enough. What impressed itself on me then and what remained with me down the years was certainly not the Victory, a land-born bore that was mute beside narrative paintings of Nelson's death, but the singularity of the site.

A flat city surrounded by water (it really is an island, although practically it's more akin to a bulbous peninsula); an architecture made in the expectation of war; a grey belligerent stateliness that the first syllable of Pompey (etymology unknown) might be referring to; endless terraces of tiny houses in a relentlessly right-angled grid – martial Mars gets the grandiosity, the cannon fodder and the dockyard navvies get two-up-two-down.

And above all this is the escarpment of Portsdown Hill, a sheer cliff that's the northern horizon, that bears the low accents of the fortresses that Palmerston, mistrustful of Napoleon III, had built in the 1850s, although in principle if not detail they might be of the 1550s. From close by they are mighty and monumental. As a child I was frightened by the deep dry moats overgrown with thorns and by the unyielding brick. But from below they are not much; they're too squat, too tight to the horizon. Portsdown Hill needed improving. And you can't beat a cross, can you? Well, you can. You could have a neon calvary with vermilion blood. But if a 500-foot-tall cross was good enough for Franco at his gross hypocritical monument to the dead of the civil war, Los Caidos, a 200-foot-tall cross would be good enough for Portsdown Hill and the Church of the Best Ever Redemption, which I have occupying one of the fortresses.

To have given Portsmouth, Pompey, a pseudonym would have been an absurd lie, for so much of the fuel of the fiction is based in topographical specificity. The founder of that church which is no more or less absurd than any other is a crippled former comedian who killed his wife in a car crash when eight times over the limit; he suffers a vision on Portsdown Hill which is peculiar to that cliff, with its views over the harbours and the straits and the sea. Places get, so to speak, equal billing with persons. The two, a maritime dystopia and its victims, are intertwined, dependent on each other, engaged in a mutual and poisonous succour.

There is a pair of insidious and complementary received ideas about the status and ramifications of place in English fiction. The first is that a provincial setting necessarily implies a provincial mentality, dreary naturalism, parochialism, desiccation, making a virtue of modesty, lack of ambition, the self-curtailment which is the sine qua non of minor art. Etc. This is not an invariably correct point of view and it's most certainly not one which anyone would dare apply to, say, an American or French work with an Iowan or Charentais setting.

The second idea seems to follow almost inevitably. It is that the further the distance – spatial or temporal – between the British writer and his or her subject, the greater the appeal it possesses. Hence the commonplace eschewals of the present (in favour of the past) and of England (in favour of abroad). I'm not entirely innocent: Pompey begins in 1945 and ends in 1976; and it is also located in, inter alia, Brussels, the Belgian Congo, Lorraine (where, north of Nancy, there is a canal village called Pompey). But it is not a historical work, and it is mainly about England, and English peccadillos, and English backwaters, which are endlessly fascinating.

It's just a matter of daring to look beneath the carpet and then of having the stomach to peel the festering underlay away. Decay is supposed to be a gaudy and animated spectacle. But look around: we're meant to have been going down the pan for as long as anyone can recall. It's not especially colourful, though, not particularly lively. Again, it needs improving; it needs heightening, stretching, bending, cutting and tucking. There is a subfusc exoticism in the sticks which can be brought out. Fictional representation has no obligation to be contaminated by the drabness (or any other quality) of its subject. It makes up its own truth, it is uninhibited by reportorial adherence to facts.

But a balance has to be achieved. Licence, unchecked, is the parent of fantasy, of the incredible. There's a mental boundary beyond which it is injudicious to trespass. Let us take, for example, oh, three Chinese extortionists, make them Triads, that's the way, then have them assault the owner of a takeaway; and put them to live in a ménage à trois in a '30s semi in a blossom-filled avenue in the suburb of Cosham. Of course, such people did exist, and were invited to pay their debt to society in the early '80s. I cannot overstate the appeal to me of these troilists and their home and their inept small-timeness but they didn't fit, not even parenthetically.

And the peculations, scams, backhanders and all-round corruption in the city that *Private Eye* zealously reported throughout the '70s had to be binned, too: it might all have been base and sordid, but it wasn't my baseness, my sordor. Equally, there was no mileage in Portsmouth's distinction of possessing what has more than once been voted the worst modern building in England: this is Owen Luder's all-concrete (hence 'brutalist') Tricorn Centre; here was to be found, incidentally, the takeaway whose owner refused to pay protection. I happen to admire its stern and sinuous plasticity but that was neither here nor there.

Still, there were aspects of quotidian Portsmouth, other than its site, which I could use off the peg, unamended. The registration mark STR 1P was due to be granted to a Portsmouth vehicle in August 1975; but some thinker at the DVLC in Swansea spotted it and, believing it would be an incitement to naturism, forbade its issue. I am sure that had that registration come up twelve or so years later, the then minister of transport, the then Mr Cecil Parkinson, who was of the profound conviction that individual freedom was enshrined in 'cherished' number plates, would have allowed an appeal against the Welsh spoilsport. But in the mid-'70s the decision was final.

Then there's the business of OMO, the excellent washing powder which was once claimed to and no doubt still does 'add bright, bright brightness'. There's a story – unsubstantiated, needless to say, but that's no reason to shun it – that it was the practice of the wives of sailors who were away at sea to place a carton of the stuff in the front window to exploit the acrostic properties of the name, Old Man Out; a use which Lever Bros probably didn't foresee. At Tipner an unmetalled road, fenced by shuttering planks and ground elder, leads to a place that has no peer in Britain. It may also be glimpsed from hurtling vehicles on the M275 above it. This is a nautical and military scrapyard. It looks like the garrison of a decrepit private force, Sterling or Walker's arsenal: tanks, armoured cars, bomb disposal robots, anchors the height of a house, the rusting hulks of World War One submarines, perpetually smelting metal, the air littered with particulates. The place had diminished since I made part of a 'documentary' there four years ago; this, apparently, under pressure from the council which clearly failed to appreciate the entropic beauty of the city's most singular spectacle. Me, all I granted it, ruefully, was a sentence: that's the trouble with local colour, it's often the wrong hue. Its temptations are to be resisted. Indeed, it's arguable that the less one knows a place, the less one is acquainted with its minutiae, the more potently it can be imagined and recreated.

My Pompey is a city on to which I've grafted the noxious detritus of countless other cities while adhering (mostly) to cartographic actuality. I guess I could have gone to Kinshasa which, at the time I'm writing about, was still Leopoldville. But why? To check out 'the moist crateral sores on beggars' limbs', to see if 'soft-boned children bite on roots protruding from the unmade pavement', to verify that there are termitaries on the bank of the Zaire at the place where one of my characters meets his awful death? I saved the air fare and bought a map. Inventing the Congo of more than thirty years ago was a cinch beside inventing Pompey. Did the Union Minière du Haut Katanga offer mercenaries pygmy hunts as sweeteners? I've decided that it did. Did the Union Minière begin to plan the Katangese secession as early as 1956? I conjured up a Marxist historian, Pierre Normand, author of *Tshombe ou La Negritude Blanche*, to say that it did.

Any area of which one's prior knowledge is negligible is wonderfully liberating. Pig ignorance can be a boon, a truism known previously only to politicians. Haematology? No problem. I became an expert for as long as it took me to invent a new blood group and a couple of diseases, HoTLoVe and Neild's Syndrome, the latter, a delinquent strain of haemophilia, named for the nephrologist Prof. Guy Neild of London University who gave me a crash course over one course of lunch one day. Haemic malfunctions thus share the appeal of abroad and the past.

I'm presently drawn to something that will include an adopted child's search for its natural mother, euthanasia and traffic management. Those bits will be a doddle: I have no experience of any of them. It's the setting that's going to be difficult – the fluid, ungraspable, unmapped London of today.

(1993)

OH GOD

WE ALL REBEL against our parents. My mother wrote me a frilly edged postcard from Lourdes. I was at school. She was on a trip for teachers. It ended – still ends – with the words 'pity them'. The monochrome verso shows a sort of partially inhabited cave. Her school had been a convent. She hated it, hated it so much that she would detour en route from the station to her parents' house so that I should not see it. She never forgave them. They never forgave her for going to uni. Sure, her life at that school was nothing like that suffered by the girls in Peter Mullan's extraordinary film *The Magdalene Sisters* – but her dismal recollections of its punitive ethic incited me to despise Catholicism, nuns, candles, Mass.

And when I grew up I found that I still despised them. For different reasons, obviously: born of thought rather than experience. It wasn't till after she died that I disloyally walked along the road in Southampton where my mother's school was, still is.

Her son makes his pilgrimage – wrong word – to Lourdes in a spirit of equal disloyalty. My mother said, subsequently: 'Just don't go there, darling.' And, dutifully, I didn't go there. Till now. She was there with her lover, a Scot called Douglas: never met him. I was there with her ghost. I drove along the riverside hotels wondering which one, where Mum, did you adulterate your marriage?

The local word for river is *gave*. The great chef Alain Dutournier describes himself – sentimentally – as *enfant des gaves*. He lives in Paris. Smart idea. For Lourdes is a place which, if you are sentient, you will flee. I ate an OK-ish omelette. Then I set out for conversion. I'm a biddable atheist. Something or other might, just, very improbably, convince me that there is a god, or God. That something is not Lourdes: I am my mother's son. And then some. I didn't feel pity, I felt anger. I felt too that I was in North Wales. The same stone, the same coaches,

338

the same heavy mist, the same fruzz. It might as well be Betws-y-coed the day that Jim Laker took the last of his nineteen wickets and the Empire swelled with congratulation; that was the summer before Suez, the last summer when anyone might say 'What a wonderful country you have there'.

I tend to think that about France even though I am francophilephobic – but, then, so too are the fifty million Brits who live in the Dawdoink and Lottungerron. There is a France – most of France – which anglophones never seek: the industrial north-east (mine host: Jean-Marie Le Pen), Nancy (as fine a set-piece as Bath), Nantes (the subject of Barbara's most beautiful song), Nevers (fill this parenthesis). And then there is Lourdes, a town which makes one thankful for Henry VIII's greedy polygamy and our subsequent post-Protestantism.

I guess that a valetudinarian devotee of Spa in the Ardennes or Vichy in the Bourbonnais or any burgh ending in 'bains' or 'bad' would, initially, feel quite at home. Such towns, and their British equivalents – Malvern, Harrogate, Cheltenham – share a look which derives from their having been built for seasonal, temporary occupation rather than permanent habitation. Lourdes is kindred. It is entirely urban. It looks like a resort. The hotels are numberless. I once stayed a night, for want of anywhere else, in a Christian hostelry in north Norfolk. How horrible was it? How long have you got?

Lourdes is, at least, in a country which understands hospitality and gross exploitation of a site. It could have been any site in the Pyrenean foothills for, as my mother disgustedly observed, 'there are springs everywhere'. It was a poor peasant who suffered a hallucination of the Virgin. Maybe she had forgotten to take her medication. It is certainly poor peasants who frequent Lourdes. There are more *petits gens* than at a Michel Sardou concert (as self-appointed *petits gens* we attend those, my wife and I, the only *angliches* in a mire, sorry *mirepoix*, of French self-congratulation). It is however another performer that Lourdes brings to mind. Bob Dylan. When this genius cum charlatan – and the two properties are invariably linked – played Earls Court in the mid-'70s his gigs were packed out by paraplegics, quadriplegics, the chair-bound and the literally legless. He might cure. Well, why not. As Jules Lagrange has stated, he is a far better rock singer than John Keats ever was, and rock is religion-lite: one of the high priests, Eric Clapton, observed years ago that the best pop songs are electric Victorian hymns (though his lord in Blind Faith's In The Presence Of The Lord is heroin).

Lourdes's supplicants are as willingly deluded as we all are in the presence of Jagger and Jones, Lennon and Hannon. What is different is that rock has no monuments: a CD, a DVD, a tape, an LP or (for older readers) a 45 is not a building. Beside the cave where the ignorant girl experienced her 'vision', whose immemorially smoothed surface the suggestible queue to touch and kiss, there has grown a tectonic abomination, a church that is so horrible it is almost lovable. It is astonishingly irrational, bewilderingly ugly, one of the most powerful testaments to humankind's sycophancy to its most enduring fiction: the material it is built from recalls some kind of bad confectionary, liquorice chews perhaps.

The bereavement of beauty is startling. Devotional superstition triumphs over architectural felicity. Indeed, Lourdes demonstrates an obstinate contempt for the niceties of art. There is a manifest populism at work here. The tons of sacred gewgaws – in plastic, stone, wood, leather, glass, MDF, blockboard, etc. – may be laughable but are nonetheless sternly instructive. The lesson is that their message is of far greater import than their mere form or appearance. The faithless secular aesthete should be chastened. I wasn't. I came, as a dutiful son, to mock. And I left, still mocking, and with the prescribed pity.

(2005)

[SCRIPT #6]
FRAGMENTS OF AN ARBITRARY ENCYCLOPAEDIA
(MEADES IN FRANCE, PART 1, 2012)

STRIP ALONG BOTTOM OF SCREEN AT START OF SHOW:
*No check tablecloths, no 'Gallic' shrugs, no strings of onions, no
art of living in Provence, no dream homes, no boules, no ooh la la,
no accordions, no Dordoink, no cricketing expats, no Piaf, no black
polo-necks, no gnarled peasants, no picturesque bastides, no street
markets, no vive la différence.*
EXCEPT WHEN SPECIFIED THE ONSCREEN WRITTEN TITLES — LIKE CHAPTER
HEADINGS — AT THE START OF EACH SECTION ARE IN ROGER EXCOFFON'S
FONTS: BANCO, CALYPSO, CHOC, MISTRAL, ETC. SOMETIMES THESE TITLES
ARE CENTRE SCREEN; SOMETIMES THEY EXPAND OR CONTRACT; SOMETIMES
IN CORNER; SOMETIMES THEY REMAIN ON SCREEN, OTHER TIMES THEY
DISAPPEAR. ALWAYS BRIGHTLY COLOURED, ALMOST DAY-GLO. THESE TYPEFACES
— ESPECIALLY MISTRAL — ARE SEEN IN THE BACKGROUND OF SHOTS: ON
SHOPFRONTS, POSTERS, VANS AND LORRIES. THIS IS NOT DIFFICULT TO
ACHIEVE. WE DO NOT DRAW ESPECIAL ATTENTION TO THEM.
    JUMP CUTS. TILTED CAMERA. 'WRONG' FRAMING. ABRUPT CHANGES OF PACE/
COLOUR/LIGHT.
    ASTONS. CAPTIONS. SUBTITLES. PLACARDS. GRAFFITI — ALL THESE USED TO
GET OVER HISTORICAL INFORMATION/CONTEXT.
    NON-NATURALISTIC SOUND THROUGHOUT, POST-DUBBED. FLEMISH MUSIC.
GERMAN MUSIC, OOMPAH. 'LOVE TO GO A WANDERING'. BELGIAN BRASS BANDS.
SWISS YODELLING. NORTH AFRICAN MUSIC. LA FARIGOULE (MARSEILLE BRASS
AND WOODWIND BAND).
    FRANCE LOOKS DESERTED. NO PEOPLE IN SHOT IF IT CAN BE AVOIDED. THE
ONLY PEOPLE WE SEE ARE IN THE RURAL DISTANCE: HUNTERS WITH LUMINOUS
JACKETS AND WHITE VANS. GUN SHOTS. EMPTY CITY STREETS AND COUNTRY
LANES. AUTOROUTES WITH NO TRAFFIC (EASY TO ACHIEVE). FORLORN TERRAINS
VAGUES. THE SOUND OF EMPTINESS.
    TITLE: VALISE
    DAY. BRICK OR STONE WALL — MUST BE TEXTURED, ROUGH.
    V/O, ECHOIC, AS THOUGH UNDER WATER, UNEARTHLY, NOT ENTIRELY CLEAR):
    She told me this without rancour: it was merely what had happened
    to her. She hoped to die French. She did.

HOLD WALL — JUMP TO JM IN SHOT
    Every man has two countries, his own and France. This is
    habitually misattributed to Thomas Jefferson. In fact it is a line
    put into the mouth of Charlemagne in a late nineteenth-century
    play by Henri de Bornier — who was French. So it's a predictably
    chauvinistic boast, and not to be taken seriously . . .
        Save that in my case it's true. It *became* true. I frequently
    visited France as a child: Breton beach holidays, cheap hotels,
    roast horse, weird sojourns with my grandfather's elderly business
    friends, supposedly scholastic exchanges in dusty provincial
    towns. I was forever hauling a huge suitcase. *Une valise vaste.*

**DAY FOR NIGHT/LIGHT RAPIDLY CHANGES TO NIGHT. FILM PROJECTED ON TO
FACE, BODY AND WALL. IMAGES CONTORTED BY CLOTHES AND ROUGH WALL.**

LOOP: SLO-MO ALGERIA, OAS, FLN. ATROCITIES. METRO CHARONNE. HEADLINES. OCTOBER 17 BODIES IN SEINE. PAPON. EXPLOSIONS. GENERATORS WITH EXPOSED WIRES [LOOP TO USE THROUGHOUT IN DIFFERENT SIZES]

Then came 1962. That really was the year of *la valise*, the year France became *my* second country.

I was fifteen. Early in April, early one evening, *between dog and wolf* . . .

SUBTITLE: *ENTRE CHIEN ET LOUP* = LATE DUSK, WHEN A DOG AND A WOLF ARE SUPPOSEDLY INDISTINGUISHABLE

. . . I got on a ferry at Dover — and got off in a war zone.

It was the shameful thrill, of war's omnipresence and its fearful randomness, that made France my second country. On paper the war was over. The Evian accords had been signed a fortnight previously. Algeria had got its independence. For the OAS the war was not over. The armed faction of the millions of French citizens who were betrayed at Evian continued to plant bombs and attempt assassinations.

TITLE: VALLIN, EUGÈNE (1856–1922) (ART NOUVEAU LETTERING) NANCY, ART NOUVEAU BUILDINGS — RUE FELIX FAURÉ, NANCY

In the earliest years of the twentieth century the furniture maker Eugene Vallin made his first tentative steps in architecture. Nancy was a celebrated crafts centre: glassware, marquetry, cabinet making, ceramics, metal work. Nancy's art nouveau was not really that new: much of it was revivalism of the rococo and the baroque: art nouveau mostly occurs in places touched by the baroque. It was covert revivalism. These artists heeded Montaigne's counsel to the plagiarist:

HORSE WITH, PROGRESSIVELY, DAY–GLO LIME–GREEN TAIL, THEN DAY–GLO PINK MANE, THEN DARK GLASSES AND WHITE STICK. WHINNIES OF PAIN

'Behave like a horse thief: dye the tail and the mane and sometimes put out the creature's eyes' — i.e. cover your tracks.

30 RUE SERGEANT BLANDIN

For all its whimsy, art nouveau was political. The strain that developed in Nancy was in deliberate opposition to Strasbourg, then a German city whose recent buildings were in the German jugendstil. Nancy defined itself with its own version of art nouveau just as many small countries and aspirantly autonomous regions and city states did: Riga, Catalonia, Liguria. The Nancy School took as its model the English arts and crafts movement.

INTERIOR. EXCELSIOR BRASSERIE NEAR STATION

Unlike the arts and crafts it was not opposed to industrial processes — but it shared the conviction that a well–turned door handle would make the world a better place: so it was the profusion of its enamel and the excellence of its stained glass that prevented the First World War and the bloodbath at Verdun. Needless to say it was the preferred style of *the caviar left*.

SUBTITLE: *LA GAUCHE CAVIAR* = CHAMPAGNE SOCIALISM TITLE: VAUBAN, MARQUIS DE (1633–1707) BELFORT AND/OR LILLE, PALAIS DES BEAUX ARTS, MAQUETTES OF FORTRESSES

Fragments of an Arbitrary Encyclopaedia

Sébastien le Prestre, Marquis de Vauban, was an expert in porcine husbandry, an economist, the greatest of military engineers. He built or rebuilt some 300 fortresses on the country's borders, borders that were ever-expanding. The places where they are sited comprise a litany of belligerence. They are forever associable with battles . . .

BELFORT. EACH OF WORDS IN FOLLOWING PTC SPOKEN IN A DIFFERENT PLACE IN FORTRESS, DIFFERENT FRAME SIZE, DIFFERENT BACKDROP
. . . sieges, evacuations, trenches, trenchfoot, shame, victorious commemoration, mutilation, barracks, disease, humiliation, mass death.

TITLE: VAUDEMONT
EXAGGERATED WIND NOISES. THE SITE FROM A DISTANCE ACROSS THE PLAIN THEN UP ON THE VAUDÉMONT-SION HILL: THE CONVENT, THE BASILICA, THE VIEWS FROM THE HILL, THE BARRES MONUMENT
The hill south of Nancy that Sion and Vaudémont are perched on is holy or spiritual or mystical — something along those superstitious lines anyway. Rosmerta, the Gaulish goddess of fertility, a big girl, was honoured here in the fourth century BC. The Romans erected a temple to Mercury. The first Christian site was of the fifth century. The basilica was built in the 1870s. Its tower has a Mary on top of it. She's twenty-five feet tall. Another big girl. If she's capable of virgin birth why should she not have an over-eager pituitary gland?

When Alsace and northern Lorraine were returned to France after the First World War a folksy ceremony of reunification was held here. It was presided over by Maurice Barrès. The hill was by then known as La Colline Inspirée after his novel of that name which is set here and which treats the tension between Catholicism and the nationalistic stirring supposedly prompted by the animism of this place.

The reason *this* hill should have supernatural attributes and imaginary properties dumped on it is clear. It is the only hill for miles around. The topographically prodigious is routinely claimed for god when it actually belongs to the marvels of geological happenstance.

Barrès was a politician, an eventually repentant anti-Semite, an unrepentant anti-Dreyfusard, a clubbable bigot. His life was irremediably coloured by the German seizure of Alsace and northern Lorraine when he was eight years old. He was a xenophobe yet was steeped in German literature. He was a Lorraine supremecist who subscribed to a doctrine akin to *Blut und Boden — la terre et les morts.*

SUBTITLE: THE EARTH AND THE DEPARTED
I am not free to think as I wish. I can live only in relation to the dead of my race. They, and my country's soil, tell me how I should live. He believed that his native forests spoke to him.

Much that he wrote was tosh. But he wrote it captivatingly. It was not just the gullible whom he inspired but an entire generation. Vaudémont Sion was proclaimed the sacred hill of

the nation. And despite its being on the country's very margin
Barrès persuaded France that Lorraine was its spiritual heartland.
Whatever that meant.

DOMREMY. JOAN OF ARC HOUSE

The celebrity witch Joan of Arc also came from Lorraine. The month
after she was canonised, the Chamber of Deputies voted in favour
of Barrès's plan that a national day should be devoted to her.

VOICE FROM SKY:

Elles crament bien, les pucelles.

SUBTITLE: THESE VIRGINS BURN NICELY

**COLOMBEY LES DEUX EGLISES. DE GAULLE KNICKNACKS. LA BOISSERIE —
GROUNDS VISIBLE FROM DE GAULLE MUSEUM**

Charles de Gaulle, whose idea of nationhood owed much to Barrès,
chose to live at Colombey les Deux Églises in Haut Marne on
the edge of Lorraine. He had no connection with the area.
Strategically, it was between Paris and the garrison towns on the
German border.

**LORRAINE LANDSCAPE — STEPPE AND FOREST, BOTH OF THEM HARSH. FROM DE
GAULLE MUSEUM LAYER UPON LAYER OF DISTANT HILLS**

But in that part of the brain where psychology and topography meet
it was more than that. De Gaulle wrote of 'vast, raw, forlorn
horizons . . . melancholy emptiness where nothing changes. Not the
spirit. Not the place.'

WOODS NR COLOMBEY

De Gaulle, like Barrès, was a solitary who immersed himself in
the depths of woods, took succour from the signals he discerned
in nature. Like Barrès he believed that he was communing with a
higher reality.

Destiny, then, is a delusion prompted by . . .

**AUDIO: EXAGGERATED NOISES OF THE FOLLOWING SO SPEECH IS DROWNED AND
IT HAS TO BE SUBTITLED**

. . . birds' chatter, leaves gusting through forest naves,
shafts of sunlight, raucous twigs, dapple, sprinting clouds and a
multitude of unseen small animals, scurrying.

**COLOMBEY. 45-METRE-HIGH CROSS OF LORRAINE. C/U ON JM. CAMERA PULLS
OUT. CROSS IS REVEALED HIGH ABOVE THE WOODS AROUND IT**

In July 1940 the opium-smoking, job-seeking admiral Émile Muselier
sucked up to de Gaulle by proposing that the Cross of Lorraine
should be adopted as the emblem of the Free French. The two-barred
cross is of Byzantine origin.

EXAMPLE OF CRUCIFIXION WITH INRI BAR

The top bar was added in sacred iconography so that it might be
inscribed INRI — Jesus of Nazareth King of the Jews.

HERALDIC DEVICES. ILLUMINATED ILLUSTRATIONS FROM BOOKS OF HOURS

The device is to be found in the coats of arms of Hungary, Poland
and Slovakia. It reached Lorraine by a heraldic-genealogical route
that involved the dukes of Anjou.

Such a cross would be a potent symbol for the Free French

because it graphically defied the hooked cross — the swastika. De Gaulle's predictable acceptance of Muselier's proposal further augmented Lorraine's religiose nationalist mystique. Lorraine stood for France.

TITLE: VAUGEOIS

The genealogy of certain strains of nationalistic doctrine and cultural protectionism is risibly straightforward. Here is a family tree of a simplicity which matches the simplicity of the unchanging ideology that is handed down from one generation to the next.

BELFORT LION

There are no skeletons in this ancestral cupboard, no by-blows, no exogamous liaisons, no fruits of careless miscegeny, no métissage, no mongrelism, no papal nephews, no first Mrs Rochesters, no jabberers in a distant bin, no mixeté, no secret remittance men, no wayward scions with Nellcote habits, no black sheep, no black *anything*. Everything is B B B: blanc, blanc, blanc. Everything is P P P: pure, pure, pure. And perfect.

Henri Vaugeois founded Action Française in 1899. It was both a movement and a newspaper propagating that movement. It was anti-Dreyfusard, anti-Semitic, anti-corporatist, anti-democratic, anti-republican, anti-protestant, anti-masonic, xenophobic. Maurice Barrès was soon involved. So too was the demagogue Charles Maurras.

It was Catholic and monarchist. It supported Mussolini and Franco. However its enthusiasm for Hitler was tempered. Not because it disapproved of his doctrine but because it still hadn't got over the defeat by Germany at Sedan more than 60 years before. And its subsequent pro-Vichy stance derived from the odd conviction that by collaborating France would retain its identity.

Identity is a perennial concern of the far right. Its enemies are cosmopolitanism and rootlessness. Identity in this sense is a form of communitarianism which defines people by their race and inherited culture rather than by their individuality, aspirations and talents. It is a prison.

In the years after the war — the Second World war, this — one of Action Française's street hawkers was the young Jean Marie Le Pen. Later he would serve in the army in Indochina and Algeria, and enter parliament as its youngest member in 1956 under the banner of Poujadisme, which took its name from the stationer and rabble-rouser Pierre Poujade, the deafening voice of the silent majority.

There is then a direct line from the nationalism and revanchism fomented by the defeat in 1870 to the chippy paranoiac nationalism of the present day. There's a direct line from Barrès and Maurras to Jean Marie and thence Marine Le Pen.

This fringe is always with us. It is, also, always impotent. Its erectile dysfunction is due to its being a perpetual dupe.

More skilful populists incessantly portray it as a no-hoper, an enduringly lost cause, a bogey — while at the same time filching its ideas and sanitising them.

Every other strain of French politics changes, by the month, by the week. The absurd right's inflexibility is a national, and, of course, nationalistic, marvel. It stays put, stays pure, stays white — mostly. It stays wounded. Like Philoctetes: its sores will never heal.

TITLE: VEDETTE
NOISE OF EXPLOSIONS CONTINUES. ZOOM IN ON WASHING MACHINE REVOLVING. THROUGH THE GLASS, AS THOUGH THROUGH THE LOOKING GLASS, TO C/U OF ELECTRIC CARVING KNIFE, 45 RPM RECORD CHANGER, M OF MOULINEX LOGOTYPE, COLLARLESS CARDIN JACKET — SHOW ONLY PART OF THESE ITEMS > EARLY 1960S ADVERTS (COULD BE MOCKED UP) OF THE UTOPIA THAT ELECTRIC KETTLES HERALD

The mortally wounded OAS's endgame was played out against the glossy backdrop of *Les Trente Glorieuses* — which were not yet so called.

Les Trente Glorieuses. Today it's an everyday phrase. It derives from the title of a book of the late 1970s by the economist Jean Fourastié. It signifies the roughly thirty years from the liberation of France until the first oil crisis in 1973. A period of exponential industrial growth and the rush towards modernisation — both public and private.

SUPERMARKET. TRACK ALONG AISLE OF VAST HANGAR-LIKE STRUCTURE. JM ATTACHED TO DOLLY

The standard of living rose threefold during this period.

Fourastié was no doubt an ironist. The word glorious was a provocation. In his estimation, glory derives from the spread of affluence, the triumph of consumerism, medical advances, improved education, improved working conditions, improved public services, etc. I own, therefore I am. I am happier, I am healthier. Now that I drive a Simca Vedette and possess plenty of plenty the state is no longer *solely* an ideal to which I bear quasi-religious fealty and adolescent resentment. It is also a supplier with which I enjoy a quasi-commercial relationship in exchange for my taxes, my heavy taxes.

Material change does not cause moral progress. Nor does it extinguish pervasive belief systems. The human appetite for irreason was not diminished by synchromesh and transistors. Ancient superstitions and immemorial rites persisted. Material well-being does not eradicate them from the obstinately retentive paleocerebellum. Humans do not have to shed one idea of life in order to accommodate its antithesis. We are more like trees than like snakes.

ARCHIVE. A FORMAL CEREMONY. SARKOZY. THE TRICOLOUR. TROOPS IN FANCY UNIFORM, HORSES. ARCHIVE CUT UP AND FRAGMENTED

For Fourastié glory had, then, little to do with the nation state's mystical allegiance to soil, with self-regarding pomp and vaunting exhortations to patriotism. Which was just as well — for these were otherwise startlingly *in*glorious years for France, years marked by military ignominy, political chaos, political assassinations, attempted putsches, a bloodless *coup d'état*, presidential mendacity, social catastrophe, near revolution and

Fragments of an Arbitrary Encyclopaedia

war upon grinding war. For a martially preening nation France's war record in the twentieth century was dismal. Played four, lost four. *Throwing the sponge* is habit forming.

SUBTITLE: *JETER L'ÉPONGE* **= TO THROW IN THE TOWEL**
RUN AS STRIP ACROSS BOTTOM OF SCREEN OVER NEXT (DOUAUMONT) SECTION
Germany's linguistic invasion. Or French 'borrowings'. Vehme = a kangaroo court, condemnation by hearsay. Schlage = a blow, beating up a prisoner. Rafle = round-up. Putsch = coup. Bivouac = bivouac. Bunker = Blockhaus, Blockhaus = bunker, Fifre = fife, Arquebus = muzzle-loader. Vasisdas = What is that. Qu'est que c'est que ça?

TITLE: VERDUN; AUDIO: 'VERDUN' BY MICHEL SARDOU
THE OSSUARY AT DOUAUMONT. FORMAL. SINGLE SHOT — VERY LONG, VERY SLOW.
ZOOM IN ON CENTRAL AXIS. MIX THROUGH THE LONG LINES OF CROSSES.
V/O (AFTER A COUPLE OF VERSES OF THE SONG)
Verdun: the ten-month battle in 1916 claimed 300,000 lives. The name is France's greatest trigger to remembrance. It is to France what the Somme is to Britain. One difference, of course, is that Verdun is on French soil.

A second difference is that beside Lutyens's monument to the Somme's victims at Thiepval the Douaumont ossuary seems architecturally inappropriate. It is not frivolous but it lacks solemnity, gravity. It recalls the architecture of pleasure, monstrously distended. It is inimical to meditative remembrance. Its architects lacked the nerve to address the awful purpose of the monument and shamefully made light of it — they effected a 140-metre-long betrayal of the dead. One might equally suggest that they were victims of the modernist century's incapacity to devise a commemorative mode. No century ever needed one more.

MIX TO BAR LE DUC: EGLISE ST ÉTIENNE. SKELETON STATUE SHOWN IN LINGERING DETAIL
TITLE: VERLAINE
This is Paul Verlaine's Ode to Metz:
> *Fate made Metz my cradle.*
> *Raped, it remains demure, more virginal than ever.*
> *Childhood was bliss in this place — whose fortress was no Fortress, for its Commandant's weapon was a white flag.*
> *This was the proud mother I loved.*

METZ STATION
The commandant in question was François Achille Bazaine. In 1870 he threw the sponge after Metz had been besieged for two months, the animals had been eaten and typhus had become rampant. Hardly a cheese-eating surrender monkey as the bosch Donald Rumsfeld would have it.

SUBTITLE: UN PRIMATE CAPITULARD BOUFFEUR DU PUANT
When France's porous border admits the invading armies of its bad neighbour it admits, too, German buildings. The station at Metz is wholly German: the round arch, neo-Romanesque idiom forms no part of France's architectural lexicon. Barrès described it as an

immense squat meat pie. A characterisation that is more xenophobic than aesthetic.

METZ VILLE NOUVELLE AND GOVERNOR'S RESIDENCE
The station, and the burghers' palaces nearby, were built after northern Lorraine was annexed in 1870 at the end of the Franco-Prussian war. They are wholly German. These are didactic buildings, contrived as instruments of subjugation. They are architectural manifestations of Prussia's imperial power.

ÉGLISE ST QUENTIN (IN BURBS) AND METZ CATHEDRAL
The restorations of existing buildings were designed to rewrite history and render them teutonic.

TITLE: VÉROLE
JM WALKS PAST WALL WITH GRAFFITO: *Plaisir d'amour ne dure qu'un moment. Chagrin d'amour dure toute la vie.*
Vérole. An extra supplied by the *vivandière.*

1962. I had never seen graffiti. They did not exist in the tidy obedient untroubled England of the early 1960s. There was no need for them. Graffiti were the mute howls of the suppressed, the laments of the betrayed.

GRAFFITI — IN DIFFERENT HANDS. SEPARATE SHOTS IN DIFFERENT LOCATIONS. JUMP FROM ONE TO THE NEXT. '21.00: À GENOUX, C'EST L'HEURE DE LA PIPE'
AUDIO: 'LES SUCETTES' — CONTINUES OVER FOLLOWING GRAFFITI: 'ALGÉRIE FRANÇAISE', 'OAS', 'FLN', 'ICI ON NOIE LES ALGÉRIENS', 'MITTERRAND + THATCHER = SIDA'
TITLE: VERSAILLES
NIGHT. C/U ON STONE WALL. TRACK ALONG WALL. PULLS BACK TO REVEAL A CEMETERY
1962. The night I arrived at Dunkirk the OAS fighter Lieutenant Roger Degueldre was arrested in Algiers.

He was tried by a kangaroo court. Edgard de Larminat, the general who was to have presided over it, killed himself rather than sit in pre-judgement. On 5 July 1962 Algeria became officially independent. The next day, Degueldre was clumsily executed by a firing squad most of whose members deliberately missed. This judicial murder took half an hour.

EIGHT SHOTS, INCREASINGLY *SOTTO* SCREAMS, DEATH GRUNTS
The sentence was devised to appease *bien pensant* piety, to sate liberal bloodlust. After his death he was the object of black propaganda which claimed he had been *a Belgian*, a collaborator, a member of the Waffen SS Wallonian Rexist brigade. He had in fact been a member of the Resistance — as had many other members of the OAS. This, of course, did not, and does not, accord with the calumnisation of the OAS as fascist.

On the contrary: as Winston Churchill had observed twenty years previously, the quasi-fascist was Charles de Gaulle who conducted *his* resistance — and his career trajectory — from the safety of Carlton Gardens. Degueldre's grave in the Gonard cemetery at Versailles is the site of an annual ceremony attended by his

fellow outcasts who survived, men and women turned into pariahs by
de Gaulle's treachery. It routinely prompts protests. Algeria is
always with us.

TITLE: VESOUL; AUDIO: BREL SINGS 'VESOUL'
SHOT LIKE A POSTCARD. THE TOWN NAME IN THE CENTRE. EACH QUARTER
OF THE 'CARD' SHOWS A DIFFERENT SCENE. A V-BUS. THE LYCEE. A
RUNNING TRACK. CONFORAMA. SHOOT FROM BRIDGE OVER ROAD NEAR EX-URBAN
WAREHOUSES. A ROUNDABOUT, ETC. CAPTION: 'JACQUES BREL WAS BELGIAN'
TITLE: VEXATIOUS LITIGANTS
STRASBOURG JUGENDSTIL. RUE RAPP, AVE DES VOSGES, ETC. FRENCH
IMPERIAL: PLACE DE LA RÉPUBLIQUE

The conjunctions of Germany and France, of Belgium and France,
of Italy and France have produced no significant architectural
mongrels. Lille is Parisian boulevards and Flemish back streets.
Urban Strasbourg is either ur-Deutsch or hyper-Français. No fusion
has occurred. Architectural apartheid was the rule until modernism
became the norm.

STRASBOURG EURO QUARTER. EUROPEAN COURT OF HUMAN RIGHTS AND OTHER EU
BUILDINGS

The European Court of Human Rights derives ultimately from the
1789 Declaration of Rights of Mankind and Citizens. It is, of
course, misnamed.

It is the European Court of Special Pleading. We are all
showered at birth with the promise of potential entitlements.
Should that potential not be realised we can endlessly complain
here and so profit the pious shysters of the human rights industry.

No matter. The architecture of this dodgy institution on the
French-German border is indistinguishable from that of buildings
on the Polish-Lithuanian border.

Homogenisation has its benefits. Modernism has no nationalist
etiquette attached to it. It is the pan-European idiom, it does
not express difference or diversity. The constant injunction to
celebrate vibrant diversity is half-witted. It is *shared* qualities
that should be appreciated. Emphasising differences merely consigns
people to the prison of their tribe, their caste, their religion.
It creates ghettos.

MONTAGE OF NON-FRENCH NAMES: SHOP SIGNS, STREET SIGNS, POLISH FOOD
SHOPS, BULGARIAN FOOD SHOPS, IRISH PUBS, ARAB GROCERS, MALIAN PHONE
SHOPS

When not being invaded, France opens its sieve-like borders to
anyone who wants to be French. Immigrants bring with them their
mores, language, clothes, cooking, genes, religion, poverty — and
are expected to dump the lot. They will be accepted so long as
they can somehow render themselves French — which is the noblest
fate any being or any thing can enjoy. This touching delusion
colours much of the behaviour of France - the enlightened republic
whose values are universal. Universal in this context means
French.

France's regionalism — France's *fragmentation* — is exacerbated
by its borderland absorption of Belgium, Germany, Switzerland,
Italy, Spain. What is notable is the absence of reciprocity.

France borrows from its neighbours. Approximate linguistic expressions apart, its neighbours are less prone to borrow from it. Universalism doesn't travel.

TRAD WINSTUB (E.G. CHEZ YVONNE, STRASBOURG OR AU NID DE CIGOGNE, MUTZIG) ALSACIEN — I.E. GERMANIC — DECOR

What do you call good German cooking? You call it Alsacienne cooking. Boastful, but probably true. Alsace enjoys the deserved reputation of being the most gastronomic region of the most gastronomic country on Earth. Yet there is an obvious paradox here, for its cooking is only French by appropriation. The characteristic dishes are German: *zwibele weie, presskopf, baeckoffe, flamenkuche.* And many dishes arrived from even further east with the Ashkenazi diaspora. The bias against Dreyfus was more than anti-Semitic: he was a Jew whose family had lived, until its annexation, in Alsace, a place of ambiguous loyalties that excited suspicion in deepest, unequivocal, French France.

TITLE: VICHY

The precise figures will never be known. Private fear. Public shame. The actively collaborationist and the actively resistant represented a tiny fragment of France's population. Both extreme factions attracted turncoats who moved between them — the invariable behaviour of the politically opportunistic. The myth of resistance and its counter-myth have contaminated France for well over half a century.

HAIRDRESSER'S SALON. WINDOW STICKER: 'RELOOKING STYLE 'PIPEUSE DE FRITZ' 25€'

For every shaven-headed woman, for every Laval justly executed . . .

SUBTITLE: PIERRE LAVAL, 1883-1945, PRIME MINISTER, 1942-44

. . . for every scapegoat like Brasillach condemned by a show trial . . .

SUBTITLE: ROBERT BRASILLACH, 1909-45, NOVELIST, PAMPHLETEER, CONTRIBUTOR TO L'ACTION FRANÇAISE

. . . there were numberless subsequently amnesiac officials who simply swapped one regime for another. Well-connected war criminals rose through the peacetime ranks to powerful positions under the patronage of all the presidents of the Fifth Republic, from de Gaulle to the laughably compromised François Mitterrand.

Just as their forebears were unquestioningly obedient in their submission to Hitler and to the dictates of the racial state, subsequent generations of Germans have been unquestioningly obedient in their expressions of shame and ancestral culpability.

France has suffered no such collective expiation. The occupation was another episode in France's interminable civil war: right against left, monarchism against republicanism, Catholic against the evil alliance of masons, Protestants and secularists. There was no unanimity while they were being waged. There has been little unanimity since: hence no peace and reconciliation, whatever they may be. Allegiances are, after all, provisional, forever shifting. Yesterday's enemy is today's ally. For the moment.

Fragments of an Arbitrary Encyclopaedia

The OAS included former collaborators and former resistants, men who had been deported to Dachau and Mauthausen. The black propaganda continues fifty years on. Denunciation, forged documents, satanic rumour mills. The truth is never simple: the Minister for the Colonies in the first Vichy administration was Henri Lemery. He was black.

TITLE: VIDELA
NIGHT. CROSSES IN A MILITARY CEMETERY

Roger Degueldre became a martyr of sorts. Most of the OAS's members suffered a more banal fate. Georges Bidault — another resistant, the successor to Jean Moulin as president of the Conseil National de la Resistance and twice prime minister of France — went into exile till an amnesty was declared in 1968.

Even after the amnesty many remained in Alicante. Others migrated to London where they prospered in the wholesale vegetable trade. A few ultras found their way to Argentina.

MONTAGE: THE 1978 WORLD CUP — TICKER TAPE, KEMPES, VIDELA'S HARD MOTIONLESS FACE. AUDIO: CHEERS TURNING TO SCREAMS

Nostalgic for the good old days, they taught the crafts they had practised in the dungeons of Oran and Colomb-Béchar to General Jorge Videla's state sanctioned torturers.

TITLE: VIENNA
FISHING PONDS AT METZ/VAUX. JM HOLDING BAGUETTE

The baguette is not French. Its creation myth has it was introduced to Paris from Vienna in 1830s by August Zang. In France Zang was a baker, in Vienna he was a press baron. This is a not uncommon combination of pursuits: One thinks of Rupert Murdoch, for instance.

C/U CHINESE EGG CUSTARD TARTS
TITLE: VIET MINH ('ORIENTAL' SCRIPT)

France withdraw from Indo-China after the debacle at Dien Bien Phu in 1954. It might have been yesterday. Georges Boudarel was a French teacher in Saigon who had defected in 1950 to the Viet Minh. He set about re-educating French prisoners. When they proved ineducable he tortured and starved them. He was responsible for the deaths of almost 300 of his compatriots. He was sentenced to death in absentia.

Years later one of his victims who had survived recognised him, in Paris. He was put on trial for crimes against humanity. He had embarked on a university career, teaching history — of course. Scores of academics, useful idiots, came out in his support, muttering about colonialist revisionism and a manhunt. Boudarel was freed under the terms of an amnesty.

Daniele Minne aka Djamila Amrane was also amnestied. She had planted a bomb in an Algiers café in January 1957. Four women died, forty people, again mostly women, were wounded, five children and many adults suffered amputations. She, too, pursued an academic career which culminated in a chair at Toulouse University. Her disciplines — you couldn't make this up — were women's studies and the history of decolonisation. There are countless kindred instances of the scum rising to the top.

The French state's masochistic honouring of its enemies and its sadistic contempt for *les petits gens* is engrained. This is only comprehensible when you consider that the French state, having no one to rebel against, rebels against itself. The fifth republic was declared in 1958. Its four predecessors each committed suicide in an access of self-hatred. Infantile leftism is not the exclusive lot of the left. The right, too, is of the left. Or it was: Nicolas Sarkozy's crime in the eyes of the *bien pensant* politico-media-judicial establishment is that he does not subscribe to this consensual charade.

TITLE: VISIONARY, VISIONARIES
HEAD SHOTS/PORTRAITS OF LEDOUX, GODIN, LE CORBUSIER, PASCAL
HAUSERMANN. (EACH NAME CAPTIONED IN CALYPSO FONT). LEDOUX PORTRAIT
EXPANDS TO SHUT OUT THE OTHERS. WIPE TO: ARC ET SENANS, NEAR BESANÇON
Claude Nicolas Ledoux's royal saltworks was a utopian saltworks. Utopianism is relative. It must be measured against the norms of its time. For those who lived and worked here, it was humane, salubrious and a world away from the incarceratory treadmills and noxious slums of salt works thitherto. The fragment that was built is cyclopean, aggressive, fetishistic, proto-brutalist, lowering, exhilaratingly sullen . . .

Ledoux was French. His architecture wasn't. His inspiration was Vanbrugh whose buildings he visited in 1769 and 1770. Vanbrugh is wrongly pigeon-holed as baroque, Ledoux is wrongly pigeon-holed as neoclassical. These labels are fatuous. There are no schools, only talents.

Among this city's unrealised buildings was a utopian brothel whose utopian ground plan was in the form of a utopian penis — that is, fully and permanently erect. Such architecture, whose function is advertised literally, would come to be called *architecture parlante.*

SUBTITLE: ARCHITECTURE THAT TALKS
A public toilet in the form of a syringe. A new House of Commons in the form of a suitcase full of money or a cash dispenser, that sort of thing.

FULL FRAME GODIN PORTRAIT, LE FAMILISTÈRE, GUISE, NEAR LAON
Jean-Baptiste André Godin was an oven manufacturer in Picardie. The company that bears his name still exists. So too does his Familistère, realised according to the ideas of Charles Fourier soon after the revolution.

Fourier's plan was for the world's population to be divided into groups or phalanxes of 1,800 people — autonomous, polyandrous, communal, distributist. Once society was thus organised a period of 35,000 years of harmony would follow. Obviously.

Godin's scheme was more modest. Nonetheless he housed over half a phalanx. Children were educated according to Fourier's methods — precursors of today's discredited 'child-centred learning'.

BRIEY, NEAR METZ. FULL FRAME LE CORBUSIER PORTRAIT. L'UNITÉ
D'HABITATION SEEN FROM DISTANCE RISING OUT OF WOOD.

Fragments of an Arbitrary Encyclopaedia

JM IN CORBUSIER CHAIR IN FIELD

The modernism of *les trente glorieuses* was a direct reaction
to the formal, smooth, pre-war modernism whose master was Le
Corbusier. The reaction to Le Corbusier was led by Le Corbusier.
An artist who was forever reinventing himself.

DETAILS OF L'UNITÉ. C/U OF COLOURS ON BALCONIES

He was a Swiss peasant who wanted to be a French genius. He was a
sculptor; a catastrophic theorist; collagist; aphorist; totalitarian
toady; collaborator; monk; socialite; cultural colonialist; utopian
follower of Fourier and Godin. The five *unités d'habitations* that he
designed owe much to the example of those pioneers.

Once the machine is taken for granted it no longer demands
glorification. Like Roger Excoffon's typefaces Le Corbusier's post-
war manner uses machines but doesn't worship them.

**CHAPEL AT RONCHAMP, NEAR BELFORT. OFTEN PHOTOGRAPHED, OFTEN FILMED.
SO SHOOT FROM UNFAMILIAR ANGLES. FULL FRAME. NO SKY. SUBTITLE: NOTRE
DAME DU HAUT, RONCHAMP, HAUT-SAONE**

The architecture is plastic, expressive. There are deliberately
rough edges. The materials play at primitivism. Purity of form is
suppressed. Impurity of form is more interesting.

Did I realise this in 1962? No. But it did prompt wonder and
delight. I didn't ask myself why. I didn't make a link to the
Citroën DS and to the Mistral typeface. Ronchamp was a piece of
an unmade jigsaw which, whatever it ended up looking like when
finished, would proclaim the conjunction of France and tomorrow. In
my second country the future had already arrived. Rather, *a* future
had already arrived. A future that had nothing to do with nuclear-
tooled ideological gangsters in 405-lines black-and-white. That
future did not belong to us.

**RAON L'ÉTAPE, 70 KM SE OF NANCY. MUSEUMOTEL. 'ORGANIC PODS'. SHOT TO
EXCLUDE THE OUTSIDE WORLD — AS THOUGH THE 'PODS' ARE ALL THERE ARE ON
EARTH**

The French future, on the other hand might have looked as though
it had suffered multiple amputations and would scream if it could
— but we had a share in it. It was ours. It was here. The future
existed in the present.

Pascal Hauserman who designed it was Swiss. Like Le Corbusier,
like Godard, like Rousseau, like . . .

**VOICE DROWNED. FULL SCREEN PORTRAIT OF HAUSERMANN
TITLE: VITRINE**

Les trente glorieuses saw the advent of Mobilettes, transistor
radios, kitchens. It saw the advent of bathrooms and salubrious
lavatories.

The line that Peter Nichols gave to one of his grotesques in
The Gorge became ever less applicable:

VOICE FROM SKY

Where the French fall down is in the toilet.

Soon everyone had a telly; once Britain was able to boast,
smugly, that the French also fall down on television. No longer.

And soon everyone had white goods, electric blenders, Ma Griffe and Vespas, Cacolac, Bic pens, Colibri lighters shaped like a pebble, 45 rpm discs, multichanger record players, gadgets, more gadgets, the world's most thrilling car, the Citroën DS.

People could at last go beyond *window licking*.

SUBTITLE: LÈCHE VITRINE = WINDOW LICKING = WINDOW SHOPPING
Les trente glorieuses had their own *colours*. There was a time when every other car was cobalt blue — somewhere between the colour of Gitanes and the colour of Gauloise, two of the country's predominant scents along with urine, sewers and two stroke fuel. They had their own materials — Formica, tergal, kerval.

WHEN THE NAME OF A FONT IS UNDERLINED THE FONT NAME, WRITTEN IN THAT FONT, JUMPS UP TO FILL THE SCREEN WITH JM STILL VISIBLE BEHIND IT
Nations and cultures covertly define themselves by their typefonts just as they more manifestly define themselves by the languages written in those fonts: Alte Schwabacher and Plagwitz in Germany; Euskara, Etxeak snd Kaxko in the Basque country; Corcaigh and Paternoster in Ireland. Brito and Breizh in Brittany.

These are logos — they are as political and exclusive as a uniform or a flag. They express nationalist sentiments and secessionist longings.

B&W HEAD SHOT OF EXCOFFON. SCAR PICKED OUT IN PURPLE
Les trente glorieuses had their own fonts . . . the creations of the greatest typographer of the age, Roger Excoffon.

His fonts shun folksiness. They do not evoke any sort of past. They abjure the heritage lark. He did not self-consciously set out to create types that would come to be identified with France. That just happens to be their unforeseen fate.

THIS SECTION IS NEITHER V/O NOR PTC BUT TEXT USING MISTRAL WITH INTERRUPTIONS FROM CHOC AND CALYPSO. THE TEXT FILLS THE SCREEN, IT ROLLS NOW VERTICALLY, NOW HORIZONTALLY. AUDIO: BARBARA'S '*DIS QUAND REVIENDRAS-TU?*' MISTRAL
Excoffon's creations represent a rupture in typographical practice, a break with the formality and sobriety of sans serif modernism — which was supposedly functional and which adhered to the machine ethic.

They make a further break. Most typefaces that are founded in handwriting — scribes — take as their source italic or copperplate scripts. Mistral is based on Excoffon's own hand. The sheer ingenuity is extraordinary. Other scribes do not attempt to join the letters. Excoffon devised a method of doing so, of eliding all the letters of the alphabet, of accommodating every permutation. Each letter is moulded, even contorted, so that it possesses a sort of universal joint.

Legibility, sense, is often apparent only through a letter's conjunction with its neighbours. In isolation the marks do not invariably read as letters.

CHOC
This tendency was taken further in Choc which is heavier and

which lacks ligatures. It is based not so much in handwriting
as in scrawl. Characters are pushed to the limit, abstracted.
Hieroglyphs? Ideograms? Sketches of fleeting phenomena glimpsed
out of the corner of an eye from a fast train? A bird's swoop, a
pollard, a gambolling lamb?

MISTRAL

Within a couple of years of its publication in 1953, Mistral was
being used in the most improbable circumstances — on the cover of
the Playfair cricket annual, which is where I, a child obsessed
by handwriting, first saw it. But it remains sutured to France,
a cricket-free zone. And of all the works of applied art of *les
trente glorieuses* it is the most enduring. It had no precursors.
And no successors save a few borrowings — the Radisson logo, for
instance, attempts to combine Mistral and Picasso's signature.

SHOW LOGO AND PICASSO'S SIGNATURE

Such borrowings are so patently plagiaristic that they might have
been created expressly to emphasise Mistral's uniqueness, to show
that great art leads nowhere because it is inimitable — think of
Barbara, singing now.

TITLE: VIVANDIÈRE; AUDIO. SONG SUNG DRUNKENLY

> *Gauche, droit, sabre au côté!*
> *La cantinière se laisse baiser(2X)*
> *Et en avant et en avant la cantinière!*
> *La cantinière du régiment(2X)*

NIGHT. ILL-LIT STREET. BLEAK. HIGH WALLS LIKE BARRACKS

Garrison town. Barracks town. They're all in the east. They
all look the same, like *lycées*. High walls. Toxic canals. Grim
stations. Old defences that failed to defend.

And *vivandières. Cantinières.* The women who followed the
troops, who cooked meals, supplied drink and sex. The *vivandière*
and the *cantinière* occupy a high place in the low imagination.

TITLE: VOIE SACRÉE. MAGINOT LINE FORTRESS. CIVILIAN GUN FIRE ADDED IN POST — HUNT, NOT WAR

Dunkirk, Cambrai, Cassel — where the grand old Duke of York, less
of a liability than the current bearer of that title, marched
his men to the top of the hill and then marched them down again
— Malplaquet, Namur, Sedan, Belfort, Metz, Valenciennes, Toul,
Maubeuge, Verdun . . .

These are the names of western Europe's killing fields . . .
cavalry-friendly terrain, Panzer-friendly terrain. The borders,
which are marked, evidently, by man-made defences, are bereft
of natural boundaries. No Alps. No Pyrenees. Today, the only
belligerent threat to France and its neighbours is that of
credulous sexual inadequates equipped with seventh-century
superstitions and twenty-first-century weapons. They are occluded
yet visible, graspable as poison mercury. Their species of war and
its riposte leaves no built legacy.

Old war was concrete, graphic. Twenty years ago the battlefields
beside *la Voie Sacrée* — Barrès's coinage — were still evidently
battlefields. But all that could be pillaged has been pillaged:

barbed wire, small arms, shrapnel shards, gun barrels, even tanks
— tanks! This is the work of impious traders in relics.

AREA AROUND DOUAUMONT. MONTAGE OF MEMORIALS (EVERY TOWN AND VILLAGE HAS ONE). ECHOING REPEATED VOICE: 'MORT POUR LA FRANCE'

Half-buried rusted tanks were greater incitements to remember than
the multitude of memorials, which institutionalise death, glorify
squalor.

France remembers without god. A secular state honours its dead
citizens without invocation to threadbare fictions, without the
intercession of the primitive mind's most banal whimsy.

TITLE: VOLOGNE
WATERFALLS, STONE BRIDGES. W OF GUEBWILLER. RIVER GURGLING RUSHING WATER — GETS LOUDER. PASTORAL MUSIC

The Vologne rises near Gérardmer and flows into the Moselle. Soon
after Gregory Villemin was born in August 1980 his parents began
to receive anonymous letters threatening the child. On the evening
of 16 October 1984 his body was found in the river at Lepanges sur
Vologne where the family lived. His legs and arms had been tied. His
father's cousin was arrested, charged then freed. A few months later
he was shot dead by Gregory's father. Gregory's mother was arrested,
charged then freed. The sometime collaborator and amnestied OAS
fighter turned graphologist, Colonel Antoine Argoud, was employed to
determine the identity of the letter-writer. He failed.

L'affaire Gregory became France's most notorious murder of the
latter twentieth century. It is still with us. No one has been
convicted — save, in the court of public opinion, several branches
of the police, the magistrature, the press, and all Vosges
peasants. Who are, it goes without saying, inbred.

TITLE: VOLTAIRE
MOCK-UP: TACKY POSTER FOR 'THE ALSATIAN AND LORRAINE' (BIG-HAIRED PORN STAR WITH GERMAN SHEPHERD)

Alsace—Lorraine. They are spoken in the same breath as though they
are more than geographically akin. But the gulf between Strasbourg
and Nancy is immeasurably greater than a hundred miles. Strasbourg
is a border city. Nancy, most decidedly, is not.

It strives to represent itself as French France — not in some
folkloric way. There is nothing *franchouillard* about it, at least
not about its architecture. It is the city that most physically
embodies the Enlightenment. Its only rival in this regard is
Bordeaux which expanded through trade: wine, of course, and
slavery — towards which the Enlightenment was unenlightenedly
tolerant: Nancy is the fruit of a nobler impulse.

It was near here, at Lunéville, that Stanislas Leszczyński
held his court in exile after he had abdicated from the throne
of Poland — abdicated for the second time, which looks like
carelessness. He was nonetheless a lucky man. An assassination
attempt was stymied by Count Poniatowski, forebear of Michel
Poniatowski, Giscard's gofer.

CAPTION: 'IN 1978 GISCARD DESPATCHED PONIATOWSKI TO TEHERAN TO ASK THE SHAH WHETHER HE WANTED THE DST TO ASSASSINATE AYATOLLAH KHOMEINI,

Fragments of an Arbitrary Encyclopaedia

THEN IN EXILE IN THE PARIS SUBURB OF NEAUPHLE LE CHÂTEAU. THE SHAH SPITEFULLY DECLINED THE OFFER AND SO BROUGHT A NEW KIND OF TERROR BOTH TO THE COUNTRY THAT HE WAS ABOUT TO FLEE AND TO THE WORLD'

His daughter, who had the misfortune to inherit his looks, married Louis XV, who disliked him but nonetheless granted him the duchy of Lorraine.

He devoted himself to receiving such figures as Montesquieu, Voltaire and Émilie du Châtelet whom he fed with the pâtisserie which he himself made. He is supposed to have invented the rum baba and the madeleine: more likely, he amended extant Polish recipes then unfamiliar in France.

He wrote several books of epicurean philosophy. He counselled happiness, optimism, virtuous hedonism, good fellowship, philanthropy and self-regard.

For him happiness was owning a dwarf, Bébé Nicolas, whom he had bought as a child from his peasant parents. Bébé was sometimes charming, sometimes intemperate, always exhibitionist, always illiterate, always jealous of dwarves smaller than he was.

SARKOZY IN LATE EIGHTEENTH-CENTURY GARB BESIDE A GROTTO. OR WITH DOG, AS IN PAINTINGS AT LUNÉVILLE MUSEUM AND LORRAINE MUSEUM, NANCY. PLACE STANISLAS AND ADJOINING STREETS: AXIAL PERSPECTIVES, IRONWORK, FOUNTAINS, STATUARY, ARCHITECTURAL DETAILS

His greatest work was Nancy, a Versailles for the people.

Stanislas's architect Emmanuel Héré spent his entire life in Nancy. His buildings are magnificent, and they are vectors of happiness. Stanislas evidently wished to outdo his enthusiastically adulterous son-in-law. France's baroque was usually sober: that's an unavoidable contradiction. The France of the Enlightenment seldom displayed an appetite for exterior exuberance. That was reserved for within, for rococo decoration. The exterior of Versailles is vast yet slight — it lacks a scale commensurate with its sheer size. The garden front is low. It is simply not big enough. Voltaire described it as 'a lavish trinket: the masses are dazzled, the learned derisive'.

Héré's work may be on French soil but it is stylistically indebted to Bavaria, to Italy, to Stanislas's Poland — where all major projects were undertaken by Italians. These were countries temperamentally inclined towards the baroque's essential irrationality, countries where Héré's work would not have seemed so atypical. Or so dated — it belongs to the fashion of half a century before it was designed. So what? The worth of art has nothing to do with novelty, with being ahead of the game.

TITLE: VOSGES
ROAD TO BALLON D'ALSACE. ROAD THROUGH VINEYARDS. GUEBWILLER TERRACES

1962. The porousness of France's borders could hardly have been more clearly or more potently or more humanly demonstrated to me. We drove in slo-mo past a bucolic wedding in the garden of a half-timbered inn with steep gables and dormer windows and garlands. The guests merrily raised their glasses to us in response to the parping of the car's horn. We arrived at a farm high on the eastern side of the Vosges.

357

JM OUTSIDE FARM BUILDINGS AMONG VINES. NOISES OFF: A MEAL. PEOPLE EATING. THEY ARE TALKING IN ALSACE PATOIS. AFTERMATH OF A MEAL. WOODEN TABLE. C/U PLATES WITH REMNANTS OF HARE STEW. C/U BOTTLE OF MIRABELLE AND GLASSES. THEN CAMERA MOVES OUT AND WE BEGIN TO SEE THAT THE OBJECTS ON THE TABLE FORM A ROUGH MAP. A LINE OF SALT REPRESENTS THE RHINE. VERY OLD HANDS FOR THE NONAGENARIAN. A PLATOON OF *BIERSTEINS* FOR GERMANY

I remember the sumptuous lunch, one of the greatest lunches of my life: hare simmered in wine, spices and bitter chocolate, the sauce thickened by its blood, buttery noodles.

I remember tasting eau de vie de mirabelle for the first time in my life.

I remember a silver thread in the far distance — the Rhine, and beyond it, on its right bank, an indistinct spectre, a once troubling spectre. The Black Forest. Germany.

I remember the sprightly nonagenarian great-grandmother whose seventy-year-old daughter had cooked the hare.

I remember her telling me that this was the house where she had been born. She had never moved from it. Yet she had changed nationality four times in her life. French, German, French, German, French. She told me this without rancour: it was merely what had happened to her. She hoped to die French. She did.

TITLE: *LA VALISE OU LE CERCEUIL*
LORRAINE, UNDULATING LANDSCAPE

'Every man has two countries. His own and France.'

Wrong. Some men have no country. The displaced of Algeria found this out in 1962. Their choice, famously, was *la valise ou le cerceuil.* The suitcase or the coffin. Expatriation or death. Hundreds of thousands left. Hundreds of thousands were murdered. The authors of this overlooked genocide were the terrorists of the FLN — and their new friend Charles de Gaulle.

AUDIO: JEAN-PAX MÉFRET, *'LE PAYS QUI N'EXISTE PLUS'*

END

7

DEAD, MOSTLY

Prodigious Lives

ZAHA HADID (BORN 1950)

ZAHA HADID'S PRACTICE occupies an 1870s building designed by the London School Board's architect E. R. Robson who, typically of his profession, was unquestionably formulaic. Still, his was a sound enough formula. Today the high, plain, light rooms are crammed to bursting with her 200 or so employees. Though they are of every conceivable race, they are linked by their youth, their sombre clothes, their intense concentration. They gaze at their screens, studiously, astonishingly silently. There is little sound other than the click of keyboards and a low murmur from earphones. They don't talk to each other. It is as though they are engaged in a particularly exigent exam. It feels more like a school than a former school. And it feels more like a factory than a school. If there is such a thing as a physical manifestation of the dubious concept called the knowledge economy this is it. This is a site of digital *industry*.

'What is exciting,' says Zaha, 'is the link between computing and fabrication. The computer doesn't do the work. There is a similar thing to doing it by hand . . .'

'The computer is a tool,' I agree.

'No. No, it's not . . .'

What then?

The workers on the factory floor – an accurate if tactless way of putting it – are, she says, 'Connected by digital knowledge . . . They have very different interests from twenty years ago.'

Sure. But this does not make immediate sense. It is a matter to return to, that will become clear(ish) in time.

*

Her austerely elegant apartment is a sort of gallery of her painting and spectacularly lissom furniture. It's a monument to Zaha the public architect rather than Zaha the private woman. It occupies a chunk of an otherwise forgettable block ten minutes' walk away from the practice. The route from one to the other might almost have been confected as an illustration of the abruptness of urban mutation. Here is ur-London. There are stock bricks and red terracotta, pompous warehouses, run-down factories, Victorian philanthropists' prison-like tenements, grim toytown cottages, high mute walls, a labyrinth of alleys, off-the-peg late-Georgian terraces, neglected pockets of mid-twentieth century utopianism, apologetic infills, ambiguous plots of waste ground. It is neither rough nor pretty. It does however have sinewy character. It may be 'ordinary' but it possesses a serendipitous fascination and is undeniably diverse. The daily stroll through this canyon of variety is surely attractive to an artist whose aesthetic is doggedly catholic, each of whose buildings seems unsatisfied with being merely one building. If she is offended by the suggestion that constant exposure to a manifestly ordinary and thoroughly typical part of London might, however indirectly, impinge on her work she doesn't show it.

But she is faintly bemused. It is as though such a possibility had never occurred to her.

This is absolutely not the sort of proposition that gets mooted within the world of Big Time Architecture which Hadid has inhabited all her adult life, for many years as a perpetually promising aspirant, a 'paper architect' who got very little built but nonetheless won the Pritzker Prize – which begs the questions of whether architecture is divisible from building, of where the fiction of design stops and the actuality of structure starts. Today she is this tiny powerful milieu's most singular star, and its only woman, its only Zaha. So distinctive a name is useful. It's a fortuity which grants Zaha effortless entry to the glitzy cadre of the mononomial: Elvis, Arletty, Taki, Sting. The first architect since Mies (van der Rohe) to be so blessed.

*

Architecture is the most public of endeavours yet, it must be said, it is a smugly hermetic world. Architects, architectural critics and theorists, and the architectural press, are cosily conjoined by an ingrown, verruca-like jargon which derives from the cretinous end of American academe: 'Emerging from the now-concluding work on single-surface organisations, animated form, data-scapes, and box-in-box organisations are investigations into the critical consequences of complex vector networks of movement and specularity . . .'

What is being referred to are merely methods of making buildings. This is the cant of pseudo-science – self-referential, inelegant, obfuscatingly exclusive: it attempts to elevate architecture yet makes a mockery of it. Zaha, however, has the chutzpah to defend it. She claims to be not much of a reader of anything other than magazines so the coarseness of the prose doesn't offend her. The point she makes is that this is the *lingua franca* of intercontinental architecture. A sort of esperantist pidgin propagated by the world's major architectural schools, the majority of which happen to be notionally anglophone yet whose pupils and teachers come from a host of countries, and the world's major architectural practices which are international and polyglot. When she talks about architecture, about urbanism, about the continuing exemplary importance of London's Architectural Association, where she studied after a childhood in Baghdad, boarding school near London and reading mathematics at Beirut university, she uses this pidgin, almost falls into it, constellates it with syntactical mishaps.

'You know, space is an interesting endeavour . . . you create an interesting . . . the impact you have on the cityscape. The whole life of a city can be in single block . . . Break the block, yeh? Make it porous . . . Organisational patterns which imply a new geometry . . . The idea of extrusion . . . One thing always critical was idea of ground, how to carve the ground, layering, fragmentation . . .'

Perhaps being 'connected by digital knowledge' is a *faute de mieux*-ish means of circumventing the problems inherent in a polyglot workforce given that verbal expression plays only a minor part in architectural creation. The gulf between clumsy, approximate jargon and precise, virtuoso design is chasmic. And it hauls in its wake some important ramifications. Despite its practitioners' fastidious, perhaps delusional protestations that it is a creative and scientific endeavour, architecture is a very big business which is involved in the creation and sale of one-off objects rather than multiples: it is a trade which mostly deals in the bespoke. Now, one consequence of being 'connected by digital knowledge' is an enforced internationalism – and a complementary loss of localism – at the highest tier. So take, for example, the Basque provinces where Santiago Calatrava has built Bilbao's airport, where Frank Gehry has famously built that city's Guggenheim Museum, where Rafael Moneo has built the better Kursaal at San Sebastian and where Zaha has no fewer than three projects: a new quarter of Bilbao; a sleek, partially buried train station in Durango and governmental offices in Vittoria. This region, whose paranoiac sense of itself and of its blood-drenched specialness need hardly be emphasised, is becoming a testing ground for exercises in a globalised aesthetic entirely at odds with its vernacular idioms of distended chalets and Hausmanian pomp. Zaha is enthusiastic about this sort of dissonance. She is opposed to new buildings which nod allusively – she would say deferentially – to their ancient neighbours. She regards such buildings as sops to populism.

'It would be interesting to do a large project without looking backwards.'

How large?

She grins: 'A city. A city! Without looking backwards. Vernacular building . . . it's like minimalism.' (She probably means neo-vernacular building.) 'People can handle minimalism, vernacular. It doesn't *disturb* them.'

Of course, Hadidopolis, the dreamed city, would, paradoxically, be less disturbing, less astonishing than a single building by her in an already established environment where the clash of idioms is potentially deafening.

'They still talk about contextual. Ha!'

The 'they' she is referring to are her bugbear, the (now rather old) New Urbanists, the begetters of crass, kitschy, retro-developments such as Seaside and Disney's Celebration, both of them in Florida. Her distaste for their twee, anti-modernist escapism is total.

In Zaha's lexicon contextual might be synonymous with *compromised*, which is the last word that could be applied to her own work. Bloody-minded, unaccommodating, serious, joyful, emotionally expressive, intellectually engaging: these are more apt. Yet, no matter what she says, each of her buildings is sensitive to its context. Being sensitive does not mean being passive. It is not a question of taking a cue from the immediate surroundings but of making an appropriate intervention which *changes* those surroundings, which creates a new place and better space. Even the most cursory scrutiny of the 25 projects she has either completed or which are currently under construction reveals an exceptional versatility and a multitude of responses. She has eschewed the temptation to develop 'the signature' that afflicts high-end architects and so prompts the unexceptionable accusation that Libeskind or Calatrava or Gehry merely plonk down the same lump of product time and again across the globe. She has style all right, but not *a* style.

The Rosenthal Contemporary Art Centre in Cincinnati is blocky, grounded, cubistic, rectilinear; its geometry may be scrambled but it's predominantly orthogonal. It is unrecognisable as being by the same hand as, say, the Phaeno Science Centre in Wolfsburg which is taut, dynamic, horizontal and looking to make a quick getaway. The Museum of Transport on the south bank of the Clyde in Glasgow has a silhouette which might be a child's depiction of a city's skyline. One cable railway station at Innsbruck is sleek and reptilean; a second is mycomorphic; a third is a homage to a species of bird that never existed. There are hard forms, constructivist dislocations, and there are soft forms, Tanguy-like fondants. Sometimes she seems to be working in steel, other times in butter, here she is chiselling wood, there she is twisting chocolate. A university building on the Barcelona waterfront recalls a negligently shuffled pack of cards. The A55 motorway's descent into Marseille from the north is one of the most thrilling in Europe. It will be further enhanced by the CMA-CGM container company HQ in the cleft where raised carriageways bifurcate. This 140-metre-high tower, in the enormous Euromediterranée ZAC (*zone d'aménagement concerté*), will be the highest in the burgeoning city. It is a perhaps reproachful complement to the effortful wackiness of neighbouring projects like Massimiliano Fuksas's Euromed Center:

the tower is as stately and elegant as a duchess's ball gown and again very different to anything else she has done.

How do she and her collaborators, chief among them Patrik Schumacher, manage to avoid the besetting architectural tic of self-plagiarism?

'Don't draw on to computer. Don't draw and then put it on to computer . . . I have five screens in front of me . . . Different projects . . . You work on developing, oh, a table while at same time you're developing master plans. It's like you have different information coming from different directions. Like photography. Out of focus . . . then you zoom in. I'll have a sketch – it'll take a few times before it takes . . . Sometimes a few years. You see, not everything can be used, not every idea can be used right then . . . But nothing is lost. Nothing.'

So a shape or form devised initially for a piece of furniture may be fed a course of steroids and become a building?

'No. That's not what I'm saying. Doesn't work like that.'

I rather suspect that Zaha has the ancient, habitual, superstitious fear that to discover how her processes work would be to jeopardise them.

*

The idea that London comprises a series of 'villages' – an estate agent's vulgar conceit – goes lazily unchallenged. Villages are small, hick, inward-looking. London is not. London pioneered sprawl, it was a horse-drawn precursor of Los Angeles. It is a city of stylistic collisions and astonishing juxtapositions. Which might be reckoned to make it susceptible to imaginative and unorthodox architectural interventions. There is, after all, no classical homogeneity to rupture, no defining idiom which must be adhered to.

Yet Zaha Hadid – an architect who is nothing if not imaginative, nothing if not unorthodox, who is fêted throughout the world as, ugly word, a *star*chitect – still does not have a single building to her name in London despite having lived and worked here for three and a half decades. There are, to be sure, schemes – the 2012 Olympic Aquatic Centre, and a building for the Architecture Foundation in Southwark; but the former's budget is being persistently called into question and pared and the other has not progressed since it was first mooted several years ago.

It would be coyly disingenuous to feign surprise at this absence of a work by her in her adopted home. A catalogue of circumstances militates against her.

She is extraordinarily engaging but equally obstinate. She has never pretended to be anything other than an artist. An artist moreover of a particularly dogged sort, one who has kept alive, or revived, the unfashionable notion of the avant-garde. And who has created her own fashion rather than blindly following the herd like, oh, 99 per cent of architects.

She is, evidently, not English; her sensibility is not English; her lack of timidity is not English; her earnestness is not English; nor her resolute ambition.

Then there is the question of her sex.

*

Architecture is globally dominated by men to a degree that no remotely kindred endeavour is. This has always been the case. The history of architecture can be written, often has been written, with no mention of women save, perhaps, of monarchs, aristocratic grandees, philanthropists: patrons not makers. The psychophysiological contention that women are less adept than men at three-dimensional thought doesn't begin to account for their acutely disproportionate position in British architecture. Only 10 per cent of practising architects in the UK are women. However, the percentage of *qualified* women architects is vastly higher. But women drop-out at an alarming rate. So alarming that the former RIBA president George Ferguson commissioned an investigative study.

He need hardly have bothered. Its conclusions were thoroughly predictable: low salaries and long hours (which equally afflict men), lack of preferment and office machismo (which probably don't). The outstanding woman architect of the generation before Zaha's, Georgie Wolton, opted for a (successful) career as a landscape artist having designed just one major building, a studio block in the north London suburb of Holloway. Sarah Wigglesworth (whose most celebrated building is also in that suburb), Amanda Levete and Cecile Brisac are London architects currently producing work of the highest order, much of it outside Britain, in cultures where there exists less bias against women. The volume and prestige of commissions received by such practitioners as Manuelle Gautrand in France or Tilla Theus in Switzerland is unthinkable in Britain.

*

Of course, the British bias is not merely against architects who happen to be women. It is against architects who happen to be architects.

The rule rather than the exception dictates that British architects who aspire to anything more than polite synthetic-modern apartment buildings or self-effacing, production-line offices have to prove themselves abroad. That is where creative reputations are made. This has been the case since the early 1970s when public confidence in architecture plummeted and architects came to be regarded as licensed vandals committing a sort of aesthetic *trahison des clercs*.

'No! Later, it was . . .' Zaha corrects me, 'it was 1975, '6. Definitely.' By which time she had been at the AA for four years. It is telling of that institution's carapace of ivory exclusivity that the popular antipathy towards the discipline it taught took so long to breach its walls.

She is certain of the date. For that was when, incredulous and indignant, she witnessed the transformation, the near-apostasy, of some of her dogmatically modernist teachers. 'Between one term and the next,' she says, Léon Krier became a former modernist, literally a post-modernist. Krier lurched, in the bipolar way that fundamentalists will, from preaching the rhetoric of imaginative, technologically based rationalism to becoming a groupie of the then still incarcerated Nazi war criminal Albert Speer, an architect whose formidable banality was matched only by the megalomaniac scale of his (mostly, thankfully) unbuilt projects. Krier would, frighteningly, go on to become the Prince of Wales's architectural advisor, and thence the brain (if that's the word) behind such *volkisch* excresences of the 'New Urbanism' as Poundbury, the cottagey slum of the future disgracefully dumped on a greenfield site on the edge of Dorchester.

'By 1978 he is god of historicism . . . You know – that attitude

367

that you can't go forward without looking back, that's the historicist position, post-modern position.' It's a position she deplores. That is to put it mildly. Zaha seems to consider postmodernism a sort of betrayal. Which may be going a bit far. Surely, I suggest, the question is not taxonomical, not what style a specific building belongs to – post-modern or any other – but whether it is good or bad. She appears not to hear. She asks for more tea. She snuffles. She has a cold.

But then I too would develop a cold if someone had put to me a proposition that impertinently questions the very core of my aesthetic. She is contemptuous of the sort of relativism which even hints that the often infantile, mostly eager-to-please idiom of the Thatcher years is serious architecture. She is, perhaps, right. Accessibility merely means lowest common denominator populism, commercial opportunism, the subjugation of the creator by market researchers and of originality by second-guessing what the 'people' will find acceptable. Zaha has been fighting all her professional life against the architecture of the market place, struggling to assert the paramountcy of the artist, i.e. of herself, of an uncompromised vision. She had to bide her time a long while.

She was the victim of a shift in taste. She could, chameleon-like, have followed Krier and many of her AA contemporaries and near contemporaries who discovered themselves suddenly sympathetic to upside-down diocletian windows, playground colours, Bluto columns, oafish pediments: the components of a new architectural 'language'. On the other hand there were those who invented with aplomb.

*

She tells me she doesn't want to talk about other architects' work before I have even broached the matter. Happily she isn't as good as her word. An architect with a detailed knowledge of architectural and urbanistic history is, astonishingly, a rarity. The living and the dead constellate her discourse. They are not the figures one might expect. Despite the status she has achieved she still, implicitly, considers herself an underdog rather than an overdog. There is something heartening and generous about the way she enthuses about the work of Douglas Stephen, an unacknowledged genius who designed less than a dozen buildings in a lifetime of scrupulously high standards and absolute integrity. She is enthusiastic about the Italian rationalist Aldo Rossi whom she describes as forgotten. Forgotten by whom? I wonder. 'Forgotten,' she insists. I point out that his rationalism was hardly all-encompassing and that whenever he was in London he would go to gaze at the clunkily historicist War Office in Whitehall. She smiles, as though to acknowledge the disparity between the architect and the man. She admires Rodney Gordon, maybe the greatest of the British brutalists, a sculptor in concrete whose finest buildings (the Tricorn in Portsmouth, the Trinity in Gateshead) have been or are about to be demolished.

Would we burn a Bacon? Take a hammer to a Gormley? No. But in Britain architecture is peculiarly expendable. She despairs of British taste. And of British short-termism.

Which is expressed in two ways. Buildings, notably those of the 1950s and 1960s, are wantonly torn down before they have been allowed the chance to come back into fashion. This, of course, is not exclusive to Britain. Even in France, which has a much greater appreciation of modernism, Claude Parent's space-age shopping centres at Reims and Sens have been disfigured.

We rue the loss of high Victorian buildings of the 1860s. Why will future generations not rue the loss of those made in the 1960s during another of those rare periods when British architecture abandoned its habitual timidity?

Secondly, buildings used to outlive humans, not least because the process of construction was so long and laborious that permanence was a desirable aim. Today's corporate presumption is that a building's duration will be hardly longer than a few decades. Its lifespan is in inverse proportion to our own continually stretching sentence. One consequence of this disposable building syndrome is that quick delivery and low cost are valued above all other considerations. Much architecture is, then, increasingly concerned with the provision of what are effectively temporary structures. Zaha has an unfashionable distaste for such ephemerality. She must like any architect worry about what will become of her buildings. One of her earliest completed projects, a pavilion for the study of landscape at Weil am Rhein on the German-Swiss border, is already looking tatty as a sink estate, and the nearby fire station she built for the furniture manufacturer Vitra's factory was considered inappropriate for that role and has been turned into a museum of chairs.

A consequence of short-termism is standardisation.

'London is becoming more and more *even*. I don't like current work here. I'm not against new projects . . . Obviously I'm not . . . But there's no planning here. No critique about what is coming next . . . There is a responsibility on the city . . . to impose – not, not, ah, *rules* but . . . quality. The state should invest in architecture like in Spain, Holland. But the dynamic here, it's all corporate . . .'

Again, it always has been. Aesthetic dirigism is as alien to Britain as economic dirigism. Public building is the exception: the long third quarter of the twentieth century – the years of abundant social housing, the new universities and their architecturally enlightened chancellors, hospitals, theatres and libraries – were atypical.

'Yup,' she sighs and shakes her head, 'London: city of lost opportunities.'

That's largely because London lacks the sort of patrons she needs: wilful, vain, philanthropically inclined plutocrats with a taste for self-advertisement, endowment and high art museums rather than for football stadia. Collecting buildings is a very expensive hobby. There is no Getty, Guggenheim, Whitney, Vanderbilt or Rosenthal.

She doesn't seem embittered but, rather, wearily resigned. As well she might be for while London is unquestionably enjoying a building boom it is equally suffering a blandness boom. PFI schemes do not encourage audacity. Indeed they are infected with an almost totalitarian conviction that architecture should be useful rather than beautiful or striking or marvellous. And most architects duly oblige for they know who calls the tune. It is as though they pride themselves on the design of risk-free buildings whose primary attribute is that no one will notice them, thus no one will take offence at them. (They are wrong. Blandness on a massive scale *is* offensive: just look at Southwark Street, across the river from the City of London, where the prolific commercial practice Allies and Morrison has committed some sort of crime against streetscape which Zaha loyally refuses to condemn.)

Why then does she base herself in a city which, if not professionally antagonistic to her, has been hardly welcoming?

'I was teaching here.'

But she was also teaching at Columbia, Harvard, Yale, Hamburg, Vienna . . .

'Vienna has the same problems as London.'

What are they?

'It's historic city.'

But many of the cities in which she has buildings under construction are equally historic. Naples, Madrid, Strasbourg, Barcelona, Seville. And as for Rome . . .

'I'm in London because best civil engineers in the world are here.'

Civil or structural engineers are unquestionably the scientists without whom architects would not exist. But, given the internationalism of both architects and engineers, it is a truly bizarre reason.

One is inclined to suspect that it's a professional disguise which masks a private inclination.

'I don't know if I'll ever do a big project in London . . . But I do have a take on the city.'

That take is as much a flâneur's as an architect's. As an architect over twenty years ago she undertook a research project which envisaged a linear city down the Lea Valley and another around the Royal Docks. The latter has come to pass, but in typically London manner – piecemeal, uncoordinated, scrappy, unambitious. And the Lea Valley is being cleared, *cleansed*, to host the Olympic Games, a trophy coveted by emerging tyrannies, tinpot totalitarians and third world dictatorships. Tactfully, and atypically for so opinionated a woman, she refuses to diverge from the party line and mutters some right-on stuff about the games' 'legacy'.

Maybe she believes it. Maybe not.

I wonder, because Zaha the flâneur has an immense appetite for a very different London, an insatiable curiosity which she reveals only obliquely. She palpably appreciates the very oddities and quirks of the area which the Olympic site will occupy, the atmospheric *terrain vague* of abandonment, dereliction and toxic canalisation.

*

When she talks about anything other than architecture she employs an urbane vocabulary, a flourishing grammar and the definite and indefinite articles. She is funny, and fun. About how London has changed socially: 'The kids cannot believe it when I tell them about the King's Road in those days, cannot believe it.' She is eloquent about parties, friends, flu remedies, clothes (she wears black though she professes to pine after the days of colour), a tardy florist, a driver whose limited comprehension of satnav prompts him to put in 'crescent' rather than the name of the crescent. Her word power expands miraculously.

It might be conjectured that a different part of the brain is activated, that architecture is confined to a ghetto that is actually cut off from language, that is pre-verbal or extra-verbal. She is neither dyslexic nor left-handed – two conditions which afflict a number of extravagantly gifted architects.

The awkward struggle to describe the products of her capacious imagination is hampered by her disinclination to employ simile. But simile, though it might clarify, would undermine her achievement. To compare her work to something already existing would be to detract from it. For me to state that her buildings are *like something* – frozen napkins or origami in a hurry or squeezed-out ointment tubes or a carnival dame swaying in a frock or a flock of starlings cartwheeling like iron filings subjected to a magnet or baroque drapery – is explanatory shorthand. It is not to debase them. Far from it. But I didn't make them. They are admirable for a load of reasons.

Her work derives, she says, not from observation of extant architecture. Nor from formalism. She claims to take nothing from organic morphology. No ammonites, no sharks, no petals. It all begins with painting, with pure abstraction.

But a few moments later she changes her mind. She contradicts herself and attributes her inspiration to landscape, topography, sedimentology, geological patterns . . . Indeed one of her pieces of furniture is called 'Moraine' and there is an unmistakable acknowledgment of a badlands roster of folds, prisms, hoodoos and organ pipes, and to the shifting shapes of dunes and drifts. Such architects as the Finn Lars Sonck, the Catalan Antoni Gaudí and the German Gottfried Boehm have represented rock formations with differing degrees of naturalism. Zaha goes further. Buildings are static objects. Throughout the twentieth century architects vainly attempted to imply that structures were on the move, to invest them with speed, one of the essential properties of modernity but one which is, alas, necessarily absent even in borax buildings that are streamlined or googie ones which borrow the imagery of airplanes or rockets. Much of Zaha's work implies a different sort of speed – the slow passing of millennia, the gradual attrition of wind, the grind of the sea on stones, the way rain turns chalk into pinnacles and spires. There is a scent of erosion, of time's inexorability, of future fragmentation, of mortality.

(2008)

NORMAN FOSTER (BORN 1935)

I CANNOT REMEMBER who the opposition was, let alone the result, but it was back when Ipswich Town was a successful football team. Autumn 1975, I guess. Dusty Hughes persuaded me to accompany him to a home game. What I do remember about that day is seeing a work of architecture that astonished and perplexed me: it was the Willis Faber & Dumas building (as it then was) by Mr Norman Foster (as he then was). Not that I knew the name of the extraordinary structure or its author. How come I had never heard of this prodigy of 1930s modernism? The building was beguiling – faintly sinister, sinuous, sleek as a patent leather shoe, fetishistic. I thought Dusty was having me on when he said it had only recently been completed.

Given the ubiquity of Foster and his imitators today, it is difficult to recall how the Ipswich building defied the British mainstream. That mainstream's defining characteristics were a nervy mannerism, an apologetic half-heartedness, a sycophantic eagerness to please, a desire to create buildings that would not be noticed. The bristling confidence of the 1960s had dissipated.

The late Philip Johnson rather waspishly described Foster as 'the last modern architect'. He could not have been more wrong. That epithet ignores modernism's heterogeneity and countless splits. Foster thirty years ago was among the earliest neo-modern architects. Like Paul Hamilton, Douglas Stephen and Georgie Wolton, he looked back forty or fifty years to the era of 'heroic' modernism, to that utopian period when clean lines and abundant glass were going to change the world. Foster's East Anglian landmark is gleaming black by day, transparent at night. Its most evident debt is not to Mies van der Rohe; the smoothness and spareness derive more obviously from the Daily Express building in Ancoats, in Foster's native Manchester. This was the most thrilling of the three buildings that Owen Williams designed for the newspaper and the most progressive work of its time in that city. It is perhaps significant that Williams, patronised by the Sainsbury family forty years before Foster, was an engineer as well as an architect. He disdained the arbitrary division of the two endeavours – a division caused by the Victorian eagerness to establish architecture as a profession whose task was to decorate 'mere' engineering.

Foster's allusion to this icon of his childhood is, if you like, an exercise in nostalgia, an expression of a longing for a far-off past, for an ideal seldom glimpsed in this country outside the pages of *The Eagle*, where Frank Hampson's drawings for Dan Dare bore an extraordinary kinship to Owen Williams's work. It is as though, belatedly, Foster gave us yesterday's tomorrow. This is hardly the act of a cold, robotic, technophiliac, kit-fixated ultra-functionalist.

We should not allow Foster's preoccupation with manifest structural engineering and the exposure of nuts and bolts to delude us into believing his buildings are programmed to be aesthetically unengaging and emotionally vacuous. The great medieval cathedrals hardly differentiate architecture and engineering yet they are capable of hugely moving us. As are such exemplars of unadorned engineering as the New Bedford River, the Garabit viaduct, the Forth Bridge and the Itaipu Dam. They excite our awe because of their might and their raw beauty, which is at the other end of the aesthetic scale from prettiness. Foster's oeuvre at its best carries an emotional freight more readily associable with candidly expressive architecture: the roofscape of Le Corbusier's l'Unité d'Habitation; Sir John Vanbrugh's mad arch at Eastbury; Claude-Nicolas Ledoux's Arc-et-Senans. One Foster dictum that is worth paying heed to is that buildings often look most exciting while they are being constructed. Why should a building's building not be exhibited? After all, process is acknowledged as a component of painting, fiction, music and so on. At, among others, nocturnal Ipswich, Stansted, the Hongkong & Shanghai Bank, the Torre de Collserola in Barcelona, the Sainsbury Centre at the University of East Anglia, the structure is defined by the way it's made. The buildings recount, admittedly obliquely, the story of their construction. Maybe this is a form of narrative architecture.

The supreme example of Foster's capacity to move, to achieve a thrilling sublimity, is the autoroute bridge across the Tarn, west of Millau. The actual river is maybe 20 metres wide. The gulf between the plateaux – the Causse Noir and the Causse du Larzac – is 2.5 kilometres. This is barren, savage country, which may have something to do with the volume of local opposition to the bridge, an opposition that has vanished now that it is finished. Millau used to suffer appalling jams because traffic had to divert from the incomplete autoroute through the town. Now it suffers jams because all of France wants to gaze at Milord's soaring masterpiece. Foster makes predictable environmentally correct noises about 'minimum intervention in the landscape'. The truth is that it's an appropriate intervention that has no obligation to fit in because it is a scape in itself, a monument to human ingenuity just as the magnificent terrain is a monument to geology and climate.

La Dépêche du Midi ran a front-page story about Foster visiting Millau for the first time. He helicoptered in for forty minutes. The bridge was almost finished. Now I know Bill Gates doesn't write the programs himself; I know Mr Kipling no longer personally makes all the cakes. I have used 'Foster' throughout this article to signify Foster and Partners and its predecessor Foster Associates. But architecture isn't software, it isn't transport, it isn't baking. The question of authorship is raised more than with any other practice: indeed this question threatens to occlude the work. The name is personalised and hierarchical, which prompts a different expectation. When the chef Paul Bocuse was asked who did the cooking at his restaurant when he was away, he replied with attractive candour: 'The same people who do it when I'm there.' There is a lesson here. But whether it's for Lord Foster or for his detractors is moot. And he certainly has his detractors.

This, after all, is the architect who was compared by Signor Piloti in *Private Eye* not to Wren in London but to Speer in Berlin. I'd dispute that. It is not Foster's grand gestures that tarnish the reputation. The fact of the buildings having acquired more or less fond nicknames indicates the popular reaction to them: 30 St Mary Axe/The Gherkin,

SECC & Clyde Auditorium/The Armadillo, Sage Centre/The Slug. No, it is the apparent conveyor-belt production of what might be called the practice's diffusion line that is so dismal. 'Minimum intervention' is not a phrase to use of the bulbous, convex Sainsbury's building at Holborn Circus. Are More London and the ungainly Tower Place and Albion Wharf – whose flashy paucity is shown up by its neighbour, which houses the practice and Foster's home – anything more than richly arid hackwork? Foster's place in history is assured. His place in the present is less secure.

(2005)

RICARDO LEGORRETA (1931–2011)

THE NOTION THAT architectural modernism is globally homogeneous is a falsehood propagated with varying degrees of sophistry, mendacity and, above all, ignorance by its opponents, who are apparently blind to the yawning gulf between the work of, say, Paul Rudolph and his pupil Richard Rogers, let alone to the manifold nuances to be found in the output of a single artist. There were and are many modernisms.

That of Ricardo Legorreta is sensual, elemental, colourful, spiritually exhilarating. The combination of confident urbanity and cyclopean primitivism that informs many of his creations is, of course, a trope as ancient as architecture itself. Legorreta was wedded, in a relaxed non-doctrinaire way, to the proposition that his architecture should evoke the ancient, the mysterious, the unsayable. His modernism was inhibited neither by rationalism nor by functionalism. His work, to an extent unmatched by any other Mexican, was a belated architectural expression of *indigenismo*, the assimilationist ideology wrought in the early twentieth century which aspired to build a peculiarly Mexican society, culture and identity through various forms of métissage – mixed race marriages, art that combined pre-Hispanic sources and the western tradition, literature founded in folk myth, education that granted equality to vernacular history, etc. This was a well-intentioned conceit which in today's liberal orthodoxy is predictably calumnised as a form of patronage, just as orientalism is. Nonetheless it has been pervasive enough to prevent Mexico becoming, in the manner of Argentina, a European state which happened to have strayed across the Atlantic.

A cosily 'revolutionary' strain of pueblo writing flourished alongside over-mediated 'heroic' social realist painting, notably of murals: wicked conquistadors, rapacious landowners, noble savages defined by their stoic victimhood. Architecture, more obliquely representational, lagged behind. Save for an occasional glance north towards the United States it clung to its European models – revival upon revival, art nouveau, Le Corbusier's orthogonal modernism which was massively influential in Mexico City. It was to this last that Luis Barragan, the most celebrated Mexican architect of the mid-twentieth century, had initially been attracted. But he soon tired, as Le Corbusier himself did, of icy, mechanistic, monochromatic purity. By the time that Ricardo Legorreta, more than a generation his junior, worked with him he had evolved an idiom which tempered Corbusian abstraction with borrowings from the Mediterranean gardens and landscaping of the Franco-German illustrator Ferdinand Bac and from Mexican vernacular: notably from adobe buildings and from the ostentatious colours of fabrics, decorative pottery and religious gewgaws. Barragan was a perfectionist who generally worked on an intimate scale. He was, also, costive with his talent, fussy about who he worked for (it was often himself).

Ricardo Legorreta – perhaps more worldly, certainly more businesslike – took Barragan's methods, which were akin to those of the arts and crafts, and applied them to large projects. Remarkably, he did so without adulterating them. His early work is Barragan writ large. But he soon began to develop a lexicon of monumental devices that owed nothing to Barragan, that were occasioned by the sheer size of his commissions. The Chrysler plant west of Mexico City at Toluca (1964) has at its heart two truncated cones which disguise services and a conference theatre; they might be amputee cooling towers. In subsequent works such as the Camino Real Hotel in Mexico City, completed in 1968, he went further and incorporated allusions to the very forms of the great Aztec and Mayan temples. This was a far cry from the cosmetic appropriation of ornament which enjoyed a vogue in the 1920s and '30s. A subsequent hotel for the same company on the Pacific coast near Puerto Vallarta rises above the canopy like the ghostly ochre ruin of a sacred side. Hidden on its roof is a system of aqueducts feeding swimming pools shaded by pseudo-acacias. Flowing water – in the form of mini-weirs, spouts, fountains and even controlled interior floods – is a perennial motif, as ubiquitous as it is in Moorish buildings and gardens. He made frequent use of epic flights of stairs, stepped profiles, *pare-brises*, manipulations of sun and shade that border on legerdemain and walls of a seemingly exaggerated blankness: they go on and on, unrelieved, overstatements of a void. The effect is mesmerising.

The Hollywood house he designed for the Mexican actor Ricardo Montalban appears as a complex of sand-coloured cubes without apertures, a fortress against the smog. This was the first of many commissions he carried out in California from the mid-'80s on. It is a curiosity of his career that even when he had achieved global fame among architects and their patrons he diffidently neglected to succumb to the temptation to suffer jet lag in order to impose monuments on every country. He did not even open a Los Angeles office: he went no further than buying a sports car and keeping it in a friend's garage there. It was not till his youngest son Victor joined the practice just over a decade ago that Legorreta+Legorreta began to build in Cairo, Qatar, Israel, Japan, etc. (Father and son had once sworn never to work together.)

Ricardo Legorreta's work is full of surprises. The Metropolitan Cathedral in Managua, for instance, was designed in the early '90s with cavalier bloody-mindedness: it is an exercise in the then despised raw concrete brutalism of a generation earlier. It is a matter of regret that so little of his work is to be found in Europe. His first work on this continent is also a surprise for the unwary. Zandra Rhodes's museum and workshop which Legorreta designed in collaboration with Alan Camp is a Southwark oddity. In a part of London now dominated by Renzo Piano's megalomaniac Shard and where 'regeneration' has usually meant another lump of glazed insipidity by such masters as Allies and Morrison, this made-over warehouse of the 1950s stands out like a dislocated jewel. Like nearly everything Legorreta did it causes those coming up on it to at least smile. His is the architecture of optimism and enjoyment.

(2012)

RODNEY GORDON (1933–2008)

THE EPITHET 'YOUNG architect' is habitually applied to any architect under the age of fifty. Few practitioners get a break before they are middle-aged. Rodney Gordon was one of those who did. He was still in his twenties when he designed the first major building of his astonishing and truncated oeuvre. By the time his exact contemporary Richard Rogers achieved fame at the age of thirty-eight by winning the competition for the Pompidou Centre in 1971, Gordon's career was waning fast.

It was a career unlike any other in the past half-century of English architecture: it happened the wrong way round. He designed nothing less than a handful of masterpieces – then it was virtual silence. And his art has been treated with a contemptuous shoddiness commensurate only with its brilliance.

That, however, is hardly surprising, given that Gordon was a brutalist, probably the greatest (as well as unquestionably the youngest) of the English brutalists and thus a ready target for indolent bien-pensants whose antipathy to the architecture of the 1960s is as drearily predictable, as dismally unseeing, as was their parents' and grandparents' to that of the 1860s. These people fail to differentiate between the many strains of modernism and, more importantly, between what was good and what bad. Nor, in their arrogance, do they realise that tastes change. Today brutalism is admired by a new generation of aesthetes as opposed to the clichéd, knee-jerk calumnisation of 'concrete monstrosity', as John Betjeman and Osbert Lancaster were to 'Victorian monstrosity'.

The word brutalism was coined by the architectural theorist Reyner Banham. It is a bilingual pun on the French béton brut (raw concrete) and *art brut* (Dubuffet's word for outsider art) and the all too plain English word brutal. If only Professor Banham had failed to commit it to paper and had dreamt up a less loaded term, the fate of buildings in this idiom might have been happier, for their opponents, apprised only of the English component, would not have had the ammunition of what seems like a nomenclatural admission of culpable aggression.

On the other hand they might still have abhorred it, for brutalism committed the grossest of sins in English eyes. It abjured the picturesque in favour of the sublime. It scorned prettiness. 'It put on,' as John Vanbrugh, a brutalist *avant la lettre*, had it, 'a masculine show'. A show which did not preclude a strangely butch delicacy, a steely effeminacy. Gordon might have worked in concrete but he made it sing. His buildings were articulated rather than monolithic. More than any other English brutalist he had looked at constructivism. Gordon's professed aim was to create an architecture that was 'raw, dramatic, sculptural'. At the Tricorn in Portsmouth and Trinity Square in Gateshead he succeeded on a vast scale, unparalleled in Europe. These buildings were indeed extraordinarily sculptural, their silhouettes were audacious and poetic, jagged and rhetorical. They were thrilling structures that seem to be forces of nature, like fortresses in Castille which grow from the earth.

But they were shopping centres with car parks, mere shopping centres with car parks. The Derwent Tower (aka the Dunston Rocket), also in Gateshead, was council housing, mere council housing. Gordon was an artist who believed that the everyday should be touched by the exceptional, that usually banal building types ought to be as well made as cathedrals and so rendered rather less banal. But there exists to this day a preposterous architectural hierarchy which has nothing to do with the quality of the building and everything to do with its use. In the 1960s new university buildings, post-Vatican II churches, and theatres were respectable commissions. Commercial buildings and speculative developments were not.

Gordon possessed every architectural gift save those that matter most: he was no businessman and was a poor self-publicist. Hence his difficult partnership with Owen Luder – 'a businessman', in the words of a former employee, 'who happened to have architectural qualifications'. Luder had a talent for architectural politics: he was twice president of the RIBA and even considered offering himself as a parliamentary candidate; regrettably he could not decide which party deserved him.

Gordon and Luder needed each other as much as they disliked each other. Gordon acknowledged that he was reliant on Luder's persuasive salesmanship and tireless schmoozing to get the commissions that enabled him to design with something close to carte blanche. He later wrote, gleefully biting the hand that fed: 'If it meant selling my grandmother or working for Owen Luder, provided I was free to implement my ideas, I would do it . . . and without a second thought I did.'

Luder, five years Gordon's senior, was a hard-nosed, industrious grafter from the Old Kent Road who, with great determination, had worked his way through the Brixton School of Building and the Regent Street Poly: he knew that the process of architecture was a matter of committees and compromises, budgets and restraints, of horse trading and connecting well, of tactfully balancing the conflicting interests of planners and clients. Rodney Gordon could hardly have been more different. He was exotic, convinced of his brilliance, feckless, charming, impatient, tirelessly hedonistic, uncompromising. And he considered architecture an art.

He was born in Wanstead, East London, in 1933, to non-observant Jewish parents. (He too was secular but was a determined, sometimes belligerent, champion of Israel.) His father was Polish-Russian, his mother Chilean. They were well off. They moved to High Wycombe and then, in 1945, to Chelsea. They spent a lot of time at the races.

Their only son was academically precocious and went up to University College Hospital Medical School at sixteen in 1949. Two years later, inspired by the Festival of Britain, he enrolled at the Hammersmith School of Building, where he was taught by the émigré German modernist Arthur Korn, who had been in partnership with Erich Mendelsohn. Korn captivated Gordon and encouraged him to apply to the Architectural Association, then, as now, a hothouse.

Gordon graduated from the AA in 1957 and went to work for the London County Council architects department, a sort of de facto postgraduate school through which scores of subsequently celebrated architects had passed. It was during this period that he designed the now listed Michael Faraday Memorial at Elephant and Castle, an ingenious cloak over an electricity substation which has mystified millions of passers-by.

He also designed a predominantly timber house (again listed) for himself and his wife at Walton-on-Thames, though he would live there for less than five years before his marriage broke up and he returned to his natural habitat, Chelsea, just in time for Swinging London, where he devoted much energy to fast cars and starlets.

At the very end of the 1950s he had hooked up with Luder and had gone into creative overdrive. First there was Eros House in Catford. Ian Nairn wrote: 'A monster sat down in Catford, and just what the place needed . . . a staircase tower which is either afflicted with an astounding set of visual distortions or is actually leaning . . . the most craggy and uncompromising of new London buildings turns out to be full of firm gentleness.'

Then came the Tricorn. Nairn again: 'At last there is something to shout about in Portsmouth.' He was, of course, right. But Portsmouth did not see it that way. And nor did a sinister constituency of philistine *tricoteuses*. Over and again the Tricorn was adjudged the ugliest building in Britain in crass polls organised by populists seeking favour with 'the people' who would, no doubt, given the chance, equally deem *Ulysses* unreadable, the Second Vienna School unlistenable and Resnais unwatchable. The Trident Centre in Gateshead was a similarly magnificent structure, even more outrageously expressionistic at street level and exhilaratingly primitivist when seen from across the Tyne.

Gordon again used a gamut of subtly non-orthogonal angles to achieve the distortions Nairn had noted at Catford, acts of architectural legerdemain of the highest order. It was, however, no more commercially successful than its Portsmouth counterpart. By 1970, when his friend Mike Hodges came to use the Trident car park as a location in *Get Carter*, Gordon had left Luder. He questioned whether he really was a partner rather than an employee, he was infuriated that much of the Trident's detailing had been subcontracted to journeymen and, even though their authorship was an open secret among architects, he resented Luder's ascription of his designs to the Owen Luder Partnership.

Throughout the 1970s Gordon collaborated with the Abbott Howard Partnership in a climate of growing mistrust for assertive architecture. His reckless combination of overwork and the pursuit of pleasure caught up with him and he suffered a heart attack at forty-two. He subsequently formed Tripos Architects with Ray Baum, but professional life without someone like Luder was difficult. He had neither the patience nor the persistence to woo clients who bored him. Tripos built a factory for Ford at Dagenham, some R&D facilities and labs at Bishop's Stortford, and an office development at Richmond on Thames. Gordon's only important post-Luder work, however, was a spectacularly outré metal-clad tour de force at the bottom of St James's Street which was, predictably, derided in the early 1980s but now appears novel and inspired.

He drove Porsches, rode Ducattis and Bonnevilles and, latterly, despite an amputation and a prosthetic leg, a wonky motor scooter. He flew gliders and skied keenly. He was instrumental in the foundation of the Uphill Ski Club which teaches the physically handicapped to ski.

The Tricorn was demolished in 2004. The Commission for Architecture and the Built Environment and English Heritage declined to recommend that it be listed. Listing was also denied to the now boarded-up Trident. Its fate was decided by 'consultation' with 'the people' and with Tesco. It was destroyed. The future of the Derwent Tower is also in doubt.

This great unsung original was curiously sanguine about the municipally ordained disappearance of his works. He had had his day making them, seeing his visions translated into potently plastic actuality. His admirers properly compare their destruction to the burning of books.

(2008)

IAN NAIRN (1930–83)

IAN NAIRN WAS born in 1930 at Bedford. Which as Pevsner noted is one of the rare British towns – Inverness is another – that makes something of its river. And a mile from his parents' house in Milton is the headquarters of the Panacea Society which guards Joanna Southcott's Box – to be opened in case of a national emergency. Nonetheless despite such excitements he called it 'the most characterless of English county towns' – which suggests that this indefatigable traveller never went to Trowbridge. He can hardly have remembered it from infancy because his father, an engineering draughtsman, lost his job at Cardington in the wake of the R101 disaster which occurred little more than a month after the son's birth.

They moved to Frimley – which really is characterless and best known for the Lakeside Country Club where, as you'll be aware, the Lakeside World Darts Professional Darts Championship celebrated its twenty-fifth anniversary earlier this month. Its best buildings, a group of houses by the young Laurie Abbot posing as the love-child of Gotfried Böhm and Pancho Guedes, were not, of course, built when he moved there, and nor were they built when Nairn wrote the jaundiced Frimley entry in the first edition of the Surrey *BoE* – 'few old houses, most of those killed with kindness'.

Frimley is one of an inchoate, eliding cluster of places which also includes Farnborough, Camberley, Deepcut, Aldershot, Sandhurst, Pibright, Bisley. The names are telling. This is the UK's most populous military concentration. The Farnborough buildings associated with the Empress Eugénie are fascinating, John Shaw's Wellington College is a tour de force and there's a jolly white corrugated iron church at Blackdown of which Nairn wrote nostalgically: 'It used to be painted red and orange and looked exactly like a toy church.' But these are exceptions – the rule is dismal. You can count on the army to render its surroundings banal, ugly and gracelessly tidy. The germ of the idea of Subtopia was, then, nurtured in a martial rather than civilian hothouse.

Nairn lived there till he went to Birmingham University to read maths when he had just turned seventeen. After graduation he did national service as a pilot in the RAF and then accepted a commission. He married at the age of twenty-one. It was often the only way in that grim interlude between the promiscuity of wartime and the invention of sexual intercourse. A former teacher of mine remarked: 'Those long engagements used to play absolute havoc with the urethral tract.'

When he quit the RAF he pestered the *Architectural Review* to publish him and eventually it did. In the summer of 1955 he wrote what remains the single most famous edition in that magazine's 113-year history. This was *Outrage*, a furious, far from polite attack on Subtopia – that is, on mediocre spec building, on drabness, town planners, traffic planners, crummy public spaces, deficient townscape, signs everywhere, imaginative paucity, and the creeping homogenisation of Britain.

But it was not, yet, an attack on architects, whom he still regarded as liable to provide the solution – or *a* solution – to Britain's post-war malaise. The date is important: building licences had only been abolished six months previously at the end of 1954, after a decade during which very little construction had been achieved and architects had been underemployed. He had no reason to doubt that the host of architects waiting for their opportunity would work for the common good. He had an almost romantic conviction in art's potency.

Outrage was not unprecedented. Nairn allied himself to a tradition which stretched from George Cruickshank and William Cobbett to Anthony Bertram and Clough Williams Ellis. But the bloody-mindedness was unprecedented during the years of the welfarist consensus. This was aestheticism of a sort. But it was not the aestheticism of the aesthete class – of Kenneth Clark or James Lees Milne, Sacheverell Sitwell or John Pope-Hennessy – it was redbrick aestheticism, non-U aestheticism, tap-room aestheticism. *Outrage*, it must be remembered, was published a year after *Lucky Jim*, a year before the first performance of *Look Back In Anger*. Nairn belonged to a literary generation that abhorred prissiness, that took a pride in being bolshie, obstreperous and unsettling, in upsetting the applecart, in trampling on the old cultural solaces – Osborne, Peter Nichols, Kenneth Tynan, Harold Pinter, Geoffrey Hill, Robin Cook, B. S. Johnson, Simon Raven, John Arden and J. G. Ballard, who shared many of Nairn's preoccupations.

Outrage either captured or, more likely, created a widespread feeling of glum indignation. The Duke of Edinburgh announced: 'We have a new word – Subtopia.' It was taken up by further unlikely champions such as the *Daily Mirror* – then, admittedly, a literate paper. It made Nairn's name, at the age of twenty-four.

He died of cirrhosis of the liver aged fifty-two in 1983. A mere six short years later, in 1989, this was noted by someone at the BBC – no doubt because Peter Gasson had just published an annotated updated version of *Nairn's London*. The BBC asked me to introduce a short season of his programmes. While they were being transmitted I ran into an architectural historian who always brings to mind Evelyn Waugh's disparagement of Stephen Spender: 'To see him fumbling with our rich and delicate language is to experience all the horror of seeing a Sèvres vase in the hands of a chimpanzee.' This particular chimpanzee was keen to put me right. 'The problem with Nairn,' he instructed me, 'was that he was always trying to go for pseudo-poetical effects.' Had he omitted the *trying* and the *pseudo* he might have been on to something. For Nairn *was* a poet. The fact that he was a journalist doesn't obviate that. Flann O'Brien was a journalist and so were Ezra Pound and Maurice Richardson and James Fenton and Clive James. There is a long tradition of journalists who do not write journalese. Nairn was an anarchic pedagogue who lurched from drayman's mandarin to saloon-bar vernacular. He increasingly embraced obliquity and ambiguity. He was increasingly impatient with fashions, movements, schools – and their petty propagandists.

The essayist – and journalist – Ivan Rioufol wrote recently that: 'I believe in what I see. The doctrinaire see what they believe in.' Nairn might have said, probably did say, the same. He was the least doctrinaire of writers. He was no more a modernist than he was an antiquarian. Or he was as much a modernist as he was a nostalgic. He was not parti pris. He was not dogmatic though he was, as I say, pedagogic, and often pedantic, as Len Deighton pejoratively noted, though what's so wrong with pedantry? It is preferable to imprecision.

Nairn was, then, unpredictable and so, in the eyes of the totalitarian modernist architectural establishment of the 1960s, he became untrustworthy, even treacherous – always a good thing for a writer to be. He was loyal only to his own sensibility, to his own intellect. The architectural profession – like all professions a conspiracy against the layman – expects architectural journalism to be a form of slavering PR. Or at least to cosy up to it, share its dubious values and the self-aggrandising, bogusly scientific patois which is taught at the Architectural Association. That establishment is not at ease with a belligerent writer like Nairn who never took anyone at their own estimate. The old fool Lionel Brett, Lord Esher, a mediocre town planner and hack architect whose book *A Broken Wave: The Rebuilding of England* is a classic of self-exculpatory denial, remonstrated with Nairn when he wrote an *Observer* article in February 1966 entitled *Stop The Architects*. That headline was improbably his, but the first line most certainly was his: 'The outstanding and appalling fact about modern British architecture is that it is just not good enough.' Esher, and the self-important, self-pitying constituency he represented as president of the RIBA, came to consider Nairn a traitor, an apostate, a heretic. People like Esher had believed that he was one of them. He wasn't.

Here, little more than a decade after *Outrage,* he was making another noise, an evidently unwelcome noise so far as the architectural profession was concerned. Nairn was, like nearly all writers, but unlike nearly all architects, an autodidact: architects are pathetically unable to accept the criticism of non-architects. What *qualifications* to castigate architects has Mr Nairn, asked one indignant practitioner in a letter to the *Observer*. It should further be borne in mind that the majority of the architects whose work he derided was a generation older than he was – which further incensed these people. His contempt for them is patently manifest in *Britain's Changing Towns*, published in 1967, a year after *Nairn's London*, but nearly all written before. It comprises sixteen short essays on places as varied as Glasgow and Norwich, Llanidloes and Sheffield. They had been written for *The Listener* in 1960 and 1964, when he still retained a vestige of hope. For the book, almost undoubtedly a cash-in on the success of *London*, Nairn revisited the places and appended a hasty postscript to each essay. Most of these postscripts are monuments to disillusion, to suppressed rage at the harm inflicted by comprehensive redevelopment – the regeneration of its day – to bewilderment at what he perceives as slipshod or unimaginative or crass, to regret at so many opportunities missed. Sometimes, of course, he got it wrong: he confidently asserts that Bicknell and Hamilton's signal box next to New Street station in Birmingham 'is likely to look absurd in ten years' and in Derry he suggests – in 1967 of all years – that sectarian tension has lessened since 1960.

He was making a warning noise that within a few years would become a clamorous screech. Nairn more or less single-handedly set in motion a bandwagon which he failed to control. His approach was nuanced. That of the figures who jumped on the bandwagon was not. They were doctrinaire. Christopher Booker, Bruce Allsopp, Simon Jenkins, the Prince of Wales and so on – ad nauseam, ad vomitum: the failure of modernism became a given, the standard-issue received idea, the conventional wisdom, the knee-jerk reaction.

Nairn was not blind to modernism. He was however temperamentally unsuited to it – I mean, he lacked the credulous gene, he was incapable of remaining a believer, and modernism was a faith, *de rigueur* at the *Architectural Review* where it was a thought crime to consider it a style. It demanded blinkered optimism, and he was anything but an optimist. He was an Augustinian not a Pelagian. His career is a chart of increasing disenchantment, of an ever-growing realisation that he had been sold a pup, that he had believed in yet another god that failed – though so long as he was connected to the *Architectural Review* he seems to have been a believer of convenience, paying lip service to the faith.

'There's nothing wrong with modern architecture,' he said, 'only with *bad* modern architecture.' Most of it was, and remains, bad. Nothing peculiar about this – so is most of the architecture of any epoch bad, the precious eighteenth century included – indeed Nairn would say the precious eighteenth century especially. The modern works he admired were of all schools. Modernism is a broad church. He certainly responded most favourably to those unorthodox buildings which echoed the work of the architects who occupied the highest places in his personal pantheon: Butterfield, Teulon, Nash, Vanbrugh – and Ledoux who might be said to have apprenticed himself to Vanbrugh's ghost. Hence his championing Owen Luder's or, rather, Rodney Gordon's masculine shows at Catford, Gateshead and Portsmouth of which he wrote 'the rhino has got into the marketplace'. He didn't attribute them to Gordon because he avoided social contact with architects and wasn't privy to their gossip. Again he was going against the grain, discerning worth in the architecture of commerce which was habitually – and wrongly – found wanting beside public works: the new universities, civic theatres and so on. He was generally frugal in his approbation of such works.

Jim Cadbury Brown's Royal College of Art was an exception – Nairn oddly refers to Cadbury Brown, in his late forties when he designed it, as 'young'. Maybe he meant young at heart. In a quite different register he was also appreciative of the delicacy of Cadbury Brown's sometime collaborator Eric Lyons's Span houses and of such provincial practices as Wheeler and Sproson in Fife and Robert Paine in Canterbury. He was immune to the allure of big names and bloated reputations. He was lukewarm about Denys Lasdun's cluster blocks in Bethnal Green and the LCC's sub-Corbusian Alton Estate beside Richmond Park, he was contemptuous of Basil Spence's grandiosity in the Gorbals, he dismissed Holford's Kent University as grindingly ugly. And then there are the sacred Smithsons, who with half-witted insolence claimed that 'Lutyens perverted the course of English architecture'.

In a book published a few years ago about press, film and telly coverage of architecture someone called Kester Rattenbury wrote that 'the one occasion when I was *personally affronted* by criticism was when in one of the Sunday papers a reviewer called Ian Nairn said that the Smithsons had not built anything for ten years and it showed'. I love that '*called* Ian Nairn' – one would have expected that an architectural critic – I presume that is what this person claims to be – writing on such a subject would know who Nairn was, just as a literary critic would know who Cyril Connolly was. But no. She goes on to state, with no evidence whatsoever, that Nairn could not *understand* the Economist Building's *disciplines*. What he actually wrote was that the vertical recessions looked like 'medieval buttresses on the National Health' and he mocked the sacred couple for their vanity – they insisted that the newspaper's offices be kept tidy. Personal affront! Where do they find these people? Did it not occur to this thinker that Nairn understood the Smithsons only too well, could see through them and their elaborately obfuscating texts: those that can, build; those that can't, compose manifestos and advance theories. The Smithsons belong not to the history of architecture but to that of self-promotion – not, thankfully, wholly effective self-promotion given how little they were invited to build.

Incidentally, Nairn's attitude to Lutyens, then at the very nadir of his posthumous fortunes, was both against the grain and admirably intelligent. He distinguished between Lutyens going through the motions – at Walton on the Hill for example – and Lutyens at his best – Britannic House, Tigborne Court: note that those two buildings could not be more stylistically disparate.

Nairn's contemporary Reyner Banham – a less gifted writer with an embarrassing fondness for both beardie hep-cat jazz-talk and after-dinner speaker's bumptious archaisms, *hath wrought,* etc. – sought to *interpret* what the Smithsons and their ilk were up to. He set himself up as a sort of conduit between them and a mystified or uninterested public. He identified himself with, linked himself to the self-consciously outsiderish avant-garde. And today's avant-garde, today's self-proclaiming outsiders and rebels always become tomorrow's establishment. This applies in all areas. Had the police horses in Grosvenor Square made a better job of trampling protest-kids in 1968 we'd have been spared such creatures as Jack Straw and Peter Hain.

Nairn could not have been more different from Banham. At a time when the conventionally unconventional was fashionable and the cosy world of alienation beckoned, Nairn really was an outsider, a loner – and as we all know, loners are troubled, just as suburbs are leafy and monstrosities concrete. He had acquaintances rather than friends. Drinking overcame shyness. He was not clubbable, not a joiner. He knew that team spirit is the refuge of the second rate.

Outrage prompted Duncan Sandys, then Minister of Housing and Local Government, to found the Civic Trust – typically Nairn would have no part of it, to have done so would have compromised his independence, his passion, his anger. He never sat on a committee. He was not in the business of making friends – indeed he relished biting the hand that fed. In the part of the Surrey *Buildings of England* which Nairn did not write, Pevsner states: 'Epsom is richer perhaps in Late Stuart, Queen Anne and Georgian houses than any other place in Surrey.' In *Nairn's London*, however, we read: 'Epsom – a terrible disappointment, both in its Georgian houses and its market place.'

Again, in the North Lancashire volume of the *BoE*, Pevsner describes the Greek revival Harris Library in Preston as 'almost *unbelievably* late for the style'. In the series *Football Towns* Nairn observes: 'To *hell* with whether it's late or early in the style.' That was *six years later* – best eaten cold and all that. This is the reaction of someone who, although he might be an ambulatory encyclopaedia of architecture, is not primarily concerned with buildings as historical artefacts or with classifying them stylistically but with their essential, enduring qualities and, above all, with their place in current townscape. Townscape after all levels everything, causes all buildings to cohabit in a perpetual present – it is the result of anachronic collisions.

To hell with whether it's late . . . is also the reaction of someone shedding the orthodoxies of architectural writing, maybe indeed abjuring architectural writing altogether. Struggling out of the straitjacket of what was not then called genre.

Nairn's London belongs to no genre save its own, it is of a school of one. It is not the work of an architectural writer. It is the work of a writer full stop. The masterwork, I believe. Its qualities are literary. There is barely a page which does not contain some startling turn of phrase. And his admirers are literary rather than architectural. They include Geoffrey Hill who cites him in *King Log*, John Gross who includes two passages from *London* in the *Oxford Book of English Prose*, Michael Wharton who by his own admission plagiarised him in the last programmes he made for BBC radio before he began his half-century stint as Peter Simple at the *Daily Telegraph*. These are not, to put it mildly, people who are readily impressed.

Nairns's London is the obverse of the Buildings of England or the Shell Guides or Paul Goldberger's *New York*.

We don't go to Dorset in search of Hardy's landscapes and villages, or if we do we won't find them – because they exist only in the compact which the writer makes with his audience. Similarly, *Nairn's London* is really only to be found within the covers of the book. Sure, *a priori*, it is a guide. But that is simply the mask it presents to the world. The book is a *vade mecum* but of a perverse sort. A guide – any other guide – is predominantly reportorial, and useful. *Nairn's London* is questionably useful and it does not report. It *creates* places which did not previously exist. It's fiction without characters, or rather the décor of a fiction with just one character. The true subject of *Nairn's London*, the book's protagonist, is Nairn and his sensibility, his visual, intellectual, emotional responses to what is around him – and his expression of those responses. It is authoritative precisely because it dispenses with all pretence of impersonal objectivity. It treats buildings and places as objective correlatives, mute inanimate material things which obliquely correspond to this mood, which signify that temper, which exemplify a particular humour. He skews them with extraordinary acuity – with nothing more than a simile or a neat phrase or a sly aside or a raucous throwaway. An effect of this is to condemn them to be forever as he decreed when he took their temperature. His version of many parts of Greater London is more potent than their actuality. Which is no reason not to go, say, north-west to Hampstead – 'a bit of a joke, though many of its inhabitants are deadly serious about it' or south-west to Richmond – 'Funny, louche, subtle. A bit of Wimbledon, a bit of Windsor, a disturbing whiff of Soho or Brighton'. Or south-east to the Midlands town called Woolwich, and thence back across the river to North Woolwich and – if it's still there – to The Gallions, 'the Bedford Park Hotel transplanted to haunted mud flats – a good private place to hatch a revolution', and a place which – for obvious reasons – recalls Chesterton. He avoided anthropomorphism, just about. However, this is his description of Teulon's St Stephen's, Rosslyn Hill: 'A great hulk on a sloping site with a brooding and bulgy central tower made into a macabre Gothic dirge, moody and flashing with unexpected poetic juxtapositions . . .' This, surely, is a portrait of the artist as a ruinous church.

Why should a meditation on a city masquerading as a guidebook not be a greater work of literature than a novel. Why should gastronomic anthropology posing as a recipe book not be greater than a novel. Elizabeth David and Ian Nairn shared the same inspired publisher, Tony Godwin, a man who appreciated that the hierarchy of literary genres is as bogus as the hierarchy of building types. What is so special about the novel in general? Why do churches deserve more attention than a warehouse or shack or silo? It depends *which* novel, *which* church . . . In an era when novels are subcultural artefacts ghost-written for Jordan I guess that to understand this hierarchy we have to look back to the golden age when novels were not ghost-written – but written, by Jeffrey Archer.

Four further books were meant to follow. Guides is evidently an approximate term. Only one was written, *Nairn's Paris*. Paris is evidently a different city from London of which he possessed a cab driver's knowledge. There is no diminution in the prose. It is still as vital, as moving, as droll, as taut as in the predecessor: 'Paris meets the countryside in a dozen sleazy ways.' But he was a stranger, a curious tourist. A gleeful sense of discovery is tangible. But so too is his race against a deadline and an absence of even the most cursory research. He is taken, for instance, with Paul Tournon's modern, but not modern movement, churches – yet omits the grandest, St Esprit. He is impressed by Le Vesinet which he considered to be a proto-garden suburb. He achieves the near impossible and finds La Butte Bergeyre, where houses of every interwar style are tight against each other like in Brussels – yet he doesn't know its name and nor does he realise that it's far from atypical, it's one of about a dozen kindred developments in the hills of the nineteenth and twentieth arrondisements.

Of course, it's not all like that. He needlessly fears for the future of the Gare d'Orsay. He is enthused by the wonderful Musée des Plans Reliefs in Les Invalides which displays Vauban's maquettes. In the Bois de Vincennes he stumbles upon the bizarre remnants of the 1931 colonial exhibition – the Togo pavilion has recently been turned into a buddhist temple. He says of St Denis Cathedral that it's 'like finding Winchester Cathedral in the middle of Tottenham'. He unfavourably compares the vast remorseless social housing project at Sarcelles with the insanitary bidonvilles nearby.

But, hardly surprisingly, Paris and the assortment of distinctly un-Parisian provincial sites which bizarrely take up the second half of the book don't resonate with him the way that London and the Home Counties do. The book remains a quirky guide. It does not make the leap that *Nairn's London* does because he is not privy to subtle demographic gradations, he is unaware of the associations and implications of places, he doesn't grasp the nature of Parisian itineraries.

This obligation to treat with the unfamiliar would dog the rest of his career. He wrote no more books. Tony Godwin, who more than anyone else had discerned the particularity of Nairn's genius, had resigned from Penguin shortly before the manuscript of *Paris* was delivered. Godwin was the midwife without whom Nairn had no one to tell him to push.

For Nairn, as for many *Observer* journalists in those days, the pre-Murdoch *Sunday Times* of Harry Evans was a cynosure. It was a somewhat guileless paper that was preoccupied by trends and fashions. That is, by moral, cultural, typographical and sartorial fashions and above all by fashionable causes. Look at that skinny model – no, no, it's a starving child. Part of it was groovily slick, part was heart on sleeve, designed apparently to foster collective guilt. It was not an obvious billet for a contrarian polemicist such as Nairn. But it paid very well indeed. He probably doubled his salary when he moved there. And because there was so much money sloshing around, the section editors grossly over-commissioned. These people were known as space barons, space on the page was their currency. Much work, especially that which defied the paper's orthodoxies, was never published.

The attraction of Nairn to the paper was that by the end of the 1960s he was a big name: the best-known topographical writer of the era and an increasingly recognised presence on television at a time when even recondite shows got large audiences because – unless you count the regional opt-outs – there were only three channels, none of them run by the educationally subnormal or the excitingly post-literate.

Having lured him, the *Sunday Times* seems to have been at a loss to know what to do with a meditative melancholic with a disinclination to think positive or believe in progress's progress or take an interest in the latest thing. So he ended up writing architecturally accented travel itineraries accompanied by infuriatingly unoriented maps stretching down an entire broadsheet column. Ostend to Trieste via Salzburg – that sort of thing. His time there was not happy. He eventually accepted a generous redundancy payment. And went to the pub.

The BBC, for whom he intermittently made films from 1967 for almost a decade, was more sympathetic to him. But it was no more astute about how to get the best out of him. His telly lacks the complexity and subtlety of his writing. And that is because it *isn't* written. He'd arrive unprepared at a location and busk it. He'd jot down a few notes. There was no time for reflection or for parsing the sort of sentences which inform his prose. What he says is largely improvised – and improvisation is no friend of invention. The programmes were hastily and cheaply produced, mostly by directors and crews who more normally worked for *Nationwide*, which was all cheery local colour and characterful cardigans. The notions of measured cinematography, perspective correction, deft editing, contrapuntal rhythms, a thematic thread, an apt musical score and so on were alien to them. The programmes are clumsily shot like news reports, technically coarse, artless.

But put all that aside. Nairn had marvellously singular and unforgettable presence on screen. The gaucheness which was a social impairment was a boon in telly land. He was a person, not a personality. There was nothing slick about him, nothing remotely cool. He rails against drunks at the Munich beer festival – the wrong kind of drunks no doubt. He suggests that Danes, all Danes, should get out more and see the world. You never get the feeling that he'd pined all his life to stand in front of a camera. He was doubting and melancholic and vulnerable. He betrays a sort of emotional nakedness. His only affectation is his short 'a' – last, castle. At least I assume it's an affectation in someone who all his life, save when at university and in the RAF, lived in the south – but he longed to have come from Jarrow. It's a voice full of quiet despair. The repetitively falling cadence suggests an acute awareness of human impermanence and death's proximity. It knows, however, that even though everything's going to the dogs Canute's example must be followed. Being in his electronic presence is heartening.

Being in his actual presence was anything but. In November 1982 I had just been hired by Tina Brown, the editor of *Tatler* – and Mrs Harry Evans. I had the idea of getting Nairn to write again. Tina reported this to Harry who said I had a fat chance of doing so and referred to him as the *formerly* talented Nairn, though it was Harry's meretricious paper which had been partly responsible for quashing that talent. I rang Nairn and he said to meet him for lunch at the St George's Tavern in Belgrave Road behind Victoria Station: he lived nearby in Ecclestone Square in a flat which his wife had to clean when he was out because he had an antipathy to hoovering and dusting being done in his presence. He had also developed an antipathy to work. He hadn't done any for five years. It was easy to see why. Normally when you're in the throes of an addiction you surrender your body to businessmen from Bogota or Afghan warlords or Imperial Tobacco or rogue chemists in Minsk or Minervois peasants or Seagrams. Nairn, ever original, ever perverse, had surrendered his body to Bass Charrington.

His lunch comprised fourteen pints of beer. He looked dropsical. His always enormous head was distended, and flabby pasties of flesh lay on his collar and reached for his shoulders like aspirant epaulettes. He had grown a Brahmin cow's dewlap. He was barely coherent but thanked me for the invitation to write. He said he'd think about it. A week or so later he asked me to meet him again. He was evidently on a regime because he drank only eleven pints. He said he wanted to do something about Wren but he wasn't sure quite what. He said, again, that he'd think about it. He must have known as well as I did that he wouldn't write it, that he wouldn't ever write anything again. Less than a year later he was dead.

Robert Hamer, Peter Cook, Oliver Reed – you can see where I'm heading. No matter that Hamer wrote and directed three of the greatest films ever made in this country, no matter that Cook was for decades the funniest man in Britain – Derek and Clive, James Last on tour – no matter that Reed gave consistently memorable performances, none more memorable than the one he constantly fell over into – he was the People's Pissartist – no matter what they achieved these men are forever classified as spendthrift talents, their own worst enemies who drank it all away, who failed to fulfil their promise, who wasted their life. Nairn belongs in this company, this august company. Despite having over twenty years written eight books, one of which is an unalloyed masterpiece, having produced countless journalistic articles, having made almost twenty TV films, he has the label marked *failure* attached to him. Which is absurd. The last few years of his life may have been marked by alcoholism, depression, sloth and so on – but what went before possesses such vigour and integrity that it will endure.

Indeed it does endure. He invented a way of looking, a way of writing. And his exemplary spirit endures too. His epigoni are *not*, of course, to be found in the style-conscious pages of vibrantly diverse architectural magazines and diversely vibrant design magazines. They are to be found writing trenchantly sod-you blogs such as Bad British Architecture by the anonymous Ghost of Nairn whoever he or she is – a he I think – and Sit Down Man You're A Bloody Tragedy by the conservative Marxist Owen Hatherley. They are to be found too

among the ever-increasing battalions of soi-disant psychogeographers
– who are distinguished from plain geographers by neglecting to take
their Largactil before they release themselves into the edgelands of
Sharpness or the boondocks of Sheppey.

His work apart, this inchoate school is, as they say, his legacy.
And there is too the implicit health warning: if you're going to drink
yourself to death, do not use beer as the agent of your annihilation.
You have to ingest ten times the volume that you would of a distillate.
Stick to vodka or whisky or kirsch or gin or mirabelle or rum or
kummel or pastis or . . .

(2010)

JOHN BETJEMAN (1906–84) AND NIKOLAUS PEVSNER (1902–83)

ARCHITECTURE HAS MANY USES. Slaughter, incarceration, equine husbandry, tyrannical bullying and religious observance, manufacture, charity, man's inhumanity to children, sweetly sordid pleasures, governance, boastfulness and so on.

Now, let us suppose we are students of charcuterie hoping to get a grasp of the subject: we do not consult the pigs, we do not go to the sty.

Kindredly if we are students of architectural history . . .

No, we have a caste of writers who, one way and another, can explain or classify or at least speculate about what we see around us, the horrors and the beauties. Writers who seize on buildings and create from them a body of work.

That body of work may be the dispiriting hackneyed sycophancy of the broadsheets – what a pummelling Milord Foster's jejunum has taken down the years! It may, even worse, be the theoretical academic discourse maculated by a gross jargon that pretends to scientific exclusivity – what the French call *la langue du bois* rather than *la merde du boeuf*.

Novels, verse and films are full of buildings and places. Colin McInnes's Little Napoli in *Absolute Beginners*, Perec's apartment block, Zola's marshalling yards, Gatsby's house – 'a factual imitation of some hotel de ville in Normandy' – the chapel at Brideshead that Waugh did not admit was a factual imitation of the chapel at Madresfield.

There is too a fragment of all the writing that is *specifically* concerned with architecture which can be read as literature. That's to say it is read not because it proposes an opinion we might concur with or dissent from, not because it is pedagogically effective, not even because of what it addresses or illumines, not because of the information it imparts but because it is worth reading for its own sake: it overcomes its supposed subject. It transmutes into literature. The paramount ability required of an architectural writer, as of any other writer, is the ability to write: it is words that are the material. Being a buff about crockets or understanding how a hyperbolic paraboloid roof works are not really the point.

The verbal symbols of these plastic features, and of glass, gargoyles, regenerative icons, upside-down diocletian windows . . . The words that represent these three-dimensional features inhabit a part of the brain which determines writing's linearity, a different part of the brain from that which manipulates and orders these features in what is dubiously called reality. An architect need not be literate. John Vanbrugh and Thomas Hardy were exceptional – even if Max Gate does little to dissuade us that Hardy made the right career choice.

The architect's endeavour foments afflatus, it prompts prose and verse. This is a purpose of architecture. Certainly, an unwitting purpose and one misunderstood by the majority of architects. But one of its greater purposes.

When does writing elevate itself? The concluding words of Sir John Summerson's *Architecture in Britain 1530–1830* are 'a still lonelier child, devouring the Waverley Novels in a garden at *Herne Hill*, the garden of his father, the *sherry merchant*, Mr Ruskin'. I'm not sure why this brief passage is so moving, so elegiac. But it is. It prompts what Nabokov in *Lectures On Literature* calls the 'telltale tingle: the reader reads not with his heart, not with his brain but with his spine'. Architectural history, architectural polemic, architectural exegesis is or can be literature.

One thinks of Banham's corny beat generation hep cat finger clicking jive talk and instantly dated lexicon. Arty yet artless. It desperately aspires to be literature and it fails.

But against that we have such prodigies as Professor Mordaunt Crooks's crackling zippy prose, all furious intensity and attack and energy. It is hugely infected with what the high Victorians called go! –

a quality initially ascribed to race horses, which then seems absolutely apt given this writing's breakneck speed.

We have Nairn's muscular sentimentality, taproom elegies and determination at all costs to prove himself non-U – like a beery Anthony Burgess.

We have most recently Owen Hatherley whose exhilarating mockery of New Labour's PFI schemes and whose welfare state nostalgia are captivating rather than crude because they are delivered in writing of the surest suppleness and fluidity. And without the hectoring spartist rancour that seems to afflict his generation.

Anything that's any good creates a genre of its own.

Now:

The Living Dead At Manchester Morgue is a very creepy film of the 1970s – shot in that city, in the Peak District which stands in for the Lake District and at the former Barnes Hospital in Cheadle – properly described by the *BoE* as Gothic, grim and debased. The pubs are sinister. The yokels are psychopathic. And agrichemicals create zombies.

Appointment in Liverpool. A film about a woman's attempt to avenge the murder of her father at Heysel stadium.

May Morning is a horror film set in an Oxford of atavistic ceremonies, debilitatingly drunken students and murderous rites.

Death Watch is a dystopian sci-fi film whose protagonist played by Harvey Keitel has a camera implanted in his head. Set in a sort of post-apocalyptic city: Glasgow.

London Night is a more or less hallucinatory depiction of a single night in the early 1950s.

Passing Time is set in a city called Bleston in the early 1950s. The novel's narrator becomes obsessed with, among other things, a rose window.

What these disparate works have in common is that though they are set in Britain they are the work of directors, screenwriters and novelists who are Spanish, Italian, French, German.

The Britain they represent, or create, is recognisable but faintly skewed, oblique, distorted. It's like a photograph which has been reversed – which is possibly the way we see places we know in dreams, before they are corrected by the brain.

Banal aspects of everyday life which are self-evidently taken for granted are lent emphasis. That which appears normal to us is marvelled at, puzzled over and shown to be very strange when seen through eyes that are not culturally attuned. In Bleston, an analogue of Manchester, the French narrator of Michel Butor's *L'Emploi du Temps* is served a slice of cake. He asks what it is called. Sponge. How very appropriate he thinks. Places we habitually overlook are invested with exoticism: Arnaud Cathine's *La Disparition de Richard Taylor* is partly set in Minehead, partly in Sheerness. The whole of Uwe Johnnson's *An Unfathomable Ship* is set in the latter. The Oxford of *May Morning* – Alba Pagana, Pagan Dawn – is a city of debilitatingly drunken students and barbaric rituals. The French essayist Denis Grozdanovich who was once that country's junior lawn tennis champion and subsequently its squash champion took up real tennis in middle age. He told me that he knew Britain through the location of its real tennis courts and had played at all of them. I muttered that he can't have seen much of the country because there are only a handful of them. Twenty-seven, he replied triumphantly – and listed them.

This curiosity about what we are liable to be incurious about is a priceless quality.

A decade ago the apparently terminally English Timothy Mowl, whose forebears have presumably been throttling Johnny Foreigner since the dawn of time, wondered in his brief discursive book on the relationship of Pevsner and Betjeman whether Pevsner 'should, with no English social background, have been encouraged so quickly to a position where he could exert an unwise, and, in a very real sense, an "alien" influence?'

As I said when I reviewed the book, the sentiment is ugly in its unabashed xenophobia. Further, the position in question was that of prolific journalist, assistant editor of the *Architectural Review* (which he had 'quietly worked his way into', read: *jewed* his way into) and the author of the nascent *Buildings of England* – the last position being one that he himself created. Again, the accusation that Pevsner was responsible 'by his prestige and tacit approval' for the development and shape of British architecture is as absurd as Mowl's contention that the site of the Attlee government's ground-nut fiasco was the Gold Coast. Tanganyika, actually.

Why was Pevsner so culpable when, for instance, P. Morton Shand, Siegfried Giedion, Geoffrey Boumphrey, Anthony Bertram, John Gloag and Oliver Hill were not? It is easy with hindsight to attribute to Pevsner a much greater influence than he actually had. Writers enjoy flattering themselves that they and their predecessors have power. It's mostly a delusion but is, of course, immeasurable. Ruskin was the most persuasive critic of his age but Ruskinian Gothic is the exception rather than the rule even in Belfast and Bristol, the cities where it was most enthusiastically adopted.

No man since Ruskin, whose inspired earnestness he shared, has changed this country's attitude to architecture more than Pevsner. We owe him (and his collaborators) an immeasurable debt: he opened our eyes, he still opens our eyes. But he doesn't make us see. There is a difference. He was not, and never pretended to be, an evocative writer. He describes, and leaves us to do the on-site ocular work. His approach was that of a scientific taxonomist. He was not merely an architectural modernist but a literary modernist. From Joyce through Faulkner to Nabokov (a lepidopterist) and Robbe-Grillet (an agronomist), modernist description was preoccupied with occlusion, with an accretion of detail that inhibits our summoning the object in our inner cinema. This is a device that aspires to neutrality, a neutrality that Pevsner seldom ruptures – but when he does . . . His one-liners and taut dismissals and laudatory clauses are all the more potent for their infrequency. The discipline with which he rationed himself is admirable. And, of course, it goes against the 'amateur spirit', the heart-on-sleeve expression of enthusiasm. There can be no doubt that he was an enthusiast for the buildings of his adopted country (which he loved more than it loved him) – why would he have put himself through such a gruelling assault course had he been otherwise? He was also an amateur in the now archaic meaning of that word.

What he wasn't was an amateur in the sense of dilettante. Builders, masons, surveyors, architects, engineers, carvers, ironworkers – the overwhelming majority of those who have professed these callings down the centuries that Pevsner devoured have indeed been professionals. Yet the study of their work was largely undertaken by passionate dilettantes who sought to pass off torpid imprecision as 'impressionism'.

If we consult a work of, say, mycological or avian reference, we expect exactitude rather than bumbling guff wrought with a pen dipped in lard.

So why should buildings be different from fungi and birds? The easy reply is that buildings are humankind's rather than 'natural' creations. But it is, equally, wrongheaded: if humankind can create the Zwinger and Blenheim and the Stupinigi, it should be acknowledged capable of creating an exact way of classifying them.

I use those examples of the baroque because the one aperçu I've enjoyed this past week is to do with Pevsner's taste, and his initial specialism was the baroque. It is necessary to amend Borges's famous dictum that 'every writer creates his own precursors', which is a neat way of saying that because E was influenced by A, B, C and D we necessarily come to see A, B, C and D in a fresh configuration and discern a connection between them, even if that connection is only E.

Mowl's witting polarisation of Betjeman and Pevsner suggests an unusually bravura approach to biography. Candida Lycett Green told him that towards the end of their life her father and Pevsner got on harmoniously. Mowl wrote: 'I prefer to emphasise their conflict as that was more real.' Real! Real is a dodgy word at the best of times. In this case it apparently means something along the lines of 'more exciting' or 'makes for better copy'. Roundhead v cavalier. Romanticism v classicist (modernism was classical). Trad v Modern. Celtic v Rangers: Pevsner is evidently Protestant Ranger. The lure of the adversarial. The lure of the taxonomical – which Pevsner himself succumbed to over and again.

We should celebrate their kindred qualities. What they had in common. We should not condemn them to a sort of posthumous separate development. What they had in common was a fierce curiosity about what was not yet called the built environment. Most people did not share that curiosity. They were in a tiny minority. The fact that that minority has grown is in large measure due to them and their example.

There's a certain English mindset which regards integration and assimilation as presumptuous. This mindset produced in one generation the sort of anti-Semitism that was laughed away as 'casual' or 'social'. In another generation – today's – that mindset has produced the fetish of faith schools and cultural diversity – which is always vibrant and is actually a kind of tribalism. A kind of separate development. And is not cultural – but ethnic.

We have free will. We may be born of Jewish or Christian or Muslim *parents* – but we ourselves are not born Jewish or Christian or Muslim. With all due respect to the late Michael Jackson and the pelmeted orange women of northern England we cannot change our ethnicity. But we can change everything else, unless we are in a racial state where racial laws devised by Kings College School Wimbledon's most celebrated old boy, Richard Walther Darré, decreed that if you had a single Jewish ancestor after 1750 you were still Jewish.

We can change our culture, our language, our nationality, our religion. Pevsner did so. The lot, for different reasons. This should be cause for celebration. Anyone who can write between two and three thousand words a day in a language which is not native to him deserves only admiration. Equally, anyone who rids himself of his parents' superstitions, who eschews daft diets, must accept that he is no longer reliant on or part of a particular community.

The two or three thousand words per day were the words, mostly written in B&Bs, which helped to slowly open our eyes to what was around us, to what we took for granted. This is not to deny the writers who preceded him. But his systematic determination and the eventual comprehensiveness of the Buildings of England are universally unparalleled. George Steiner declared that we can only understand our own culture when we know another. Pevsner allows us to make a short cut. We perceive our own architectural ethos through a sensibility formed in a tradition that *was* not ours – but which *is* ours now, it has become ours, for over three or four decades he created that great a shift, he benignly contaminated us with that sensibility. Through the process of defining Englishness he himself became English – it was an appropriation for him.

But architecture has for thousands of years been international: commerce and empire. Think of how much of Greece there is in the Athens of the even further north, Elgin. Yet Pevsner over and again finds national characteristics in buildings and art. 'One thing is certain. Klimt was eminently Austrian, being radical yet never heavily serious.' One has to suppose then that Adolf Hitler was not eminently Austrian.

What *we read* bears a sheen of impersonality. What *he wrote* is anything but. It is every bit as subjective, biased, partial and emotionally engaged as Betjeman's verse and prose and films – but covertly so. Well, fairly covertly.

I'm old enough to remember Victoria Street before it was done over. The first publication I wrote for, *Books and Bookmen*, had its offices in Artillery Mansions – which is merely 'dull, plain, all over gothic'. Across the street however stood a building whose pituitary gland had failed to inhibit its growth. It belonged to an architectural freak show. It was immense and lovably distended. And then, of course, it was gone. Not that Pevsner can have rued its demolition. 'The crowning monstrosity of Westminster, Windsor House, 1881–3 by F. T. Pilkington, a nightmare of megalomaniac decoration, with huge (three-storeyed) entrance and huge corbels under the bay windows. Built of Yorkshire sandstone, like the Army and Navy Hotel.' That *like* should have been *as*: Windsor House had originally been the Army and Navy Hotel. The German 'So' means both like and as – which makes one wonder whether he made notes in German or was thinking in German . . .

He described the same architect's Winchester House in Old Broad Street as 'that gargantuan folly' and, not altogether accurately, as 'Continental baroque'. In fact to judge from photos – this was one I never saw – it rather resembles the lavish Second Empire buildings to the south of Parc Monceau. And indeed that is how, in the Edinburgh volume of the *Buildings of Scotland*, Colin McWilliam characterised what he called Pilkington's later style, when addressing the startling confection that is now Daniel Stewart's and Melville College's boarding house. Pilkington's wilful obstinacy and perverse originality is apparent whatever idiom he essayed. In Pevsner's eyes he is proof that 'Butterfield was not alone in his curious yearning for ugliness'. Quite. He had plenty of company.

Bassett Keeling's St George on Campden Hill has 'An atrocious front with a narthex or short columns'.

St Michael's College outside Tenbury Wells: 'Woodyer's idiosyncrasies come out more strongly (than in the neighbouring church) especially in the crazily steep narrow dormers.' He was even less impressed by the pump rooms which gave Tenbury its suffix: 'Gothicky or Chinesey fair stuff, i.e. without seriousness or taste.'

Joseph Peacock's St Simon in Chelsea is 'Overcrowded with gross gothic motifs in the triumphantly bad taste of the mid-Victorian years . . .' However, he qualifies that with: 'which is so often a relief after too much pedantic correctness.'

And Teulon is predictably 'hamfisted, he indulges in sensational display, he is crude, heavy, insensitive, noisy . . . and impressive'.

Joe Orton wondered in his diaries – What's all this about artists being sensitive? Pevsner seems to have been fascinated and bemused by the wild men of the 1860s, which was also the decade of the sensation novel. But he shies away from their work as he does from any that may be touched by kitsch or excessive mannerism. Unlike Betjeman he'd prefer not to even entertain the proposition that there is nothing so exhilarating as philistine vulgarity.

As with the modern gothic of the 1860s, so with the brutalism and neo-expressionism of the 1960s. He never makes a specific link but he berates both in the same terms.

Dadi Surti who must have been unique in having worked for both Albert Richardson and Erno Goldfinger, designed a Barnardo's school in Kidderminster. 'It shows the architectural fashion of the '60s at its most outré. Dagger-shaped pre-cast concrete members produce a restlessness made even less acceptable by the crude contrast of black and white.' Unhappily this excrescence has been demolished to make way for an estate of insipid executive houses – bad taste gives way to no taste. Surti himself has fared rather better. He has a prolific practice in Karachi specialising in chemical factories and zoroastrian fire temples. As one does.

In the introduction to *Chester* (1970/71) Pevsner notes with apparent relief that 'the brutalists have not yet invaded Cheshire'. This was a man who believed passionately in the idea of progress yet wished progress had been petrified at some point in the 1930s. He mostly abhorred what post-progress, the future, brought. With the celebrated exception of Jacobson's St Catherine's College Oxford, the work he witnessed being made during the last 20 years of his life was a disappointment when it wasn't an affront. He believed that there existed an architecture that was pure. He believed that perfection might be attained. This is a matter of faith, of faith undaunted by a lifetime's empirical observation of the contrary.

It's a faith John Betjeman did not share. His writing, on the contrary, is haunted by a sharp recognition of our imperfection and of our works' imperfections. And of death's ubiquity. And of melancholy creeping up unseen. As a writer on architecture he is seldom Pevsner's peer. But, different thing altogether, as a writer about places he enjoyed powers of evocation that Pevsner could not match and did not attempt to match. As a writer full stop, a limner of the human condition, Betjeman suffers from the damning accusation of having composed 'light verse'. Light verse was never so grave. The anti-intellectualism he adopted – along with Osbert Lancaster, John Piper, Philip Larkin, Kingsley Amis and so on – should not fool us as it may have fooled Pevsner. It is itself an intellectual posture. Nor should we be taken in by the teddy bears and straw hats, the camp Anglicanism and toothy bumbling, and the pretence that he longed to be a member of the Royston round table or the Baldock WI.

He was a pro. A pro what, though? A pro Betjeman – or Betjemen. There's the telly topographer who says in a throwaway aside that 'Nowhere in England is dull, not even on a wet day'.

There's the music-hall turn, every bit as end of pier as Mick Jagger who, as a young man, tried without success to snog Candida Betjeman.

There's the ambulatory encyclopaedia of forgotten houses, ruinous chapels, disgraced sgraffito artists of the '90s.

There's the man who knew someone who had been bilked by Baron Corvo and someone else who had visted Swinburne at The Pines.

There was the purposeful flâneur, the man who resolved to know everyone.

Yet – and this is seldom the case – the multiple roles, if that's what they were, are parts of a harmonious whole. They coalesce, they are consistent, they infect each other. He was proof of Buffon's dictum: *Le style, c'est l'homme même.* To separate the hymner of buildings from the laureate of branch lines is to somewhat diminish him, for his topographical prose and his memorable performances in front of camera are indissolubly linked to his verse. This is not to say that the three are interchangeable. On the contrary, whatever he did was technically specific to the medium. Nonetheless the astute eye, the heightened sensibility and the desperation to avoid boredom are constant. The same goes for his subjects which are, in a way, his *inventions*, so persuasive and transforming is his vision. He treats architects dead long before he was born with the proprietorial familiarity that certain late Victorian novelists treat their characters. When he first wrote of, say, Cuthbert Brodrick or Harvey Lonsdale Elmes in the 1940s they could, for all anyone knew, have been his creations – most of his audience would never previously have heard of them, such was the mid-twentieth century's ignorant antipathy to the architecture of the mid-nineteenth century.

Betjeman's interest in these then forgotten figures was for many years taken to be at best obstinately teasing, at worse perversely aberrant. Pevsner's wasn't: earnestness has its place. In *Metroland*, in 1973, Eddie Mirzoeff filmed him at Grimsdyke. His immersion in the period with which he was by then popularly identified lends him the air of a revenant from the milieu of the Royal Academician Reginald Goodall who commissioned it, Norman Shaw who designed it and W. S. Gilbert who drowned in its pond: rhododendrons, the Old English style, a proximity to Harrow, a wasteful death – did he not invent this too?

Places rather than mere architecture were his greatest forte: architecture, a component of places, was too limiting. Betjeman was excited by the humble, by the everyday, by the allegedly meretricious, by preposterous kitsch, by collisions of the bathetic and the sumptuous.

He brought an aesthete's sensibility to bear on found objects which better behaved or less professionally opportunistic aesthetes would shy away from, shrieking. He wrote in 1965 to Laura Waugh about Compton Acres at Parkstone: 'a series of gardens of such unexampled and elaborate hideousness . . . that Evelyn will want to put pen to paper again, and that wonderful gift he has for bringing out the startling and alarming and funny in the trivial will be spurred into renewed activity.' He was, needless to say, describing his own gift.

Compton Acres, a twee creation of the 1920s, happily still exists, still *simpers*. It belongs to the same gamut of taste as Limited Edition Porcelain Masterpiece Collection™ Nuptial Souvenir Figurines of Sir Elton John and Mr David Furnish. But such no-holds-barred bereavements of taste in large-scale endeavours like landscape or architecture are rarer today than 40 years ago when Betjeman discovered those gardens. He too believed that 'there is nothing more exhilarating than philistine vulgarity'. But philistine vulgarity is not what it was. Today it is 'quoted', 'ironically'. It has lost its innocence. We live in a more self-conscious and thus more self-curtailing age. The precepts of 'design' are relentlessly disseminated and ubiquitously adhered to because we are, thankfully, *visually literate*. Even retail parks, an epithet Betjeman would have relished, are infested by synthetic modernism. Our Man In Malmesbury snooping (there is no other word) as he once did in arched alleys and down hidden paths where back gardens give on to allotments would no doubt find a town Responsibly Attuned To Its Heritage, where lean-to roofs in corrugated iron (the material of the future in 1900) have been replaced by historically appropriate tiles and where Crittall windows no longer illumine Jacobean cottages. Bricolage and extemporisation have been superseded by kits from B&Q: assemble it yourself is not the same as bodge it yourself, with scrounged materials intended for other purposes.

But then England is wealthier. What wealth brings, evidently, is the spread of tidiness. A derelict cottage or ivy engulfed farmhouse is a rare sight. Buildings are saved, certainly, but at what cost? They are no more allowed to wither than are octogenarian former starlets. It is improbable that Betjeman foresaw – or wished to foresee – the future absurdities of the conservation movement which he personified and popularised to the point where it is a cultural and social crime to support the demolition of, say, Georgian hackwork. He was however prescient about the fate of provincial town centres. In the introduction to *First and Last Loves,* he observes that: 'when the suburbanite leaves Wembley for Wells he finds that the High Street there is just like home.' That was sixty years ago. During which years successive governments have failed to correct the retail corporatism which has scarred this country. In general, though, the fabric which he recorded in that book and showed in his films is – astonishingly and hearteningly – extant, at the moment. The society of Cheltenham – subject of perhaps his finest essay – may have changed. The old India hands and the military widows have all gone now, the way of Adam Lindsay Gordon. (Did Betjeman know that the Cheltonian Lindsay (Gordon) Anderson, to whom he wrote a fan letter after seeing *If...*, was a distant kinsman of that ill-starred poet?) But the terraces and caryatids have never looked better. The stucco gleams with new money. Tivoli blushes brightest pink towards summer dusk. Even the debased High Street is, with the exception of the dire Cavendish House, recoverable beneath the fascias and face lifts.

What, though, is forever beyond recovery is an England that was not yet built over, *regenerated*. An England which Betjeman had little cause to pay attention to, an often unremarkable, marginal England which is going, which is not noticed and will not be rued till it is gone. And whose qualities may not be fully appreciated until it finds its peculiar rhapsodist.

(2010)

FIVE ARCHITECTS

Douglas Stephen (1923–91), Georgie Wolton (1937–?),
Frederick Pilkington (1832–98), Sextus Dyball (1832–?),
Gino Coppedè (1886–1927)

THE PROPOSITION THAT cities are free museums is self-evidently true.
Truistic, even. The streets, towers, offices, monuments, fountains,
spires, alleys, bridges, parks, boulevards, palaces, churches, apartments,
exchanges, barracks and so on, are the exhibits in the most engrossing
shows on earth.

It's all happenstantial, of course. Cities were not founded as
museums, but as sites of commerce, industry, worship, belligerence
and so on. They are, however, institutionalised and exploited as
museums. This process reduces a city to a collection of sites. Or, rather,
parallel and competing collections. Thus: imperial Rome, renaissance
Rome, baroque Rome, fascist Rome (for specialists only). These are
the Romes we are likely to be apprised of, which have staked their
place in history, which we tick off in our guidebooks, which we visit
because they are sanctioned, commended. We are letting ourself down,
depriving ourself, failing in our autodidacticism and in our duties as
cultural sheep if we neglect to visit the Coliseum or Piazza Navona.
Just as we would be were we to overlook the great masterpieces
of painting in a closed gallery. One effect of ever more widespread
technologies of reproduction is that we inhabit a world that is not
merely cartographically but photographically mapped. We know
what we're going to see before we get there. There are few surprises
left. Which is not to say there are none.

But we can hardly search for them: surprises are not susceptible
to being tracked down. They are more likely to be discovered by
chance, by serendipity. And serendipity is, in this instance, achieved
by ambling, idling, strolling in a spirit of curiosity, of hope rather

than expectation. Had it not been for this base appetite for *farniente*, I guess I'd never have encountered most of the somewhat marginal architects included here. When I say marginal I intend no deprecation, no patronisation, no faint praise. Far from it. I would suggest, rather, that they are artists who never, for a variety of reasons, achieved the recognition they deserved in the course of their career and that such contemporary indifference has duly afflicted posterity's opinion. It is, evidently, hard to hold an opinion about something of whose existence one is ignorant.

It was a photograph by Bill Brandt that drew me to Campden Hill in Kensington in my late teens: it showed a heavy and exhilaratingly coarse terrace of grand late-Victorian houses on a slope. Needless to say, my eye did not lend it the sullen mystery that Brandt's lens had. I walked on. A group of very different buildings caught my eye. Sheer, white, neat, crisp, cubistic – and thrilling. They belonged, surely, to the 1930s. Yet there was something about the details – The Mount was the first of Douglas Stephen's buildings I saw. It was completed in 1965, and was entirely out of step with its time: it was at odds with both the fey Festival style and with the sculptural brutalism that was the conventional reaction to the Festival style (and which Stephen had essayed). But Stephen belonged to no school. That, I suspect, is why he is overlooked. The Mount retains its extraordinary freshness. So does his David Murray John Tower in Swindon, that town's most (only?) striking building, a mini-skyscraper that has affinities to a design of Frank Hampson's for Dan Dare, Pilot of the Future.

The same conviction – that I had chanced upon an unrecorded tour de force of the 1930s – overcame me when I first encountered Cliff Road Studios in Lower Holloway. Again, wrong. This beautiful complex was built in the late 1960s to the designs of Georgie Wolton – whose other work includes, so far as I can ascertain, one house and the garden of The River Cafe in Hammersmith. A one-hit wonder then? Maybe. But what a hit.

A case could be made for Stephen and Wolton being at the head of the very earliest vanguard of that tendency that burgeoned in the early 1990s, and that continues to this day, that tirelessly reworks the forms of early modernism to the point where they are hardly distinguishable from their models. When modernism is revived it mysteriously doesn't count as revivalism or pastiche. When the gothic is reinvented, it is automatically revivalist. Even Frederick Pilkington's gothic is thus classified, although it is difficult to imagine an architect more wholly striving for originality. The high Victorian generation to which Pilkington belonged was preoccupied with the creation of a new architecture. Pilkington succeeded in this. Only the untutored could confuse a pre-Raphaelite canvas with one from before the time of Raphael.

Similarly the proportion of Victorian building that can be mistaken by a sentient spectator for one of the medieval period or, indeed, any other past era, is slight. Nonetheless, a kind of false antiquarian prejudice is directed at the Victorians. The nineteenth-century architecture that is valued today is that atypical kitsch that looked backwards, that was most zealously fake, most doggedly mimetic: twee, luddite, cosy, merrylie Englishe. The arts and crafts movement was all high-minded prattle and good intentions, and it led directly to the despoliation of Britain throughout the twentieth century. The most exciting work of that most exciting decade and a half, 1858–73, is bolshie, bloody-minded and perverse. And nobody was bolshier, etc., than Frederick Pilkington (though he may not have so considered himself). Nobody was less concerned about the canons of good taste. In *The Buildings of England*, the greatest of his few London buildings, Windsor House in Victoria Street, is described as 'the crowning monstrosity of Westminster, a nightmare of megalomaniac decoration'. Who could disagree? But those words strike me as plaudits. It was, alas, demolished in 1971.

Edinburgh – where Pilkington's practice was based – still abounds in his ferocious, bizarrely scaled, restlessly detailed clinics and villas, tenements and churches. After the good manners and uniformity of the Georgian new towns they are marvellously tonic – enduring monuments to the awkward squad's vigour and sod-you-ism. These are qualities that are equally manifest in the deepest south London suburb of Upper Norwood. The removal of the Crystal Palace from

Hyde Park and its re-erection at Sydenham prompted an untrammelled building boom. Villas of all shapes took advantage of the views over the London Basin and Kent. The wildest of them were designed by one Sextus Dyball, a local builder to whom all devotees of inspired ugliness and sinister gracelessness will be forever indebted. Did Dyball know what he was doing? Did he have any idea of the sensations that his creations would provoke? It is improbable that he was out to shock. Rather, it seems that he, like Pilkington, subscribed to an aesthetic system that, although not far from us feels chronologically, seems unfathomably distant – a vestige of an alien civilisation.

Of course, one facet of these architects' oddness is the location of their work. What is odd in Edinburgh or London may be almost normal in New England or Paris where nineteenth-century modernism was more candid about its intentions, less inclined to disguise industrial materials. Pilkington's Dean Park House (now part of Stewart's Melville College) is an exercise in the fabulously ornate idiom of the Second Empire. Dyball greedily took from everywhere save the conventional sites of English inspiration: the suburbs of Boston and Paris and Genoa spring to mind. Was he a traveller or was he simply a voracious gaper at illustrated books?

Speaking of Genoa: in the early 1990s, I was working in Rome. Cushy billet, as we used to say in The Catering Corps. Save in one regard. Early every evening I would walk for a couple of hours, mapless. A visual diet of ancient masterpieces becomes as sating as, say, a gastronomic diet of foie gras. One yearns not for banality but for excellence of a different sort. It was one October dusk when I found myself north-west of the Borghese Gardens in a narrow street that mutated into an urban hallucination. There were arches, gravity-defying cantilevers, apartment blocks sprouting gargoyles of a type that rewrote the physiognomy of gargoyles, tiles over everything, high relief fruit and veg. No surface was bereft of encrustations. There was something of Gaudí in the air, something of Escher, something of those turn-of-the-century orientalists such as Pierre Loti and Willi Pogany. Something: but not a lot. This place – and there were several streets and squares and circuses – was entirely unexpected. And it was almost unlike any other: almost. I recalled walking a few years previously in a kindredly off-the-map part of Genoa and coming upon a street that was surely some relation. It was. The architect was the Florentine Gino Coppedè. He worked first in Genoa and then in Rome, where the streets lined with his extraordinary structures are collectively known as the Quartiere Coppedè. They were, astonishingly, built in the 1920s. They appear to be at least two decades earlier and are therefore the opposite of the latest thing. Coppedè was at the end of the art nouveau tradition, but so divorced from it that he inhabits a world of one. He led nowhere. There were no disciples. His work exists in a fantastical cul-de-sac. History, real history, is composed of such disconnected fragments. The retrospectively imposed continuum is a soothing fib.

(2004)

INDEX

Index

Index

Index

by, 183; Hadid on, 369; Nairn on, 391; overview, 381–5; Tricorn Centre, 145–6, 273, 335, 369, 382, 384, 385, 391; Trident Centre, 384–5; Trinity Square, 369, 382

The Gorge (Nichols), 353

Gosport, 146

gothic architecture, 159, 223

gothic revival, 58, 159–60, 190, 194, 198, 323–4

Goudhurst, 310

Gough, Piers, 110, 176

graffiti, 348

Grain, 260–1, 262

Le Grand Bleu, 233

Grass, Günter, 89

Graz, 182

Great Exhibition (1851), 39

Greta Bridge, 314

Grimsdyke, 411

Grimshaw, Atkinson, 45

Grolland, 63

Gross, John, 394

Grossman, Loyd, 249

Grosvenor Square protests (1968), 392–3

Grozdanovich, Denis, 404

Grutas Parkas (Stalin World), 117, 133

Guedes, Pancho, 143

Guernsey, 69

Guildford, 148

Guthrie, Woody, 135

Hadid, Zaha, 182, 360–73

Hague, William, 157

Haig, Axel, 93

Haileybury College, 153

Hain, Peter, 393

Hall, Benjamin, 194

Hamburg, 69, 91

Hamburg-Altona, 182

Hamer, Robert, 41, 261, 399

Hamilton, Patrick, 41

Hamilton, Paul, 374

Hampson, Frank, 62, 181, 375

Hampstead Garden Suburb, 76, 103, 126

Hanseatic League, 90–1, 223

Hardcore exhibition, 274

Hardy, Thomas, 12, 234, 402

Harnham Hill, 302, 325–6

Harris *see* Lewis and Harris

Harrogate, 339

Harrow, 202

Harry, Prince, 117

Harty Ferry, 265

Hartley Wintney, 190, 198

Hatfield, 311

Hatherley, Owen, 179, 399, 403

hats, flat, 276

Hauserman, Pascal, 353

Hawksmoor, Nicholas, 50, 58, 152

Hawkstone, 152

Hay-on-Wye, 326

Hayes Common, 203

Hearts and Flowers (film), 96

Hebridean Brewery, 291

Heinrich, King (Heinrich the Fowler), 67

Hell's Angels, 107

Helvin, Marie, 250

Hemingway, Ernest, 148

Hemingway, Wayne, 148–9

Henley, W. E., 191–2

Henley Court, 207

Henze, Hans Werner, 283

Héré, Emmanuel, 357

Herkomer, Hubert von, 194

Herrsching, 67

Herzog and de Meuron, 183

Heseltine, Michael, 33, 105, 173, 210, 217, 221

Hess, Rudolf, 63, 65

Heysel disaster (1985), 220

highland clearances, 294

Hilder, Rowland, 269, 310

Hill, Geoffrey, 394

Hill, Oliver, 405

Hillingdon Centre, 171–3

Himmler, Heinrich, 65–6, 67, 69, 131

Hitchcock, Alfred, 244

Hitler, Adolf: and architecture, 66; and Berchtesgaden, 68–9; and eternity, 67; fans nowadays, 117; *Mein Kampf*, 62; origins, 63, 128; and swimming, 131

Index

Index

SUBSCRIBERS

UNBOUND IS A new kind of publishing house. Our books are funded directly by readers. This was a very popular idea during the late eighteenth and early nineteenth century. Now we have revived it for the Internet age. It allows authors to write the books they really want to write and readers to participate more actively in the publishing process. The names listed below are of readers who have pledged their support and made this book happen. If you'd like to join them, visit www.unbound.co.uk

Alex Aitken
Jonathan Allen
Nathan Allen
Edward Anderson
Rob Annable
Garry Arkle
Helen Arney
Adrian Arratoon
John J. Arundell
Lee Ashcroft
Brian G. Ashton
Philip Aspin
Nicola Asseliti
Michael Atkinson
Naomi Atkinson
Adrian Atterbury

Victoria Baines
Jack Baker
Simon Balkham

Stephen Ball
Maria Raich Balp
Phil Barnard
Richard Barnett
Timothy Barnett
Justin Barrett
Andrew Baxter
Nick Baxter
Mark Beecham
Adam Bell
Jonathan Bell
Anthony Belshaw
Andrew George Bennett
David Bennett
Anne Berkeley
David Bertram
Tony Biggins
Matt Bishop
Paul Bissett
Matthew Blake

Elizabeth Blanche
Simon Blinkho
Mark Blower
Tom Boardman
Kevin Boniface
Owen Booth
Ian Borthwick
Karl Bovenizer
Justin Bovington
Siobhan Bowman
Michael Bownas
Alex Boyd
Chris Brace
Jean Brennan
A. J. Brimmell
Jonathan Brooker
Ross Brown
Anthony Brown
David Brumby
Shona Main & Nick Bruno
Tristan Bryant
Richard Budgey
Oloff Burger
Richard Burgess
Mark Burnley
Mike Butcher
Daniel Byrne

Julia King & Greg Callus
Peter Campbell
Stuart Canning
Xander Cansell
Stewart Carr
Barry Caruth
Alastair Cassell

Will Chadwick
Philip Champion
Simon Champion
Lucy Christopher
Jonathan Churchman-Davies
Matthew Churchward
Richard Clack
Christopher Clarke
Emma Cleasby
Jonathan & Rachel Clinch
James Clive-Matthews
Richard Clouston
Alison Coleman
Paul Cook
Antony Coombs
Paul Cooper
Oliver Cope
Alan Cowan
Alison Cowan
Chris Cowan
Gina Cowen
Richard Cox
Jonathan Cox
Colin Crawford
Eleanor Crawford
John Crawford
Elizabeth Crew
Mark Crump
Paul Cuff
Nigel Curtin
Jennifer Curtis
Tom Cusack
Robin ap Cynan

Peter Dalling

Subscribers

Kimon Daltas
Valerie Daly
Rishi Dastidar
Bruce Davenport
Tony Davis
Jon Davison
Robert Dawes
Steve Day
deadmanjones
Cathy De'Freitas
David Dear
Edward Dearberg
Samuel Deards
Paul Dempsey
Joseph Denby
Neil Denny
Harry Dienes
Mar Dixon
Les Dodd
Miche Doherty
Christine Donaldson
Kevin Donnellon
Peter Dorling
Paul Dormand
Rob Downes
Robert Doyle
Lawrence T. Doyle
Andrew Driver
N. S. Drofiak
David Drucker
Charles Dundas
Lawrence Dunn
Siona Dunn

Chris Eastburn

Mark Eatherington
John Eley
Tony Elphick & Christopher
   Elphick
Max Erhard
Nicholas O. L. Evans
Mark Ewing

Simon Faircliff
Brian Fearon
Norman Ferguson
Charles Fernyhough
Mari Fforde
Mike Fitzgerald
Tony & Hannah Fitzpatrick
Steven Fletcher
Piers Fletcher
Ben Fletcher-Watson
Simon Flisher
Eoin Flynn
Conor Forbes
N. W. Ford
Natalia Forrest
Euan Forrest
Mike Foster
Jonathan Foyle
Justin Freeman
James Fry

Jim Galbraith
Robyn Gallagher
E. Garcia
Amro Gebreel
Jonathan Gibbs
Paul Gibbs

Holly Gilbert
Adam Gilhespy
Ian Gilhespy
Mark Gittins
Martin Gladdish
David Glover
Lee Goddard
Richard Goddard
Alistair Gordon
Charlie Gould
David Graham
Christopher Gray
Karen Gray
Daniel Green
Andrew Gregg
Sarah Greig
Jane Grey
Matthew Grice
Alex Grimwood
Paul Grindley
John Grindrod
Thomas Groat

Matthew Hamblin
James Hamilton
Geoffrey Hands
Robert Hanks
David Harford
Eric Hargrave
Stephen Hargreaves
Susie Harries
A. F. Harrold
Guy Haslam
David Hawkins
Phil Hayes

Ada Haygarth
Marc Haynes
Miles Hayward
Chris Heathcote
Narin Hengrung
Deborah Herron
Chris Hewitt
Denny Hilton
Robert Hinchliffe
Matthew Hoggarth
Charles Holland
George Holland
Kevin Holmes
Hannah Hood
Petry Hoogenboom
Jason Hope
Richard Hope
Robert Hopkins
Roger Horberry
Giulietta Horner
Tom Hostler
Chris Hough
Simon Howson-Green
Matt Huggins
Francis Hughes
Matthew J. Hughes
Ben Hughes
Kevin Hunt
Barry Hurst
John Hutchinson
Steuart Hutchinson Blue

Alex Ingram
Jo Ireland
Mark Ireland

Subscribers

Ian Irvine
Richard Irvine
Rob Irving
Johari Ismail

Simon James
Simon N. James
Helen Jerome
Steve Jesper
Dennis Jinks
Ian Jones
Michael-Luke Jones
Philip Jones
Sean Jones
Daniel Jordan

Susanne Kahle
Keith Kahn-Harris
Andrew Katz
Dali Kaur
Nick Kell
Neil Kelleher
Andrew Kelly
Hilary Kemp
James Kemp
Aidan Kendrick
Brian Kennedy
Rolfe Kentish
Robert Kentridge
Matthew Keyte
Dan Kieran
Jill Kieran
Deb King
David Koehne
Ali Kold

Grègoire Kretz
Grzegorz Krzeszowiec

Robin Lacey
Stephen Lafferty
Dominic Laine
Nick Lamb
Rob Lamond
Chris Lamyman
Gary Lang
Alex Latter
Thomas Latter
Simon Lawrence
David Lawson
David Layland
Liz Leach
Alex Lee
Jack Lenox
Dan Lentell
Oliver Levy
Maxine Linnell
Marcos de Lima
Lis Llewellyn
Karen Loasby
Stephen Luckhurst
Karl Ludvigsen

Mark Mabberley
Denise McAlonan
Jamie MacDonald
Rory McErlean
Ewan & Sally McFadden
Mo McFarland
Ken McGannon
Darren McGuicken

Robert McHattie
Christian McHugh
Ian McIntyre
Stuart McKears
A. J. McKenna
Bridget McKenzie
Martin McMahon
Hannah Mackinlay
John Maher
Steven Maher
James Maidment
Neil Major
Chris Mallows
John Malone
David Manns
Tom Marshall
Matthew Marsom
Daryl Martin
Clint's Mate
Karl Maton
Richard Mayston
Gillian Menzies
James Mewis
Chris Meyer
Gia Milinovich
Bruce Millar
Jessica Millar
Martin Millar
Jon Milloini
John Mitchinson
Gordon Moar
Ciaran Mooney
Dean J. Morbey
Anthony Morgan
James Morgan

Richard Morgan
Simon Morgan
Stewart Morgan
Paul Morrison
Mark Mount
Jim Mourilyan
Manjeev Muker
Michael Müller
Gus Murray-Smith
Philippa Myers

Jon & Katherine Nagl
Mark Napier
Craig Naples
Stuart Nathan
Robert Needs
Donald Neil
Andrew Neve
Richie Nice
Philip Nijman
Phil Nolan

Rory O'Gara
Andrew O'Neill
Mark O'Neill
Michael Ohajuru
Paul Oliver
Stuart Orford
Alan Outten
Catherine Owen
David Owen

Neil Pace
Jack Page
Michael Paley

Subscribers

Oliver Palmer
Ian Park
Julia Parker
Matthew Parker
David Parkes
David J. Parkes
Tim Parkinson
Mark Parmenter
Steve Parnell
Charles Parry
Jane Parry
Gareth Parsons
Thomas Paternoster-Howe
Tara Pattenden
Michael Paul
Jon Pearson
Hugh Perry
Gary Phillips
Reuben Philp
Catherine Pickersgill
Greg Pickersgill
Colin Pierce
Jonathan Pierpoint
Adam Pikett
Liz Pollard
Justin Pollard
Nick Poole
David Porter
Russell Porter
Gail Poulton
Neil Powling

Daniel Quinn

Chris Raettig

Matt Railton
Glenn Rainey
Ben Ralph
Simon Ramsden
Keith Ray
Nicholas Redding
Darren Redgrave
David Reeves
Margaret Reeves
Sebastian Reinolds
Tony Relph
Alan Rew
Patrick Reynolds
Louise Rice
Peter Richards
Caroline Richardson
Dan Richardson
Neil Richardson
Trevor Rickard
Martin Riley
Ewen Roberts
John Robinson
Michael Robinson
David Robling
Ian Robson
Olly Rogers
Jenny Rollo
Christopher Rountree
Tom Rowland
Daniel Ruff
Grant Russell
Kevin Russell
Paul Rutherford

Andrew Sampson

Jerry Sargent
Donal Sarsfield
Julia Savage
Matt Savage
Patrick Schabus
Richard Scorer
Alex Scott
Michael Scott
Paul Scott
Ralph Scott
Richard Scorer
Ian Seckington
John Self
Kodzo Selormey
Tracy Sexton
Andrew Shearer
Oliver Shorten
Ahmed Siddiqui
Anthony Silverman
John Simmons
Mark Simons
Sarah Simpkin
Deborah Skelton
Alan Smith
Alex Smith
Andy Smith
Malcolm Smith
Nigel Smith
Terri Smith
Rob Spain
Andrew Spencer
Mark Spendley
Ralph Spurrier
Mark Deal & Martina Stansbie
Andy Stanton

Daniel Durling & Jack Stephens
Peter Stone
Michael Storey
Ruth Stourton
Philip Stout
Reuben Straker
William Streek
Adam Stuart
Tom Sutcliffe
Rory Sutherland
Matthew Sweet

Karen Tanguy
Christopher Tate
Donna Taylor
Matthew T. P. Taylor
Charles Testrake
Alan Thomas
Kerrie Thomas
Richard Thomas
Tim Thomas
Paul Thompson
David Thornton
James Thorp
Adam Throup
John Tiernan
Monica Timms
Nick Timms
Rohan Torkildsen
Alison Tristram
Andrew Tulloch
Peter Tunney
Paula Turner
Patricia Tutt

Subscribers

Stephen Vahrman
Ian Vince
José Vizcaíno
Henry Volans
Henry de Vroome

Francis Walden
Frank Walker
Johnny Walker
Richard Walkington
James Wallis
Dave Walsha
Anne Ward
Miranda Ward
Sam Ward
Jonathan Warner
Tim Warren
Nick Watts
Gary Savage & Natalie Watts
Paul Webb
David Weir
Daniel Weir Esq.
Philip Weller
Matthew Wells
Peter Welsh
Jamie Westcombe
Tim Whaley
Paul Whippey
Donald Whitaker
Paul White
Sandra White
Stephen White
Andrew Whitehurst

Quentin Alder & Sarah
    Whittingham
Andrew Wiggins
Rebecca Wigmore
Patrick Wildgust
Kate Wiles
Will Wiles
Philip Wilkinson
Rachel Williams
Robert Williams
Tracy Williams
Dennis Williamson
Sam Wilson
Philip Wilson
Peter Wilson
Ally Winford
Richard Wingate
Gaynor Wingham
Larissa Wodtke
Glenn Wood
Nicholas Wood
Steve Woodward
Adrian Woolrich-Burt
Paul Workman
Tim Wright
Michael Wynn
Anne Wynne

Simon York

Stefan Zasada
Zach & Helene
Peter van der Zwan

A NOTE ON THE TYPEFACES

THE ESSAYS, AND lectures in this book are set in Sabon, an old-style serif typeface created by the German-born typographer and designer Jan Tschichold in 1964. Sabon is closely based on the sixteenth-century Garamond font, and was designed in response to the complaints of a group of German printers who were struggling with the need for a traditional font that would work equally well when set by hand or by machine. They also insisted that the new typeface should be narrower than their existing typefaces in order to save space and money. Tschichold went one better. Sabon has the unusual distinction that the roman, italic and bold characters all occupy the same width when typeset, making it easier for compositors to estimate the length of a text. Tschichold was imprisoned by the Nazis as a communist early in his career but escaped to Britain in 1933, where he set the design standards for Penguin Books.

The film scripts are set in Courier, which was the standard typewriter font for decades before being reproduced on computers. Designed for IBM in 1955 by Howard 'Bud' Kettler, the company failed to take out any copyright protection on the design, so made no profit on its rapid spread. Howard Kettler had been a small-town newspaper man and had owned his own print shop before working for IBM for more than thirty-five years. He believed font design should be invisible to the reader, who should be conscious only of the words on the page. Courier's working title was 'Messenger', but Kettler thought better of it. 'A letter can be just an ordinary messenger, or it can be the courier, which radiates dignity, prestige and stability'. For decades the standard typeface of officialdom, it is also the industry standard for film scripts and screenplays.